ADVANCES IN MULTI-PHOTON PROCESSES AND SPECTROSCOPY

ADVANCES IN MULTI-PHOTON PROCESSES AND SPECTROSCOPY

Volume 14

QUANTUM CONTROL OF MOLECULAR REACTION DYNAMICS:
Proceedings of the US–Japan Workshop

Honolulu, USA 12 – 15 December 1999

Edited by

R J Gordon
University of Illinois at Chicago, USA

Y Fujimura
Tohoku University, JAPAN

World Scientific
Singapore • New Jersey • London • Hong Kong

Published by

World Scientific Publishing Co. Pte. Ltd.

P O Box 128, Farrer Road, Singapore 912805

USA office: Suite 1B, 1060 Main Street, River Edge, NJ 07661

UK office: 57 Shelton Street, Covent Garden, London WC2H 9HE

British Library Cataloguing-in-Publication Data
A catalogue record for this book is available from the British Library.

ADVANCES IN MULTI-PHOTON PROCESSES AND SPECTROSCOPY — Vol. 14

ISBN 981-02-4520-3

This book is printed on acid-free paper.

Printed in Singapore by World Scientific Printers

PREFACE

The field of coherent control may be broadly defined as the utilization of the quantum mechanical properties of matter to control the motion of electrons and nuclei. Included within this definition are the control of chemical reactions, control of the motion of charge carriers in solids, alignment of molecules in gases, liquids, and thin films, and trapping and manipulation of the center of mass motion of atoms and molecules.

Although first experiments in the control of atomic and molecular motion were performed in the early seventies, *coherent* control of chemical reactions can be traced to the mid-eighties, when Paul Brumer and Moshe Shapiro initiated the use of quantum mechanical interference, and David Tannor and Stuart Rice initiated the use of wave packets to control chemical processes. The field remained largely theoretical until the early nineties, when experiments by Dan Elliot and Robert Gordon implemented the Brumer-Shapiro theory, and Graham Fleming, Kent Wilson, Gustav Gerber, and Philip Bucksbaum, among others, used the wave packet method. Coherent control of a branching ratio was first demonstrated by Thomas Baumert and Gustav Gerber in 1994 using the wave packet method and by Gordon and coworkers in 1995 using the quantum mechanical interference method. Since then there has been a flood of new experimental and theoretical developments. Many key objectives, however, such as bond-selective photochemistry and the use of tailored laser pulses to control electronic and nuclear motion, are still in their infancy. The first international meeting devoted exclusively to laser control was held in Neve Ilan, Israel, in May, 1996. A second international meeting was held in Cancun in November, 1997, and the first Gordon Research Conference on Quantum Control was held in August, 1999.

In recent years Japanese scientists have made significant contributions to the field of coherent control. Some of their notable achievements include phase control of the photoionization of large polyatomic molecules in the gas phase by Kawasaki, the alignment of molecules with strong laser fields by Sakai, local control theory applied to isomerization and dissociation reactions by Fujimura, and a control theory based on the dressed state picture by Nakamura. Despite the growing Japanese research effort, there has been relatively little interaction between

Japanese and American workers in this field. In order to promote greater interaction between these communities, we convened a workshop in Honolulu in December, 1999 that was attended by thirty-one scientists from both countries. As the contributions to this volume attest, the workshop provided a forum for presenting the leading edge issues in this field, which are continuing to be addressed by researchers in both countries.

The main findings of this workshop (and the participants who presented them) may be summarized as follows:

- Learning algorithms have opened the door to controlling many processes (Bucksbaum and Stroud).
- Wave packet engineering has become a practical reality (Fujimura, Leone, Kleiman, Bucksbaum, Stroud, Misawa).
- The Hamiltonian can be altered by dressing it in a strong field (Nakamura, diMauro, Liu, Kono).
- Non-adiabatic transitions may be exploited for control (Nakamura, Ohmori).
- External degrees of freedom may be controlled (Sakai, Kawasaki, Dantus, Kasai).
- Progress is being made in understanding and controlling decoherence (Gruebele, Makri, Leone, Petek, Lozovoy, Tominaga).
- Control in condensed phases is robust (Bucksbaum, Kleiman, Pötz, Misawa, Petek, Kobayashi).
- The phase lag is a powerful tool both for controlling dynamics and for discovering properties of the continuum (Gordon, Khachatrian, Zhu, Nakajima, Lozovoy, Scully).
- Useful applications of coherent control are envisioned (Bucksbaum, Stroud, Fujimura, Scully).

This workshop was made possible by grants from the United States National Science Foundation and the Japanese Society for the Promotion of Science. We wish to acknowledge and thank these agencies for their generous support.

R. J. Gordon
Y. Fujimura

Contents

Part I. Control of Wave Packets

TIME-SHIFTING THE DYNAMICS OF Li_2 MULTISTATE ROVIBRATIONAL WAVE PACKETS BY STATE SELECTIVE COHERENT PHASE CONTROL

ZOHAR AMITAY, RADOSLAW UBERNA, AND STEPHEN R. LEONE

JILA, National Institute of Standards and Technology and University of Colorado, Department of Chemistry and Biochemistry, and Department of Physics, University of Colorado, Boulder, CO 80309-0440, USA

State-selective coherent phase control using single shaped femtosecond laser pulses is implemented to achieve forward and backward shifts in time (up to a global arbitrary phase) of the rovibrational $Li_2(E^1\Sigma_g^+)$ wave packet evolution $\Psi(t) \longrightarrow \Psi(t - t_{shift})$. Experimentally, the result is reflected as a time translation of the whole measured pump-probe transient. The wave packet is composed of eight rovibrational quantum states excited from a single selected rovibrational level (using a cw laser) of the $A^1\Sigma_u^+$ electronic state. The high degree of control over the exact coherent superposition of the wave packet states is accomplished using a pulse shaping setup incorporating a liquid crystal spatial light modulator to encode the desired phases in the pump pulse. Beyond the direct implications of this work involving the experimental implementation of coherent control, it is also relevant to the possibility of probing molecular dynamics and coherent configurations that, without phase control, would occur only at long times after the wave packet excitation. This is relevant, for example, to highly dissipative environments where the wave packet dephases on short time scales.

1 Introduction

The large number of possible applications, as well as scientific insights, that are opened up by using the coherent nature of laser radiation to control the dynamics of molecular systems have driven considerable theoretical and experimental efforts in the field of coherent control.[1−15] Many of these studies have dealt with "coherent phase control", where the altered and optimized parameter is the phase of the optical field.[1,4−8,12−15] For coherent phase control using single broadband femtosecond laser pulses, in the weak field limit, varying the relative phase between various frequency components of the driving light results in changing, and therefore controlling, the quantum interference between the states that compose the excited nonstationary wave packets and their resulting dynamics. On the experimental side, recent developments in pulse shaping techniques[16,17] have allowed the tailoring of pulses with relatively complicated phase functions, and various coherent control experiments have implemented these features.[9−15]

Being one of the simplest molecular systems, Li_2 serves as a benchmark and convenient basis for coherent control studies of molecules, both experimentally[14,15,18] and theoretically.[19,20] Its spectroscopy is well known, yet its electronic structure is sufficiently complicated to be interesting, as is the case for the highly anharmonic $E^1\Sigma_g^+$ shelf state. This state is formed with an avoided curve crossing, and thus it switches from a covalent nature on the inner turning point to an ionic one on the outer turning point. Moreover, lithium dimer is a molecular system where the rotational and vibrational dynamics occur on comparable time scales, which thus allows coherent control studies of simultaneous radial and angular (three-dimensional) molecular dynamics.

In the following, we present experimental results of state-selective phase control using single femtosecond laser pulses implemented to time-shift forward and backward (up to a global arbitrary phase) the evolution of a well-defined (eight state) rovibrational $Li_2(E^1\Sigma_g^+)$ wave packet, i.e., $\Psi(t) \longrightarrow \Psi(t - t_{shift})$, where t_{shift} is either positive or negative. The required phase functions are encoded into the exciting laser pulse using a pulse shaping setup incorporating a liquid crystal spatial light modulator. As will be discussed below, such time-shifting has implications in the context of both coherent control experimental implementation, as well as in using the control ability to study dynamics of the optically driven molecular system, which otherwise is difficult to access experimentally, for example, due to strong non linear effects or high rates of dephasing (T_1 and/or T_2 processes).

2 Experiment

The experimental setup has been described previously[14,15] and will be presented here only briefly. A sample of lithium metal is heated in a stainless-steel cell to 1023 K in an Ar buffer at ~2.7 Torr. A single mode cw dye laser (200 mW) is tuned to select a specific transition of the thermally populated Li_2 molecule from an individual $X^1\Sigma_g^+(v_X, J_X)$ state to a single rovibrational state (v_A, J_A) of the $A^1\Sigma_u^+$ electronic state. From this state, using femtosecond laser pulses, a wave packet is launched on the $E^1\Sigma_g^+$ state (pump step), followed by a time-delayed ionization of the Li_2 molecule (probe step). The pump and probe pulses are obtained from a Ti:sapphire laser system (200 kHz repetition rate), and they are of ~160 fs duration, parallel polarizations, and ~0.5 and ~1.5 μJ, respectively, implicating that, except for the period when the two fs pulses are overlapped in time, the wave-packet dynamics can be described within the weak field regime.

The experimental setup also incorporates a pulse shaping assembly, composed of two gold-coated gratings (1200 lines/mm), two cylindrical lenses (30 cm focal length), and a liquid crystal spatial light modulator (SLM). The SLM consists of two one-dimensional liquid crystal arrays placed between two sheet polarizers. Each array is composed of 128 independent pixels whose effective refractive index is computer controlled by an applied voltage, manipulating the spectral phase and/or amplitude of the various frequency components of the dispersed light to modify the amplitude and/or phase of each rovibrational component state of the excited wave-packet. The experimental wavelength resolution of our pulse shaping assembly is ~4.2 cm^{-1} per pixel, and the accuracy for an applied phase (0–360°) is better than 1°, depending on the particular phase.

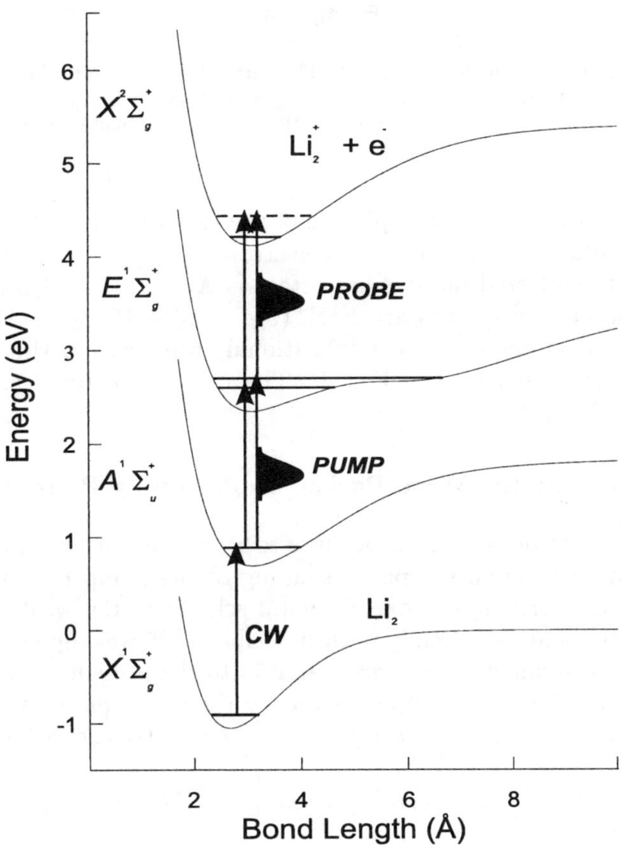

Figure 1. Relevant potential energy curves for Li_2 and Li_2^+.[22,23]

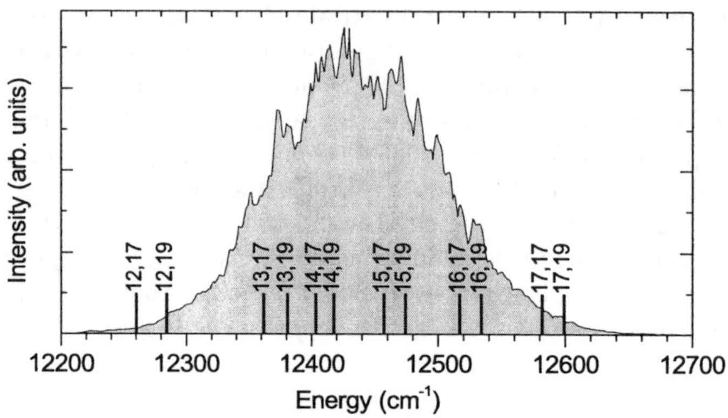

Figure 2. Spectral profile of the laser pulse. The vertical lines indicate the excitation energies of the various rovibrational states of the $E^1\Sigma_g^+$ state composing the excited wave packet. They are labeled with the corresponding vibrational and rotational quantum numbers.

Figure 1 shows a schematic picture of the experiment, including the relevant Li_2 potential curves and the ion electronic ground state. In the present experiment, the selected intermediate state is $A^1\Sigma_u^+(v_A = 14, J_A = 18)$, and the wave packet components are $E^1\Sigma_g^+(v_E = 13 - 16, J_E = 17, 19)$ (eight states). The locations of these rovibrational states within the pump pulse energy spectrum are shown in Fig. 2. The pulse is centered around 12430 cm^{-1} with a FWHM of $\sim$150 cm^{-1}.

3 Time-Shifting the Wave Packet Evolution by Phase Control

In the following we present the procedure to achieve a shift in time of the wave packet evolution through the phase shaping of the pump laser pulse. Since the theory for describing our experimental scheme in the weak field limit is well established and has already been described,[14,19] we emphasize here only the details and parameters that are essential to the present work.

The superposition state $\Psi(t)$ excited by the pump pulse (after the pulse is ended) from an initial state i is given at time t after excitation by[14]

$$\Psi(t) \propto \sum_n c_n \, \epsilon_{pump,n} \, |n\rangle \, \exp\left[-i(\frac{E_n}{\hbar}t + \Phi_{pump,n})\right] \tag{1}$$

where the index n stands for a state composed in the excited wave packet, $|n\rangle$ is its eigenfunction, and E_n represents its energy. The coefficient c_n is the matrix element of the i⟶n transition, and $\epsilon_{pump,n}$ and $\Phi_{pump,n}$ are the electric field amplitude and phase, respectively, of the pump pulse at the transition frequency $w_{ni} = E_n - E_i$. The phases $\Phi_{pump,n}$ are the physical quantities subjected to our direct experimental control through the pulse shaping setup.

In order to shift (up to a global arbitrary phase) the time evolution of the wave packet $\Psi(t)$ corresponding to the given initial phases $\Phi^{(0)}_{pump,n}$ by a constant time offset t_{shift} which is either positive or negative, i.e., $\Psi(t; \Phi^{(0)}_{pump,n}) \longrightarrow \Psi(t - t_{shift}; \Phi_{pump,n})$, the applied phases $\Phi_{pump,n}$ should depend linearly on the state energy with a slope proportional to t_{shift} (linear phase sweep):

$$\Phi_{pump,n} = \Phi^{(0)}_{pump,n} - \frac{t_{shift}}{\hbar} E_n \,. \tag{2}$$

In other words, up to a global arbitrary phase, all the superposition states of the wave packet that for $\Phi^{(0)}_{pump,n}$ would exist at time t, following the application of this linear phase sweep, would occur at time $t + t_{shift}$; a negative slope of the linear phase sweep corresponds to a shift forward ($t_{shift} > 0$), while a positive slope corresponds to a shift backward ($t_{shift} < 0$).

Experimentally, the wave packet time evolution is reflected in the pump-probe ionization signal $S_{pump-probe}(\tau)$ measured as a function of the pump-probe delay time τ:[14]

$$S_{pump-probe}(\tau) = S_{const} + \sum_{m,n;m\neq n} |A_{mn}| \cos\left(\frac{E_m - E_n}{\hbar}\tau + \Phi^{pump}_{mn} + \Phi^{*}_{mn}\right) \tag{3}$$

with

$$\Phi^{pump}_{mn} = \Phi_{pump,m} - \Phi_{pump,n} \,, \tag{4}$$

where S_{const} is a constant signal level, and the time-dependent part of the signal is composed of cosine terms corresponding to the various (m, n) quantum beats with amplitudes $|A_{mn}|$. The phase of each beat has been separated into Φ^{pump}_{mn}, which is being controlled here, and Φ^{*}_{mn}, which contains all the other contributions to the phase not under control in the current experiment. This includes the molecular phases corresponding to both the pump (bound-bound excitation and thus can induce either 0° or 180° phase) and probe (ionization) steps, as well as possible phases introduced externally by the probe laser pulse. In general, all these various phases have direct implications in the context of coherent control, and they can serve as a tool to gain insight into the structure of the ionization continuum,[8] however here they are regarded as given non-controllable parameters. Other experiments, which have been conducted in our lab, have dealt with these aspects.[21]

For $\Phi_{pump,n}$ given by Eq. (2), the phase difference Φ_{mn}^{pump} has the form

$$\Phi_{mn}^{pump} = -\frac{t_{shift}}{\hbar}(E_m - E_n) \,. \tag{5}$$

This implies that a time-shift of the wave packet evolution $\Psi(t) \longrightarrow \Psi(t - t_{shift})$ is directly reflected as a time-shift of the whole measured pump-probe transient $S_{pump-probe}(\tau) \longrightarrow S_{pump-probe}(\tau - t_{shift})$.

4 Results and Discussion

Figure 3 shows an experimental pump-probe transient (only the time-dependent part) taken without any phase manipulation up to pump-probe delay times of 80 ps, together with its corresponding FFT spectrum up to 120 cm^{-1}. The contribution from higher frequency components (120 to 256 cm^{-1}) is less than 3%. All the prominent Fourier components in the spectrum (marked with "$*$") are identified and correlated with specific pairs of states out of the wave packet states $E^1\Sigma_g^+(v_E = 13-16, J_E = 17, 19)$. This experimental identification of the energy differences between the various wave packet components is a crucial component of the present experiment, since in order to achieve the time-shifting the applied phases should follow the specific linear phase sweep of Eq. (2). The precise locations of the various rovibrational states within the pump pulse energy (see Fig. 2) were found

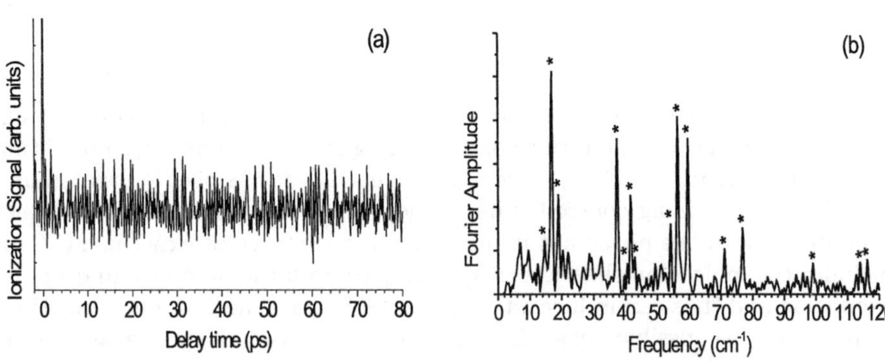

Figure 3. (a) Experimental pump-probe transient taken with non-shaped pump pulse and (b) the corresponding FFT spectrum. The Fourier components that are identified to correlate with specific pairs of states out of the wave packet states $E^1\Sigma_g^+(v_E = 13-16, J_E = 17, 19)$ are marked with "$*$" (see text).

experimentally by monitoring the amplitude of the Fourier components involving specific rovibrational states while blocking various frequencies in the shaped pump pulse.

Following the control scheme described in Sec. 3, we have demonstrated translation in time of the wave packet originally excited with the non-shaped pulse ("unshaped wave packet") by 5, 7, and 9 ps, applying linear phase sweeps over the various rovibrational states [see Eq. (2)] with a slope of -54.0, -75.6

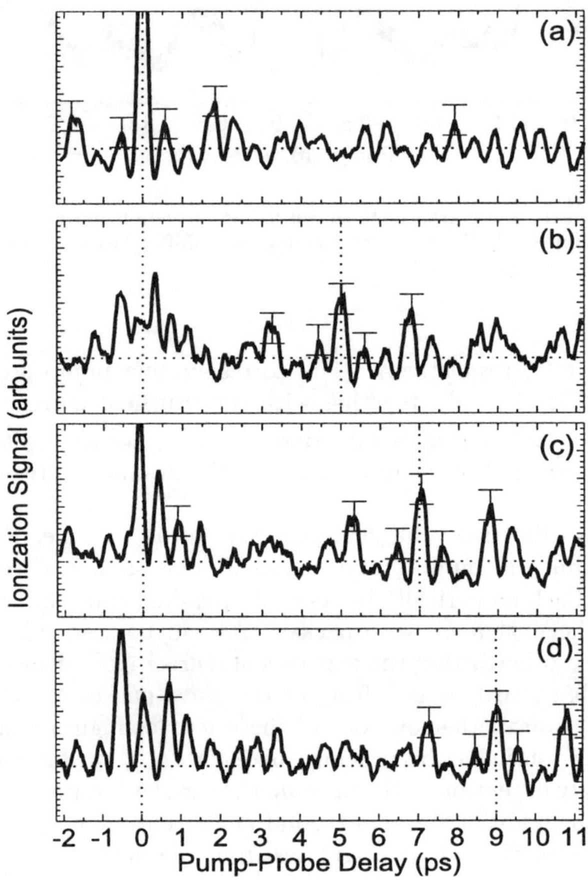

Figure 4. Experimental pump-probe transients obtained for (a) the unshaped wave packet, and for various time-shifts of it: (b) 5 ps, (c) 7 ps, and (d) 9 ps.

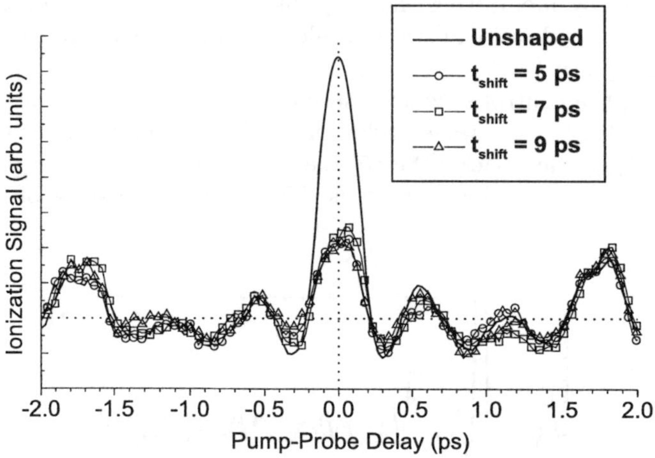

Figure 5. The experimental transients shown in Fig. 4 one on top of the other; each one is shifted back numerically by its corresponding time shift relative to the transient of the unshaped wave packet (Fig. 4(a)).

and -97.1 °/cm^{-1}, respectively. The corresponding pump-probe transients are shown in Fig. 4(b,c,d), together with the transient corresponding to the unshaped wave packet [Fig. 4(a)]. Here, $\Phi^{(0)}_{pump,n}$ stands for any chirp that exists in the unshaped pulse upon arrival in the interaction region of the heat pipe.

In order to illustrate the achieved translation in time, Fig. 5 presents all four experimental pump-probe transients one on top of the other; each one is shifted back numerically by its corresponding time shift relative to the transient of the unshaped wave packet. As can be seen, all the transients agree with each other within the experimental error bars, demonstrating both forward and backward time-shifting of the wave packet evolution from any configuration to any other one out of these four configurations. Thus, as a whole, the time-shift has been shown here for $t_{shift} = \pm 2, \pm 4, \pm 5, \pm 7$, and ± 9 ps. The only exception in the agreement between the various experimental transients is the additional signal around $\tau = 0$ that appears in the unshaped wave packet transient, which originates from the delay-time dependent (direct) multiphoton ionization of the intermediate A-state occurring when the pump and probe pulses are overlapped in time. These non linear processes are very sensitive to the power, alignment, focusing conditions, and temporal shape of the pulses and thus are less desirable for precise phase control, although such highly non linear effects are the subject of recent coherent control investigations.[12]

The actual phases applied experimentally across the pump pulse spectrum to achieve the various time shifts of the unshaped wave packet are shown in Fig. 6. The phase value at each rovibrational state is the residual from the division of the required total phase by 360°. Figure 7 displays the corresponding intensity cross-correlation profiles taken between the phase-shaped pump pulses and the original unshaped Gaussian pulse using the second-order response of a GaAsP photodiode. The cross correlation width of the unmodified pump pulse is 227 fs (FWHM) and it increases to more than 1 ps after introducing the phases depicted in Fig. 6.

One should note that the way the phases have been applied experimentally across the pump pulse spectrum (Fig. 6), i.e. in segments, each spanning over several pixels, indeed time-shifts the wave-packet dynamics, however it does *not* delay the pump pulse itself by t_{shift}. For a detailed discussion of such experimental time-shifting of the pulse itself, including the limitations arising from the pixellated nature of the SLM, the reader is referred to the papers by Weiner *et al*[16] and Wefers *et al.*[17] In the context of the present experiment, the pixellation of the SLM is not a limiting factor since each of the discrete rovibrational states composing the wave packet lies within a separate pixel, thus experimentally the applied phases can follow the needed form of Eq. (2).

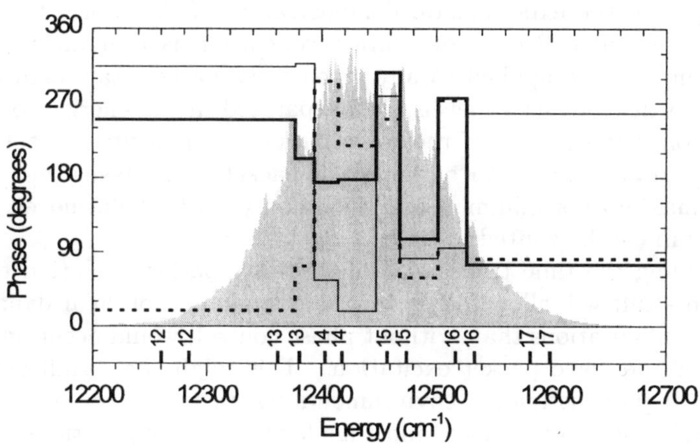

Figure 6. The phases applied experimentally across the pump pulse spectrum to achieve the time shifts of the unshaped wave packet by 5 ps (dashed line), 7 ps (thin solid line) and 9 ps (thick solid line).

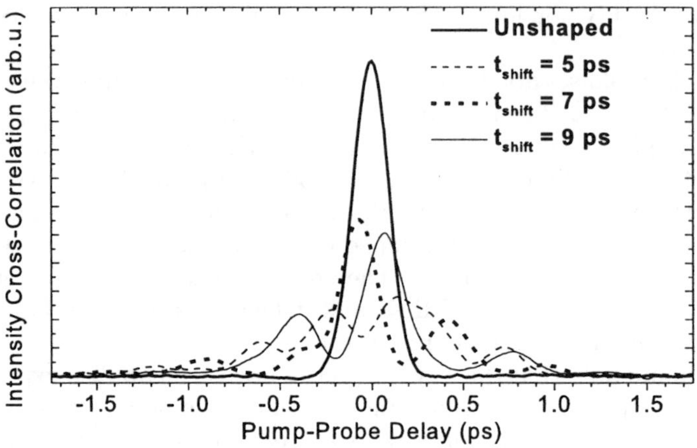

Figure 7. Intensity cross-correlation traces between the phase-shaped and the original unshaped Gaussian pulse, obtained with GaAsP photodiode. They correspond to the various phase configurations shown in Fig. 6.

5 Conclusions

Following from the experimental demonstration of this work, based on the description given in Sec. 3, one direct conclusion is that an experimental time-shifting can be applied to any given wave packet state composed of a well-defined discrete set of levels, well separated in the energy domain. For example, once the phases of the wave packet components are optimized to produce partial localization of the wave packet at a given time, and as a result to maximize or minimize the ionization yield,[15] the achieved minimum or maximum can be shifted in time.

Extending the time translation of the wave packet evolution to a much longer time-shift will allow the probing and study of molecular dynamics and coherent configurations that, without phase control, would occur only at long times after the wave packet excitation. This might be significant, for example, in highly dissipative environments where the wave packet dephases on relatively short time scales, and thus long-time configurations can not be accessed.

Future experiments will be conducted to investigate in further detail the experimental limitations, as well as the possible applications, of such time-shifting of the wave packet dynamics.

Acknowledgments

The authors wish to acknowledge the support of the National Science Foundation for this project.

References

1. *Femtosecond Chemistry*, ed. J. Manz and L. Wöste (VCH, Weinheim, 1995), and references therein.
2. M. Shapiro and P. Brumer, *J. Chem. Soc. Faraday Trans.* **97**, 1263 (1997), and references therein.
3. R.N. Zare, *Science* **279**, 1875 (1998), and references therein.
4. M.M. Wefers, H. Kawashima, and K.A. Nelson, *J. Chem. Phys.* **102**, 9133 (1995).
5. B. Kohler *et al*, *Phys. Rev. Lett.* **74**, 3360 (1995).
6. C.J. Bardeen, Q. Wang, and C.V. Shank, *Phys. Rev. Lett.* **75**, 3410 (1995).
7. B. Amstrup, G. Szabo, R.A. Sauerbrey, and A. Lörincz, *Chem. Phys.* **188**, 87 (1994).
8. J.A. Fiss, A. Khachatrian, L. Zhu, R.J. Gordon, and T. Seideman, *Faraday Discuss.* **113**, 61 (1999), and references therein.
9. D.W. Schumacher, J.H. Hoogenraad, D. Pinkos, and P.H. Bucksbaum, *Phys. Rev. A* **52**, 4719 (1995).
10. T.C. Weinacht, J. Ahn, and P.H. Bucksbaum, *Phys. Rev. Lett.* **80**, 5508 (1998).
11. T.C. Weinacht, J. Ahn, and P.H. Bucksbaum, *Nature* **397**, 233 (1999).
12. A. Assion *et al*, *Science* **282**, 919 (1998).
13. D. Meshulach and Y. Silberberg, *Nature* **396**, 239 (1998).
14. R. Uberna, M. Khalil, R.M. Williams, J.M. Papanikolas, and S.R. Leone, *J. Chem. Phys.* **108**, 9259 (1998).
15. R. Uberna, Z. Amitay, R.A. Loomis, and S.R. Leone, *Faraday Discuss.* **113**, 385 (1999).
16. A.M. Weiner, *IEEE Journal of Quantum Electronics* **28**, 908 (1992).
17. M.M. Wefers and K.A. Nelson, *J. Opt. Soc. Am. B* **12**, 1343 (1995).
18. J.M. Papanikolas, R.M. Williams, and S.R. Leone, *J. Chem. Phys.* **107**, 4172 (1997).
19. D.G. Abrashkevich, M. Shapiro, and P. Brumer, *J. Chem. Phys.* **108**, 3585 (1998).
20. L. Pesce *et al*, in preparation.
21. Z. Amitay, R. Uberna, and S.R. Leone, in preparation.

22. I. Schmidt-Mink, W. Müller, and W. Meyer, *Chem. Phys.* **92**, 263 (1985).
23. R.A. Bernheim *et al*, *J. Chem. Phys.* **87**, 861 (1987).

STRONG FIELD ULTRAFAST QUANTUM CONTROL WITH LEARNING ALGORITHMS

PHILIP H. BUCKSBAUM
Physics Department, Randall Laboratory,
University of Michigan, Ann Arbor, MI 48109-1120 USA
E-mail: phb@umich.edu

Haiku:
Coherent control
Gives us a new way to view
Quantum mechanics.

1 Introduction

Now that we can manufacture high field and ultrafast laser pulses with near-arbitrary combinations of intensity, wavelength, phase, and polarization, a question begins to emerge: Why? These may be wonderful tools for physics, but to realize their benefits we have to take the next step, and explore new physical applications. There is no single answer to this problem, no single "killer ap". In this paper I review my own group's approach to this, motivated by our experience in atomic and molecular physics, our desire to work in areas where we have an impact, and our natural tendency

Figure 1. Left: Schematic diagram of a Fourier filter. Light from an ultrafast pulse is dispersed, and then passes through a programmable amplitude and phase filter, then recombined to form any pulse shape consistent with the initial bandwidth. Right: Photograph of the Neos acousto-optic modulator

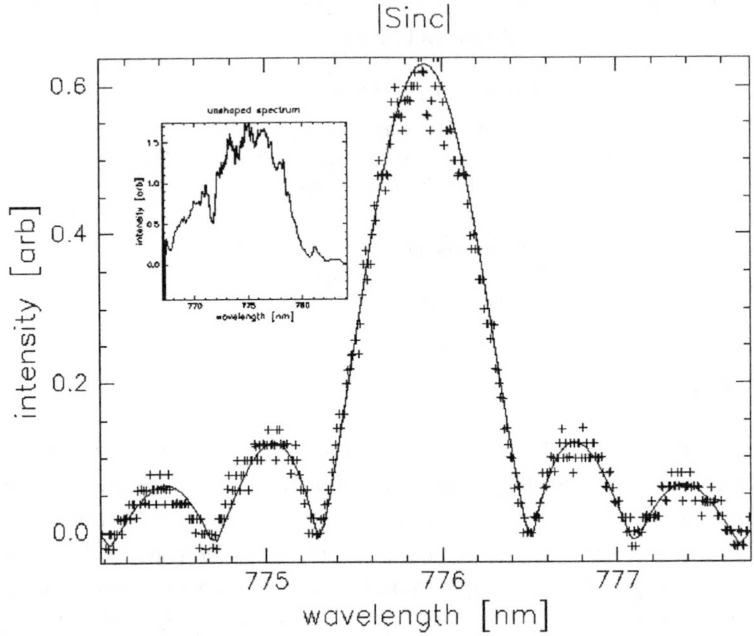

Figure 2. The inset shows the input spectrum to the pulse shaper, and the main figure shows the output after programming the filter to produce a sinx/x (sinc) function.

to work on wild and crazy ideas at least some of the time.

We produce shaped laser fields by starting with an intense short laser pulse produced via Chirped-pulse amplification (CPA). Typically, this produces several millijoules of coherent laser energy at a central wavelength of 800 nm, and with a bandwidth capable of supporting a 50-100 fsec laser pulse (several tens of nm bandwidth). The light is coherently shaped with a Fourier filter,[1] as shown in figure (1).

We use an acousto-optic filter, according to the scheme originally designed by Warren and collaborators.[2] Typical performance is shown in figure (2). The laser light can be modified further by standard nonlinear optical media to produce broad-band coherent light in the near-infrared through the near ultraviolet. We also know how to use various laser-solid and laser-atom interactions to produce far-infrared or vacuum ultraviolet coherent radiation, and the shaped optical field can be used in this case to shape the far infrared.[3]

I will discuss three applications of optical pulse shaping to quantum control: (1) Quantum information applications in atomic media; (2) Coherent control of chemistry in liquid phase; and (3) control of the motion of a crystal lattice, which has applications for x-ray scattering. These experiments have been carried out by a dedicated team of graduate students, post-docs, and senior colleagues. I particularly acknowledge the efforts of Thomas Weinacht, Jaewook Ahn, David Reis, Marcus Hertlein, Inuk Kang, Aaron Lindenberg, Roberto Merlin, and Roger Falcone. Any mistakes or omissions in this brief review are my fault, not theirs!

2 QUANTUM INFORMATION APPLICATIONS

Wave packets formed by sculpted laser pulses incident on atoms or molecules contain a lot of specific information in the form of complex quantum amplitudes, and that information can remain stored in the atom over its decoherence time. We've begun to explore efficient methods to encode and extract that information, and to increase the amount that a single atom can hold.

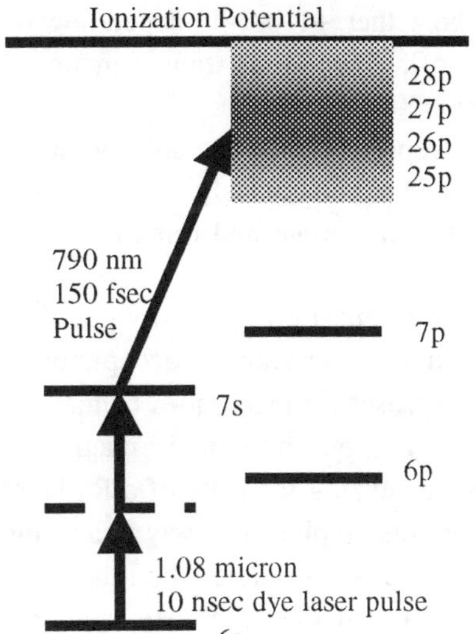

Figure 3. Scheme for creating sculpted wave packets in Rydberg states of atomic cesium.

The basic method we employ for creating wavepackets in Rydberg atomic systems is illustrated by figure 3. An atomic beam of cesium is first excited from the ground 6s state to the 7s state by a two-photon transition induced by a frequency-shifted tunable dye laser. This transition begins to saturate at intensities below the multiphoton ionization threshold. The 7s state lives for several nanoseconds, which is long enough to turn off the dye laser and turn on a coherent shaped optical field centered at 790 nm.

The sculpted light field produces sculpted wave packets in the Rydberg np series, where n ranges from 20-40. The total fluence in the field is usually adjusted to excite about 10% of the atoms to Rydberg states.

One simple way to exploit wave packets for information storage is to encode binary "1's" and "0's" as positive or negative unit amplitudes in non-degenerate *np* Rydberg energy levels of the atom. Our acousto-optic Fourier filter can address several dozen separate levels simultaneously. A shaped laser field transports probability amplitude in the 7s state to the Rydberg series, and it can be modeled as a unitary transformation on the state space consisting of these levels. For example, if each Fourier component excites its corresponding *np* level in the atom with a small amplitude ε, then the unitary transformation looks, to O(ε), like this:

$$A|\Psi>=\begin{pmatrix} 1 & -\varepsilon & \varepsilon & \cdot & \varepsilon \\ \varepsilon & 1 & 0 & \cdot & 0 \\ -\varepsilon & 0 & 1 & \cdot & 0 \\ \cdot & \cdot & \cdot & \cdot & \cdot \\ -\varepsilon & 0 & 0 & \cdot & 1 \end{pmatrix}\begin{pmatrix} 1 \\ 0 \\ 0 \\ \cdot \\ 0 \end{pmatrix}=\begin{pmatrix} 1 \\ \varepsilon \\ -\varepsilon \\ \cdot \\ -\varepsilon \end{pmatrix}\begin{matrix} 7s \\ 27p \\ 28p \\ \cdot \\ 38p \end{matrix}$$

Every Rydberg level now has the same probability *amplitude*, but the relative phase of the levels at a specified “target” time t can be either +1 or –1, depending on whether the corresponding bit of information is 1 or 0. The unitary transformation *A* in the example stores the binary number 100000000000.

In this example, a single sculpted laser pulse (represented by the unitary matrix *A*) stored information in each location of the database simultaneously. This parallel access is an important feature of quantum information storage and retrieval. Any number of bits can be flipped in a single pulse.

The next step is to retrieve the information in the most efficient way possible. The location of 1's and 0's cannot be read out by simply observing the occupation of the states using spectroscopy. This is because spectroscopic techniques cannot detect the relative phase of the probability amplitudes, but only their magnitude squared. For example, one common technique for analyzing the state of a Rydberg atom is state-selective field ionization (SSFI). Here one applies a slowly increasing electric field **F**(*t*) to the system. A Rydberg atom becomes unstable when the static field reaches the critical value where the saddle point of the combined Coulomb potential + linear applied potential reaches the bound state energy, as shown in Figure 4.

Ramped SSFI data from three different shaped wave packets are shown on the right. The integral under each peak is proportional to the *absolute value squared* of the probability amplitude for that state. Phases cannot be ascertained. Other methods for reading the state of the atom such as fluorescence spectroscopy or antistokes Raman spectroscopy have the same property, namely, they are only sensitive to the population, not the complex probability amplitude. It appears that a database where the information is stored as quantum phase is impractical, because it is unreadable.

L. Grover of Lucent Bell Laboratories has recently considered this problem, and pointed out an elegant solution. He shows that a *second* unitary transformation (we'll call it B) should be able to converting the phase information into a form that can be measured easily.[4] Grover emphasized that this second transformation is a single query of the database, which means that in only one step it is possible to search all possible storage locations in parallel. Classical digital computation must query the storage locations one at a time, so the quantum method represents a substantial potential speed-up. Another important feature is that B is *universal:* The form of B does not depend on the specific stored information, so that B can even read a register of unknown contents.

In our problem with Rydberg atoms, a convenient form for B to $O(\varepsilon)$ is

$$B(A\,|\,\Psi>) = \begin{pmatrix} 1 & -\varepsilon & -\varepsilon & \cdot & -\varepsilon \\ \varepsilon & 1 & 0 & \cdot & 0 \\ \varepsilon & 0 & 1 & \cdot & 0 \\ \cdot & \cdot & \cdot & \cdot & \cdot \\ \varepsilon & 0 & 0 & \cdot & 1 \end{pmatrix} \begin{pmatrix} 1 \\ \varepsilon \\ -\varepsilon \\ \cdot \\ -\varepsilon \end{pmatrix} = \begin{pmatrix} 1 \\ 2\varepsilon \\ 0 \\ \cdot \\ 0 \end{pmatrix}$$

This B has the property of doubling the amplitude of all "flipped" bits (i.e. 1's) and *canceling* the amplitude of all unflipped bits (the 0's). The effect is shown pictorially in Figure 5.

We performed an experimental test of this in atomic Rydberg states in cesium.[5] One or more flipped states stored in an eight-state wave packet were successfully retrieved in a single query of the data base, i.e. by superposition with the universal decoding wave packet represented by B. Two different storage/retrieval protocols were studied. In the first, we "hid" a single flipped bit in the Rydberg data register. As Figure 5 shows, the application of the retrieval query converts this database wave packet into a pure Rydberg eigenstate, which can be measured easily with SSFI.

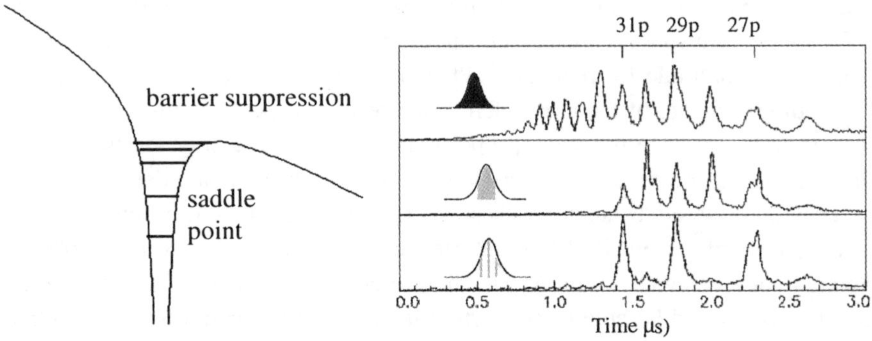

Figure 4. Left: cross sectional potential diagram for a Coulomb potential plus a linear potential from a static electric field. Right: signal collected from ramped state-selective field ionization (SSFI) for several sculpted Rydberg wave packets.

In the second situation, we showed that any number of flipped bits could be detected. Here we had to deal with the problems inherent in any quantum measurement: Each atom can only give up a single electron during ionization, and therefore

one atom only reveals one of the flipped bits with probability 1/N, where N is the total number of bits that were flipped. We therefore performed a parallel ensemble measurement using the atoms in the interaction region, where the sculpted laser pulse intersects the atomic beam. This need to perform an ensemble measurement

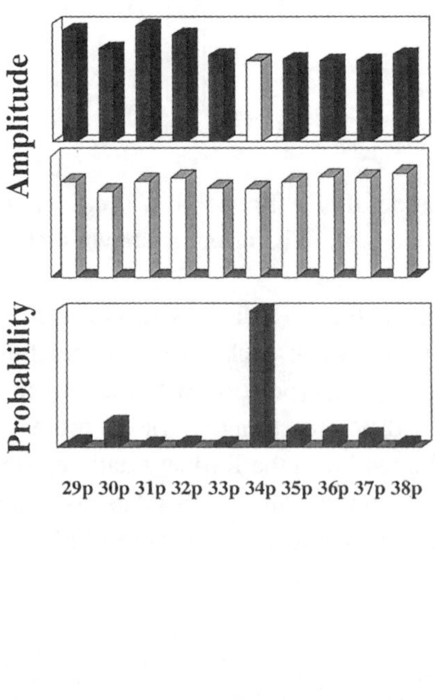

Figure 5. Schematic of data storage and retrieval in a database consisting of ten *np* states of a Rydberg atom. The top graph shows the amplitudes (bar heights) and phases (shading) of a wave packet that codes the binary number 0000010000. This middle graph shows the special wave packet that "queries the data base, and converts the data from phase to amplitude, as shown in the third graph.[5]

was also discussed by Grover.[4] We demonstrated storage of numbers as high as 2^{N-1}, with N=8 Rydberg states within the bandwidth of the sculpted laser pulse. Ensemble averaging introduces some probability for errors. In addition, there are technical problems, such as vibrations in the apparatus or amplitude fluctuations in the laser, which also contribute to errors. The results of the experiment are summarized in Figure 6.

3 QUANTUM CONTROL OF MOLECULAR MODES

The nonlinear interaction between intense lasers and condensed media is commonly used to generate new coherent light sources or probe dynamics in solids and liquids. Nonlinearities can be controlled by pulse shaping, so in principle one should be able to dial up specific pulse shapes to enhance a desired nonlinear interaction. In its most ambitious form, this becomes a general tool for bond-specific chemical dynamics. Unfortunately, the Hamiltonian for any complex molecular system is not known adequately, and without it there is no universal method to identify likely excitation pathways or derive the optimal pulse shapes. One way out of this problem was suggested in a landmark paper by Judson and Rabitz,[6] who suggested filling this gap in knowledge through learning algorithms. In this approach, the experiment runs itself by means of an intelligent feedback loop. It tries various pulse shapes, assesses their success in achieving the desired target excitation, and uses the knowledge gained in this way to improve the pulse shapes on subsequent laser shots, all without the intervention of the researcher.

We have been studying bond-selective excitation of liquids using a variant of

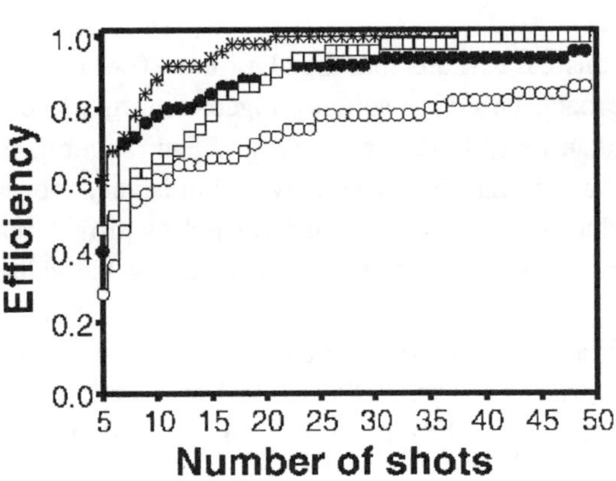

Figure 6. Results of N-bit coding with 6 state quantum register (asterisk), with 8 states (square), 2^{N-1} bit coding with 6 states (closed circle), and with 8 states (open circle). The y-axis represents the retrieval rate of programmed information extracted from first M identically prepared pulses, where M represents the number in the X-axis.[5]

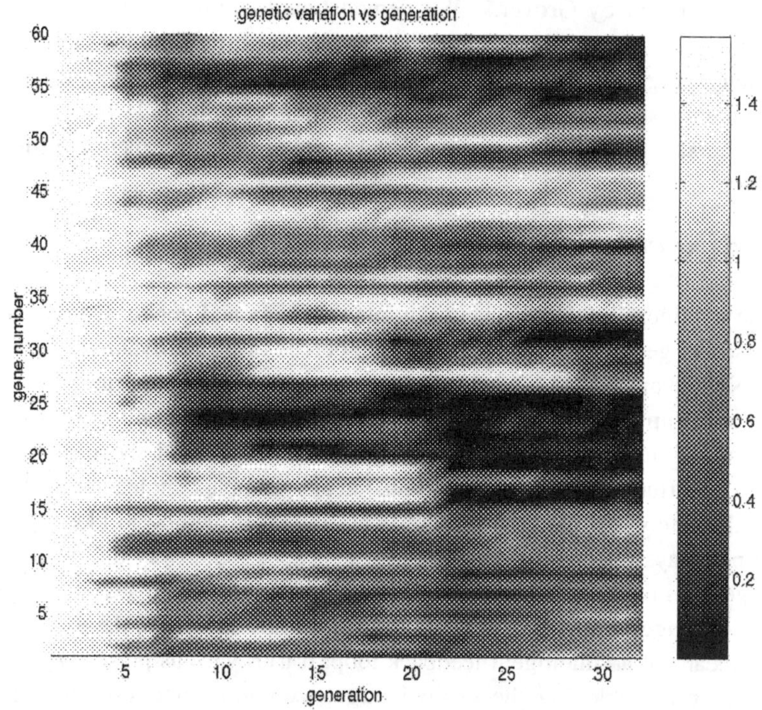

Figure 7. Gene variation versus generation during a run of the genetic algorithm. Lighter shade denotes more variation in a particular gene. Note that at the end of the algorithm, some of the genes still have a wide variance, indicating that they are not important to the fitness.

the genetic algorithm.[7] "Genetic" in this case means that the algorithm (GA) creates new pulses through a non-local approach based on splicing together traits of successful "parent" laser pulses, rather than by following a fitness gradient function, as in other evolutionary methods.[8] We have found that genetic evolution is very robust for this problem. Furthermore, the shapes of the successful laser pulses provide us with very specific hints about the excitation pathway, thus demonstrating the value of the approach.

The GA is implemented with an initial random population of pulse shapes (individuals). Each pulse shape is characterized by its "genome," or genetic code, which specifies the spectral phase in each of 30 separate wavelength bands within the pulse, and also specifies the amplitude in each of five different bands. This particular parameterization for the genome is itself subject to optimization. We chose these values by trial and error, but the GA could also be programmed to alter the stucture of its own genetic code, which would make the algorithm *adaptive*. The most important part of the algorithm is the test for the fitness of the individual pulses. In our case, each individual was tested for its ability to generate the spectrum that most resembled a target that we selected in advance. Relative success was quantified to determine the pulse fitness. Then, a new generation of pulse shapes was produced by mating different parts of the genome from pairs of the most fit individuals from the current generation. In addition, the algorithm prescribed a small probability (5%) of random genetic mutation. This basic series of processes was than iterated. The convergence and robustness of the GA solutions were measured in two different ways. First, we could monitor the variance in the genome itself. As the algorithm converges on a solution, the value of any gene that materially affects the fitness should fall into a narrow range that produces the best result. Second, we can monitor the genome for different initial conditions. If the algorithm is working, it should find a similar result. Figure 7 shows how the variance of sixty different genes changed during the progress of the algorithm in one run of the experiment. Over the course of 30 generations, some of the genes become very stable, which is an indication that those spectral phases and amplitudes are particularly important for driving the process that determines fitness.

We attempted to control the interaction between the shaped laser pulses and the

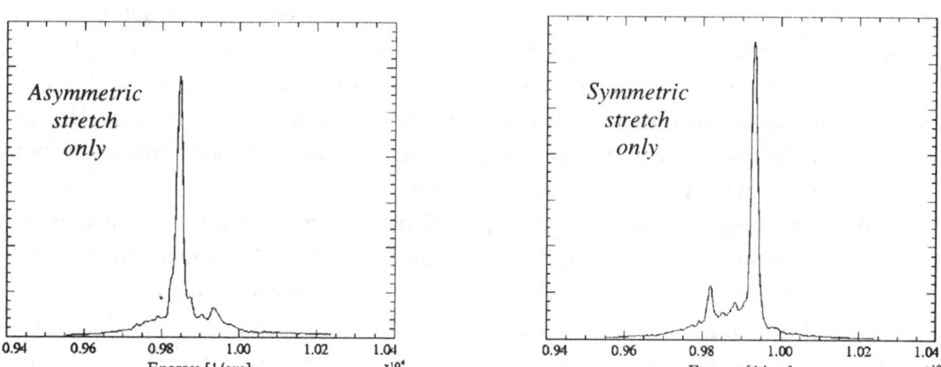

Figure 8. Selective Stokes peaks in the forward scattered light with a methanol sample. The driving laser pulse had a central frequency of 12,700 cm^{-1}.

vibrational modes of a several different multi-mode molecules in molecular liquids. The laser pulses that drive the molecular motion were intense enough so that several nonlinear processes were observed. Self phase modulation broadened the pulse as it propagated through ~1 cm of liquid. In addition, stimulated Raman scattering created spectral structure on the output pulse. In order to establish a fitness criterion that would indicate the molecule was vibrating in a particular mode, we measured the vintegrated spectral intensity of Raman Stokes peaks in the forward-scattered light. In one set of experiments we were able to find pulse shapes that were capable of selectively exciting the C-H stretch modes in methanol.

Figure 8 shows evidence for selective excitation of the symmetric and asymmetric stretch modes in methanol. The Stokes shifts for the asymmetric and symmetric modes are 2942 cm^{-1} and 2847 cm^{-1} respectively, which therefore place them far outside the laser bandwidth. By Fourier's theorem the laser pulse duration is therefore long compared to the vibrational period of either stretch mode. In this non-impulsive regime, selective Raman modes cannot be seeded directly with the laser light. Nonetheless, we found that the GA could teach the laser to excite the symmetric and asymmetric stretch modes either together or individually by changing the shape of the driving pulse.

We have considered two different mechanisms that could account for selective excitation of the C-H stretch modes in methanol. One is coupling between molecular polarizability and vibration. The molecular polarizability might effectively control vibrational excitation by providing seed light for SRS into one or more Raman-active modes. A second possible mechanism that could account for the selective excitation is quasi-impulsive coupling between the two modes. Although the coupling between different levels of each mode is definitely non-impulsive, the splitting between the two first order stokes shifts is less than the laser bandwidth, so that impulse scattering between them is possible. In the time domain, this is equivalent to saying that the laser pulse is long compared to each normal mode frequency but short compared to the beat note between the two.

We have begun to explore the implications of these two different models of mode-selection in our system. Mixtures of benzene (C_6H_6) and deuterated benzene (C_6D_6) have two Raman-active modes with a similar separation to the C-H stretches in methanol, but in the benzene mixture the modes are on ***different*** molecules: C_6H_6 (ν=992 cm^{-1}) or C_6D_6 (ν=947 cm^{-1}). GA experiments on the mixture are unable to separately control these two modes. This may suggest that mode selection is an ***intra***molecular processes.

Further insight into the selective excitation mechanism can be gained through examination of the shaped pulses that effected the control. There are some features in the solutions obtained thus far that may point to an intramolecular mechanism. Figure 9 shows the time and frequency map of the electromagnetic field pulses optimized by the GA for exciting the two C-H stretch modes in methanol. These are displayed as Husimi distributions, a means of representing wave packets that is often employed in quantum systems, but is also quite useful for these laser spectrograms. The Husimi distribution, Q(τ,ν), is calculated by transforming the optical field E(ν) according to

$$Q(t,\nu) = \int d\nu' \int dt' S(t',\nu') e^{-(t-t')^2-(\nu-\nu')^2}$$

and

$$S(t,\nu) = \int E(\nu+\nu')E^*(\nu-\nu')e^{-4\pi i\nu' t}d\nu'.$$

The electric field E(ν) was determined by combining the converged power spectra with the phase shifts applied to the AOM, which was calibrated by spectral interferometry with the unshaped reference. The optimal pulse for exciting the symmetric stretch has only one spectral lobe whereas the asymmetric stretch has two temporally and spectrally separated spectral lobes. These pulse shapes are consistent with a quasi-impulsive scattering mechanism. If the population in the two levels can be controlled through a quasi-impulsive coupling, then spectral components that are separated by the mode splitting are especially important for control. This

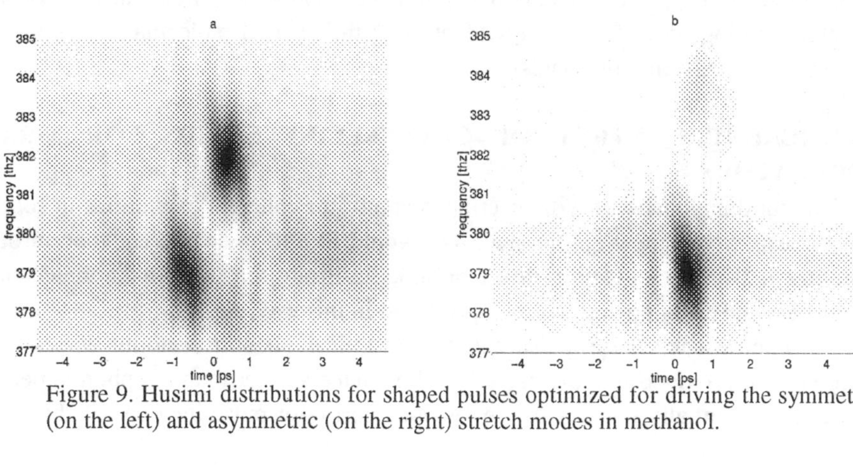

Figure 9. Husimi distributions for shaped pulses optimized for driving the symmetric (on the left) and asymmetric (on the right) stretch modes in methanol.

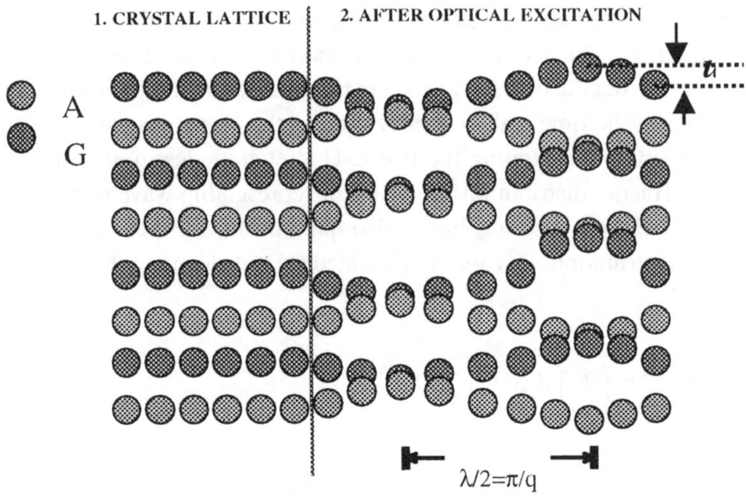

Figure 10. Schematic of optical phonon generation in a crystal of GaAs.

could explain the absence of spectral intensity at the center of the spectrum. The separation of the two spectral components in the pulse optimized for excitation of the symmetric stretch is the same as the energy separation between the two modes in the molecule.

Although the benzene experiment and the optimized pulse shapes are both consistent with a quasi-impulsive picture for the mode selective mechanism, they do not unambiguously indicate that quasi-impulsive coupling is the dominant mechanism. There are several technical details that complicate the data collection, and we are currently working on different ways of probing the excited molecular states to uncover the mode selection mechanism.

4 COHERENT CONTROL OF CRYSTALLINE SOLIDS FOR X-RAY MODULATORS

The third application I will describe is control of coherent excitations in solids. Here there is a very new application: We would like to use coherent motion of a crystalline lattice to modulate x-rays that undergo Bragg scattering from the crystal. Sub-picosecond x-ray pulses could be produced in this way.[9]

An intense short laser pulse can excite optical phonon standing waves in a Bragg crystal, provided that the pulse length is much shorter than the phonon period (typically substantially under 1 ps). A second short laser pulse incident on the same

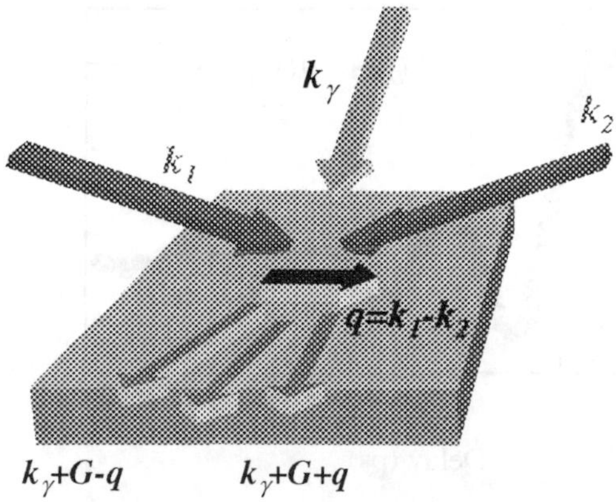

Figure 11. Schematic of the geometry for producing a coherently controlled Bragg switch in an x-ray diffraction experiment. The laser forms an optical standing wave from wave vectors $\boldsymbol{k_1}$ and $\boldsymbol{k_2}$. This produces two additional Bragg scattering conditons, at sidebands displaced by the optical standing wave vector $\boldsymbol{q=k_1-k_2}$ from the the ordinary Bragg angle.

crystal can turn off the motion after a half-integer number of phonon periods. The important parameters for this application are the lattice displacement vector $\boldsymbol{u}$ and the phonon wave vector $\boldsymbol{q}$. These are shown schematically in Figure 10. The quantities $\boldsymbol{u}$ and $\boldsymbol{q}$ can be controlled independently using the geometry shown in Figure 11. The displacement $\boldsymbol{u}$ depends on the specific Raman coupling and on the intensity of the incident laser. Creating a standing wave in the optical field can specify a specific phonon wave vector $\mathbf{q}$. This is essentially a transient superlattice that can participate in the Bragg condition. The intensity of x-rays scattered at a scattering vector $\Delta\mathbf{k}$ is given approximately by

$$I(\Delta\mathbf{k}) \propto \{\delta(\Delta\mathbf{k}+\mathbf{G}) - \mathbf{G}\bullet\mathbf{u}[\delta(\Delta\mathbf{k}+\mathbf{q}+\mathbf{G}) - \delta(\Delta\mathbf{k}-\mathbf{q}+\mathbf{G})]\}.$$

The first delta function is the ordinary Bragg condition, and the second and third represent sidebands, which are displaced by a small angle from the main Bragg peak. Coherent control of the lattice motion will therefore result in modulation of the diffracted x-ray beam. Experiments are underway to observe this now at the Advanced Photon Source at Argonne National Laboratory.

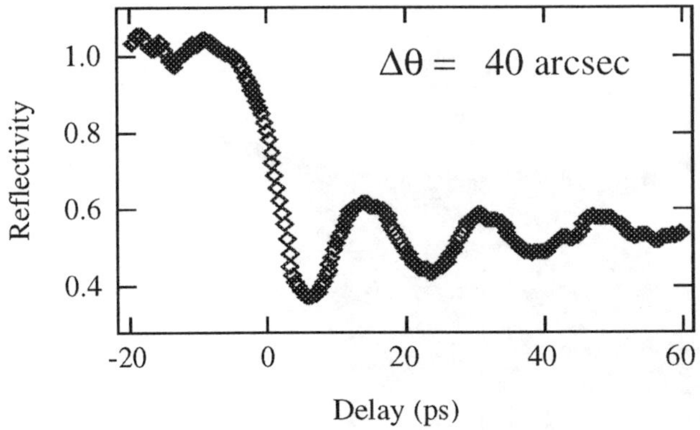

Figure 12. Bragg scattering oscillations from an acoustic phonon mode excited impusively in an InSb Bragg mirror (from Lindenberg et al.[10]).

A simpler coherent control over x-ray scattering was recently studied using *acoustic* phonon modes in a Bragg crystal, in experiments at the Advanced Light Source at Berkeley.[10] A Berkeley-LBNL-LLNL-Michigan collaboration used an x-ray streak camera to observe coherent modulation of the x-ray beam following impulsive excitation of the Bragg diffraction by a laser. Since the impulse produces a broad spectrum of acoustic modes, the Bragg condition also acted as a filter to show the modulation at a single wave vector. A line-out of the streak camera image for one wave vector is shown in Figure 12.

There are many more uses for intense programmable shaped pulses. We are trying to control molecular rotations; we are also trying to create molecules in which rotational and electron wave packets exist simultaneously. In each of these, the quantum dynamics are much richer because of the control capabilities of programmable pulse shapers and learning algorithms.

Acknowledgements: This work has been supported by the National Science Foundation, grant # 9414335. X-ray scattering work is supported in part by the Center for Ultrafast Optical Science, and the Department of Energy Division of Basic Energy Sciences.

Concluding Haiku:

Coherent control
Was teaching molecules. Now
Molecules teach us.

References:

1. A.M. Weiner, D.E.Leaird, J.S. Patel, J.R. Wullert, *Opt. Lett.* **15**, 326 (1990).
2. J. X. Tull, M. A. Dugan, and W. S. Warren, *Adv. Opt. Mag. Resonance* **20**, 1 (1997).
3. D. You, R. R. Jones, D.R. Dykaar, and P.H. Bucksbaum, *Opt. Lett.* **18**, 290 (1993); Weling, A.S., Hu, B.B., Froberg, N.M., and Auston, D.H., *Appl. Phys. Lett.* **64**, 137 (1994).
4. L.K. Grover, *Phys. Rev. Lett.* **79**, 4709 (1997).
5. J.W. Ahn, T.C. Weinact, and P.H. Bucksbaum, *Science* **287**, 463 (2000).
6. Rabitz, H.; Judson, R. *Phys. Rev. Lett.* **68**, 1500 (1992).
7. T.C. Weinacht, J.L. White, P.H. Bucksbaum, *Jour. of Phys. Chem. A* **103**, 10166 (1999).
8. Davis, L. Ed. *Handbook of Genetic Algorithms*, Van Norstrand Reinhold: New York, (1991).
9. P.H. Bucksbaum and R.D. Merlin, *Solid State Comm.* **111**, 535 (1999).
10. I. Kang, A.M. Lindenberg, R.W. Falcone, P.A. Heimann, R.W. Lee, J.S. Wark, and P.H. Bucksbaum, *Phys. Rev. Lett.* **84**, 111 (2000).

QUANTUM CONTROL OF ISOMERIZATION AND ENANTIOMER PREPARATION

Y. Fujimura, L. González[a], K. Hoki, J. Manz[a], Y. Ohtsuki, and H. Umeda
Department of Chemistry, Graduate School of Science, Tohoku University
Sendai, 980-8578 Japan
[a]Institut für Chemie - Physikalische und Theoretische Chemie,
Freie Universität Berlin
Takustrasse 3, D-14195 Berlin, Germany

An application of a feedback control theory to two types of unimolecular reactions, an isomerization of hydrogen cyanide and an enantiomer preparation of phosphinothioic acid is presented. In controlling the isomerization, we treat the quantum feedback control theory in a classical way. An expression for the optimal laser pulse is derived from the classical Hamilton's equations of motion. The amplitudes of the optimized pulse at time t are proportional to the linear momentum of a representative point of the reaction. On the other hand, the linear momentum is constructed by the nuclear wave functions evaluated by solving the time-dependent Schrödinger equation. The kinetic energy of the representative point is the controlling parameter. The results of model calculation of the isomerization HCN → HNC in a two-dimensional model show that the present control method can design pulses for the isomerization with a high quantum yield. For quantum control of the racemic modification of phosphinothioic acid, the reaction coordinate is the torsional motion of the hydrogen atom around the P-S axis of the molecule. By using the quantum feedback control theory, we obtain the reaction product of ~ 100% of a single enantiomer in the low temperature limits starting from the lowest initial state which is a coherent state representing two equivalent enantiomers with opposite chiralities. Mechanisms underlying the quantum control of the enantiomer preparation are discussed.

1 INTRODUCTION

Recently there have been considerable efforts in quantum control of chemical reaction dynamics.[1-6] Various theoretical treatments based on optimal or local control theory and on perturbative or nonperturbative methods have been developed for designing laser pulses to manipulate nuclear wave packets to the desired target under consideration.[7-26] In our previous studies,[18-20, 27-29] we developed a non-perturbative, quantum mechanical feedback control theory based on a local control method. In that treatment, the time-development of reaction dynamics was divided into short time stages in which the system can be considered as following the equation of motion of a time-invariant system. The control theory is free from the laser intensities used and is applicable to reaction dynamics in strong laser field cases as well as weak field cases. The control theory was easily applicable to simple systems consisting of one degree of freedom, but there may be some difficulty in controlling reactions in multidimensional systems.

From the general viewpoint of chemical reaction dynamics, the process of a reaction is described in terms of a representative point that moves from a reactant to a product over a transition region.[30, 31] It is natural to consider a reaction in quantum control in a similar way; i.e., quantum control is viewed as a representative point moving along the reaction coordinate in nonstationary laser fields. That is, quantum control is treated in a classical way. Quite recently, we have developed a classical, local control theory for controlling reaction dynamics.[32] The theory is based on a classical mechanical, feedback control method within a local optimization treatment presented in previous papers.[18, 27, 28] The classical treatment was extended to control wave packet dynamics by replacing the classical momentum by a quantum mechanically averaged momentum on the basis of the Ehrenfest theorem. That is, the controlled field is derived from Hamilton's equations of motion under the constraint of a small amount of laser energy. The dynamics, on the other hand, is evaluated by solving the time-dependent Schrödinger equation. In the first part of this review paper, after an introduction of an optimal feed-back control of a linear time-invariant system, we present an application of the treatment to isomerization of hydrogen cyanide.

In the second part of this paper, we present an application of the local control theory to a selective preparation of enantiomers. Enantiomer preparation is important in the preparation of substances with biochemical activity as well as in other chemical fields associated with absolute asymmetric synthesis.[33-35] Phosphinotioic acid H_2POSH is taken as a model system.[36] The mechanism of the racemic modification of H_2POSH by controlled laser pulses is discussed. Shapiro and Brumer have proposed a preparation of specific enantiomers by using symmetry breaking a specofic bond of the achiral precursor, which prepares specific enantiomers by selective photodissociation of the achiral precursor.[37] They have applied the theory to breaking a bond of HOD. The main difference between their treatment and ours is that our method is used to convert a symmetric coherent achiral state of the precursor(which has the same density as an incoherent racemate at low temperature!) into a nonsymmetric localized state representing a single enantiomer.

2 ISOMERIZATION BY A CLASSICAL CONTROL METHOD

2a Optimal feedback control of a linear time-invariant system

Consider a linear time-invariant (LTI) system whose equation of motion is given as

$$\frac{d}{dt}X(t) = AX(t) + Bu(t) \tag{1}$$

where $X(t)$ is the n-dimensional system variable vector and $u(t)$ is the m-dimensional input vector. A and B are time-independent matrices. To evaluate an optimal input vector, we define a performance index J in form of an integral of squared error (ISE),

$$J = \frac{1}{2} X(t_f)^T \Phi X(t_f) + \frac{1}{2} \int_{t_0}^{t_f} dt [X(t)^T Q(t) X(t) + u(t)^T R(t) u(t)]. \quad (2)$$

Here, the first term expresses the ISE of the system at the final time t_f, and the other terms express the ISE of both the system and the input energy between an initial time t_0 and the final time. Φ and Q are $n \times n$ real matrices, and R is an $m \times m$ real matrix. The control parameter matrices Φ and Q are related to the final status of the system at t_f and the path during $t_0 \sim t_f$, respectively. R is related to the input energy during $t_0 \sim t_f$.

The optimal input vector and system variable vector minimizing a performance index are connected to each other through $P(t)$ as,[18]

$$u(t) = -R^{-1} B^T P(t) X(t) \quad (3)$$

where $P(t)$ is an $n \times n$ symmetric positive definite matrix which satisfies the so-called Ricatti equation subject to the final condition $P(t_f) = \Phi$. $R^{-1} B^T P(t)$ is called a feedback gain matrix.[18]

A molecular system whose dynamics satisfies the time-dependent Schrödinger equation is not a LTI system. To apply the optimal feedback control described above, the equation of motion of a molecular system is approximated by that of an LTI system by dividing the time-development into short time stages.

2b Classical local control theory

Consider an n-dimensional particle motion in a time-dependent external field whose dynamics is expressed by the classical Hamilton's equations of motion,

$$\frac{d}{dt} q_i = \frac{\partial H}{\partial p_i}, \quad \frac{d}{dt} p_i = -\frac{\partial H}{\partial q_i}, \quad (i = 1,2,\dots n). \quad (4)$$

Here, $H = H(\boldsymbol{q}, \boldsymbol{p}, t)$, the total Hamiltonian, is expressed within the dipole approximation as

$$H(\boldsymbol{q},\boldsymbol{p},t) = \frac{\boldsymbol{p}^2}{2M} + V(\boldsymbol{q}) - \mu(q)\cdot E(t) \tag{5}$$

where M, $\boldsymbol{q}$ and $\boldsymbol{p}$ denote the particle mass, position and linear momentum, respectively. $V(\boldsymbol{q})$ is the potential energy, $\mu(q)$ is the dipole moment of the system, and $E(t)$ is the laser field to be optimized. Within a short time approximation, the equations of motion at the i-th propagation step starting from the point ($\boldsymbol{q}_0^{(i)}$, $\boldsymbol{p}_0^{(i)}$, $E_0^{(i)}$) at time t are expressed in terms of the ($\Delta\boldsymbol{q}_0^{(i)}$, $\Delta\boldsymbol{p}_0^{(i)}$, $\Delta E_0^{(i)}$) as

$$\frac{d}{dt}(\Delta\boldsymbol{q}^{(i)}) = \frac{\Delta\boldsymbol{p}^{(i)}}{M} \tag{6a}$$

and

$$\frac{d}{dt}(\Delta\boldsymbol{p}^{(i)}) = -\frac{\partial^2}{\partial\boldsymbol{q}^2}[V(\boldsymbol{q})-\mu(q)\cdot E^{(i)}(t)]\bigg|_{q_0}\Delta\boldsymbol{q}^{(i)} + \frac{\partial\mu}{\partial\boldsymbol{q}}\bigg|_{q_0}\Delta E^{(i)} \tag{6b}$$

where the local controlled field is given by[32]

$$\Delta E^{(i)} = -R^{(i)-1}\frac{\partial\mu}{\partial\boldsymbol{q}}\bigg|_{q_0} f_{pp}^{(i)}\,\Delta\boldsymbol{p}^{(i)}. \tag{7}$$

Here R is a control parameter of the laser field.

The performance index at the i-th short time step $J^{(i)}$ is expressed from Eq.(2) as

$$J^{(i)} = \frac{1}{2}\boldsymbol{p}^{(i)}(t_f^{(i)})^T f_{pp}^{(i)}\,\boldsymbol{p}^{(i)}(t_f^{(i)}) + \int_{t_0^{(i)}}^{t_f^{(i)}} dt\,[\,E^{(i)}(t)^T R^{(i)} E^{(i)}(t)$$
$$+\frac{1}{2}\boldsymbol{p}^{(i)}(t)^T f_{pp}^{(i)}\,\boldsymbol{p}^{(i)}(t)] \tag{8}$$

where we set ($\Delta\boldsymbol{q}_0^{(i)} = 0$, $\Delta\boldsymbol{p}_0^{(i)} = \boldsymbol{p}$, $\Delta E_0^{(i)} = 0$) since ($\boldsymbol{q}_0^{(i)}$, $\boldsymbol{p}_0^{(i)}$, $E_0^{(i)}$) can be chosen arbitrarily. The laser field is expressed as

$$E^{(i)} = -R^{(i)-1}\frac{\partial\mu}{\partial\boldsymbol{q}}\bigg|_0 f_{pp}^{(i)}\,\boldsymbol{p}^{(i)}. \tag{9}$$

Equation (8) shows that the control parameter f_{pp} is a multiplier to the kinetic energy. That is, the control is carried out through changing the kinetic energy of the representative point of the reaction system. The control field minimizes the performance index J. The control field promotes the kinetic motion of the system in the case in which J becomes smaller, and, on the other hand, suppresses the kinetic motion in the case in which J becomes larger. The control field energizes the system if f_{pp} is negative. On the other hand, the control field decreases the kinetic energy if f_{pp} is positive.

So far, we have not considered the distribution of $\boldsymbol{p}$ and that of $\boldsymbol{q}$ at the initial time. We have restricted ourselves to local control of a trajectory of a representative point starting with an initial value of $\boldsymbol{p}$ and that of $\boldsymbol{q}$. We now consider a simple case in which the control field is given by using both classical ensemble averages of the momentum $<\boldsymbol{p}>_{en}$ and coordinate $<\boldsymbol{q}>_{en}$ at each step where $<\ >_{en}$ denotes the ensemble average over the trajectories:

$$E^{(i)} = -R^{(i)-1} \left.\frac{\partial \mu}{\partial \boldsymbol{q}}\right|_{q=<q>^{(i)}_{en}} \left. f_{pp}^{(i)}\right|_{q=<q>^{(i)}_{en},p=<p>^{(i)}} <\boldsymbol{p}^{(i)}>_{en}. \tag{10}$$

The field parameter R is directly related to the magnitudes of the momentum of the system under investigation. Consider a particle in a double-well as the simplest model for a chemical reaction in which the representative point moves along the reaction path over the barrier towards the products. In the case in which the magnitude of the momentum is large enough for the representative point to return from the target well to the initial well, the system keeps oscillating between the initial well and the target well. To avoid such an oscillation, the magnitude of the f_{pp} parameter between the initial and the target well is changed. If we set a large magnitude of the f_{pp} parameter in the target well compared with those in the initial state, we can obtain a control field to stabilize the system in the target well. If the magnitude of the momentum of the system is too small at the top of the barrier, the system stops on the top of the barrier. This is because the potential is almost flat and the amplitude of the control field will be zero from Eq.(10). That is, there is no force to move the system. In this case, the ensemble of trajectories spread into two parts, one to the left and the other to the right along the potential curve. This causes a decrease in the reaction yield.

2c Extension to wave packet dynamics

The fields derived in the preceding section do not always control the quantum system because of spreading and/or bifurcation or other pure quantum behaviors of nuclear wave packets. Therefore, we need to extend the classical local feedback control method derived above to a quantum system whose wave function $|\Psi(t)\rangle$ satisfies

$$i\hbar \frac{\partial}{\partial t}|\Psi(t)\rangle = [\ \hat{H}_0 \ - \mu(q)\cdot E(t)\,]|\Psi(t)\rangle \tag{11}$$

where $\hat{H}_0$ is the system Hamiltonian.

The extension from the classical to quantum control is based on an appropriate selection of the representative points in the phase space of the system. The simplest method is just to use the quantum averaged position $<\boldsymbol{q}>$ and momentum $<\boldsymbol{p}>$ for the nuclear wave packet on the basis of the Ehrenfest theorem. Explicitly, this should be applicable to a system of a harmonic potential. However, it is expected that the simplest method is effective as long as the wave packets are not greatly separated. In this case, the control field for a quantum system at the i-th short time step is expressed as

$$E^{(i)} = -R^{(i)-1} \left.\frac{\partial \mu}{\partial \boldsymbol{q}}\right|_{q=<q>^{(i)}} \left. f_{pp}^{(i)}\right|_{q=<q>^{(i)},p=<p>^{(i)}} <\boldsymbol{p}^{(i)}>. \tag{12}$$

Here, $<>$ denotes the quantum expectation value.

2d Isomerization of HCN to HNC

Consider an isomerization of HCN to HNC by using two perpendicular polarized laser pulses as control fields. HCN serves as a model for demonstration of our control technique.[38, 39] We use a two-dimensional model in which the carbon and nitrogen atom positions are fixed and the hydrogen atom moves on the x-y plane. The potential energy surface and the dipole moment function of the reaction system were calculated by using the MCSCF method with a 6-311 G(d,p) basis set.[40] Figure 1 shows the potential energy surface and the adapted control parameter f_{pp}. In this figure, the contour around H represents the vibrational ground state of HCN. We used the same form as the control parameter f_{pp} for the x-polarized field E_X and the y-polarized field E_Y. For simplicity, we employ

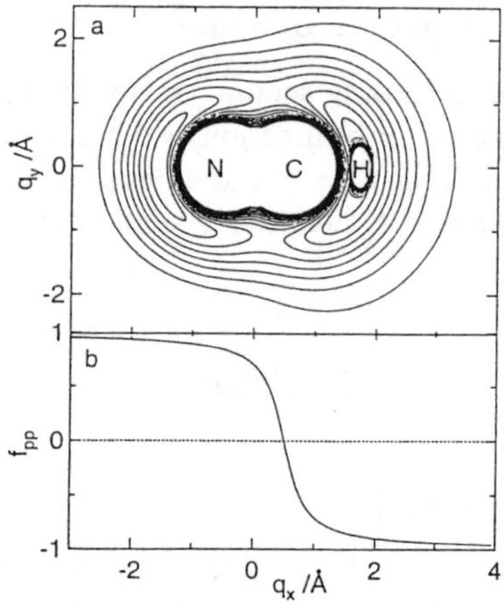

Figure 1. (a) Potential energy surface of HCN, (b) Control parameter f_{pp} as a function of reaction coordinate.

$$f_{pp}(q_x, q_y) = f_{pp}(q_x) = \frac{2}{\pi} \tan^{-1}\{s(q_x - q_b)\}, \tag{13}$$

where s=4Å^{-1} and q_b=0.5Å. The intensity parameters for E_X and E_Y were chosen to be R_X=3030 and R_Y=556 m^{2}s/V^2, respectively.

Figures 2a and 2b show the time development of the wave packet of the isomerization reaction under locally optimized laser pulses. The contours with time t describe the wave packet at that time and the dashed line denotes the trajectory of the quantum averaged position $<\boldsymbol{q}>$ of the wave packet. We can see that the trajectory runs along the reaction path and that the wave packet moves on that trajectory except for the final part of the reaction. This small difference between the trajectory $<\boldsymbol{q}>$ and the center of each wave packet indicates the validity of our simplified treatment. At the end of the control pulses as shown in Fig. 2b, the system was localized at the bottom of HNC with an excess energy of HNC bending vibration. This remaining part is the origin of the differences between the trajectory of $<\boldsymbol{q}>$ and the wave packet at the final part of the control.

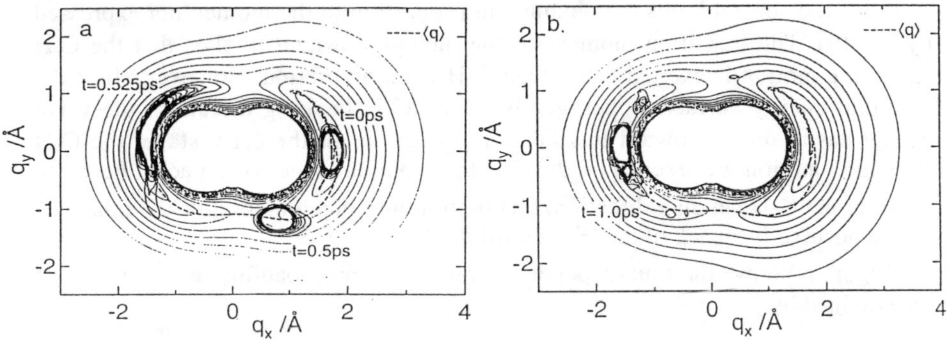

Figure 2. (a) The wave packet dynamics in two linearly polarized control pulses. The dotted line shows the trajectory. (b) The wave packet dynamics after the control pulses.

Figure 3 shows the control fields for this isomerization. The upper figure represents the x-polarized field E_X and the lower one represents the y-polarized field E_Y. In Figs.2 and 3, the laser pulse of E_Y mainly controls the wave packet except for the period in which the system just moves over the reaction barrier. This is

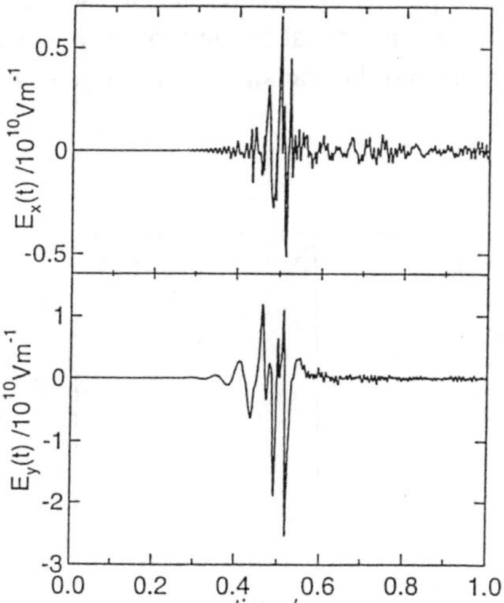

Figure 3. Field amplitudes obtained by a locally optimized control method. The pulses consist of two linearly polarized fields.

because the control fields are obtained in proportion to the momentum expressed by Eq. (9). The transition moment of the quantum system implies that the C-H stretching motion is excited faster than C-H bending motion. However, since the potential energy surface slopes gently in the C-H bending direction, the wave packet easily moves toward this bending direction. In the early stage, the C-H stretching motion was excited by the E_X field, but once the wave packet began to move in the C-H bending direction, this motion was much more quickly enhanced by the positive feedback of the E_Y control field.

Figure 4 shows the time-dependent isomerization probability as a function of time defined by

$$P(t)= \int dx \int dy \ |\Psi(t)|^2. \tag{14}$$

The integrated region is shown by a white rectangle in Fig. 2b. About 90% of the wave packet is localized in that region at the end of the control pulses. There is no oscillatory behavior in isomerization probability after the control pulses come to an end. This indicates that the wave packet remains localized in the product region of HNC.

Figure 5 shows the expectation value of the molecular Hamiltonian along the reaction coordinate. The dotted line represents the dissociation threshold of $HCN \rightarrow H + CN$. The wave packet absorbed energies while oscillating in the initial well. We can see that the energy absorbed by the system is lower than the dissociation energy even around the transition state region. As a consequence, the

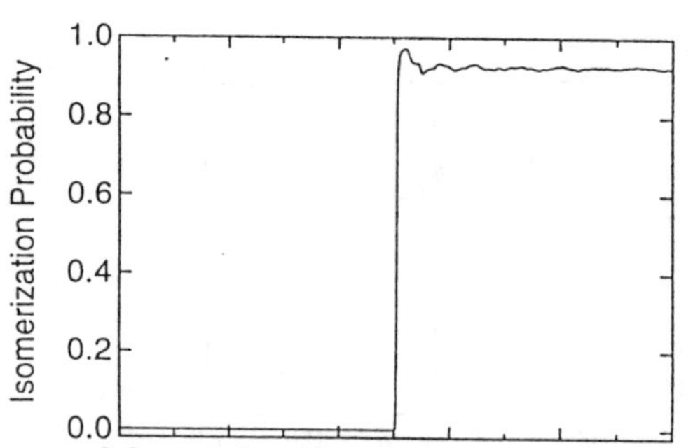

Figure 4. Isomerization probability as a function of time.

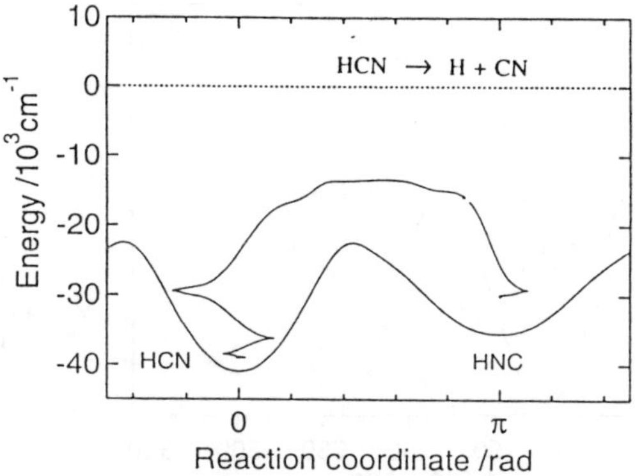

Figure 5. The trajectory of the average energy and the average position along the reaction coordinate. The broken line shows the threshold of dissociation HCN ⟶ H + CN.

dissociation is not induced by the control pulses shown in Fig. 3. Once the wave packet comes to the region in the target well, the system is stabilized to HNC by stimulated emission.

Related to the quantum control of reaction dynamics in a classical way, Schwieters and Rabitz[41] developed a classical optimal control theory based on a global optimization scheme. They applied their theory to a symmetric one-dimensional double-well system. They controlled a quantum system by introducing a cost functional with an explicit quantum-classical-difference reduction. The resulting observables were well matched between classical and quantum mechanics. The controlled wave packet dynamics avoids quantum effects such as tunneling and lowering the de Broglie wavelength. It is considerably useful for keeping quantum-classical correspondence over the whole control period, although the computational effort is quite large. On the other hand, our theory is simplified at this point. The quantum-classical correspondence was kept automatically by the compensation at each short time step to some extent. This is one of the characteristic features of our local control treatment.

3 ENANTIOMER CONTROL

3a A model system

Consider a quantum control of a selective preparation of enantiomers from a coherent initial state, with symmetric density, identical to the density of an incoherent racemate at low temperature. We used a quantum feedback control theory

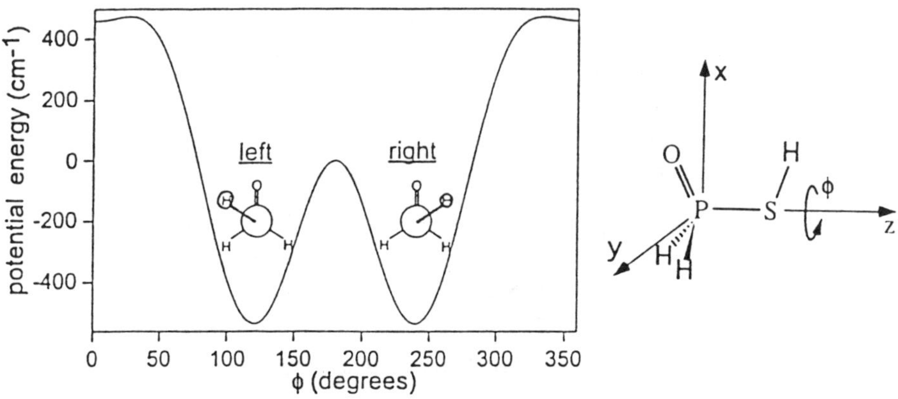

Figure 6. The axes of H_2POSH, and its one-dimensional potential energy evaluated by an *ab initio* MO method.

derived within a locally optimized procedure.[28] We took phophinothioic acid H_2POSH as a model system for the following reasons.[36] First, it should be small enough to evaluate its geometric structure and dipole moments by an *ab initio* MO method while serving as a model for more complicated biological compounds. Secondly, the two enantiomers should be separated by a potential barrier high and broad enough to prevent immediate(sub-ps) tunneling from one to the other enantiomer. Thirdly, the model molecule should have a strong dipole moment variation which makes it possible to use relatively weak laser fields to control its selective preparation of enantiomers.

Figure 6 shows the one-dimensional potential energy as a function of the torsional angle ϕ, that is, a rotation of the H-S bond around the P-S molecular axis. The cartesian coordinates of H_2POSH are also shown in Fig. 6. Here, the other degrees of freedom were frozen at the equilibrium value except for the hydrogens of the phoshino group. The potential was calculated by using the GAUSSIAN94 package of programs at MP2 level of theory [See details in Ref. 36].

Figure 7 shows the x-component of the dipole moment μ_X as a function of ϕ and the y-component μ_Y. These were also calculated using the same level of theory; μ_X is symmetric with respect to the mirror plane of symmetry; μ_Y is antisymmetric with respect to the mirror plane. The z-component of the dipole moment is irrelevant because it does not interact with the laser field.

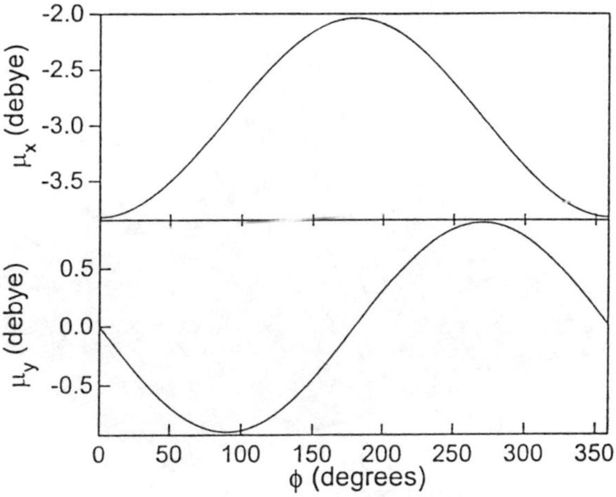

Figure 7. Dipole moments of H_2POSH evaluated by an *ab initio* MO method.

3b Selective preparation of enantiomers

Consider a selective preparation of enantiomers in the electronic ground state in the low temperature limit.[33] The initial state is taken to be the torsional ground state. The corresponding density is identical to an incoherent racemate of two H_2POSH molecules, an L-form and an R-form, as indicated in Fig.6. The tunneling time between the two enantiomers is ~ 630 ps, which was estimated from the energy difference between the lowest and the first vibrational eigenstates. That tunneling time is long enough for monitoring the preparation of the pure enantiomers spectroscopically. Wave functions of the two enantiomers can be expressed to a good approximation as a linear combination of these two eigenstates. We used the second and third vibrationally excited states as the intermediate states. Our control is a kind of pump - dump control scheme through these intermediate states.[42]

Consider the enantiomer preparation of the L-form ,which is carried out within 10 ps. Figure 8 shows the propagation of the wave packets in the presence of laser pulses, which was calculated by a local control method. We can see that the selective preparation of the pure enantiomer with yield of 100% is realized within 10 ps.

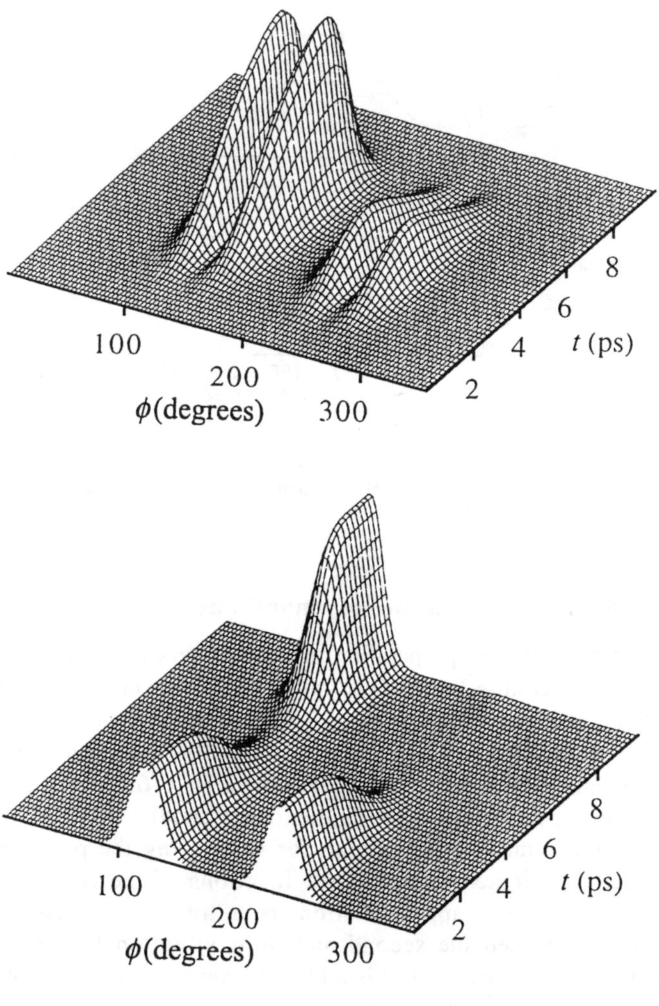

Figure 8. Propagation of wave packets by controlled laser pulses at T = 0 K. The upper figure shows the density of the wave packets in the second and third torsionally excited states. The lower figure shows the density of the wave packets constructed by the lowest and the first excited states.

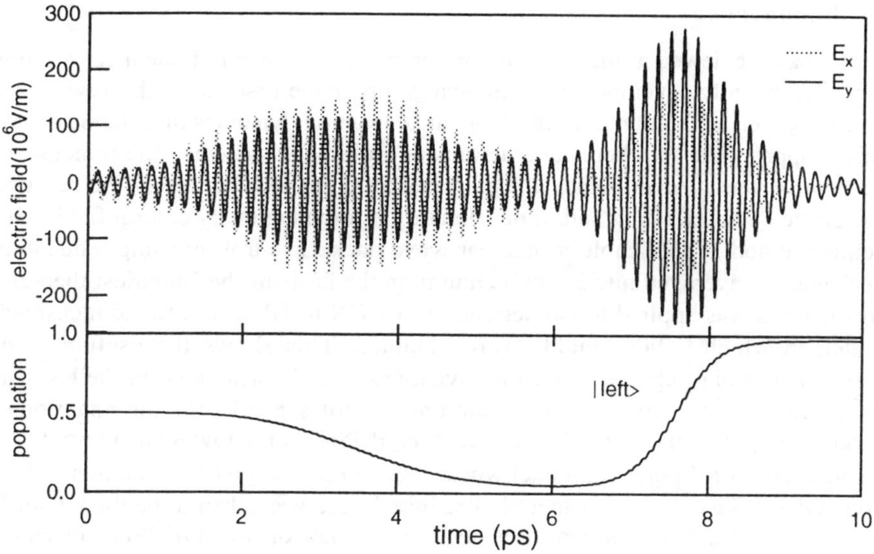

Figure 9. The upper panel shows the locally-designed optimal pulses; the lower panel shows the localization of the wave packets.

The upper panel in Fig. 9 shows the variation of the amplitudes of x and y linearly polarized electric fields, E_X and E_Y , which were obtained by the locally optimized control method. The early part of the sequence of the pulses corresponds to the pumping process, and the latter part to the dumping process. We can see the intensity differences between E_X and E_Y which reflect the strengths of the dipole moments relevant to optical transitions. We can also see that the relative phase difference between two laser fields, E_X and E_Y, is nearly constant and $\frac{\pi}{2}$ for the pumping process. This indicates that the pulses are elliptically polarized ones. On the other hand, for dumpling process, the phase difference is π, which means a linear pulse. The lower panel of Fig. 9 shows the localization of the wave packet as a function of time. The localization is defined as $|<\Phi_L|\Psi(t)>|^2$, where Φ_L is the wave function of the enantiomer of L-form. We can clearly see that almost complete localization is obtained within 10 ps.

So far we have considered the enantiomer selective preparation of L-form. If we want an enantiomer of R-form, we can use the linearly polarized laser fields, E_X and E_Y corresponding to the relative phase of $-\frac{\pi}{2}$ for the pump pulse.

4 Summary

We have reviewed a theoretical work on quantum control of chemical reaction dynamics recently developed in the Sendai group. In the first part of this review, we have presented a quantum control of chemical reaction dynamics in a classical way. This is appropriate for chemical reactions of multimode systems. The principle of the control was to manipulate a representative point moving along the reaction coordinate by accelerating and/or decelerating its momentum by control fields. Our simple method is applicable to nuclear wave packet control by using a quantum mechanically averaged nuclear momentum on the basis of the Ehrenfest theorem. This method was applied to isomerization of HCN to HNC in a two-dimensional model, in which ~90% yields were obtained. This shows the usefulness of quantum control of chemical reaction dynamics in a classical way. In the last part of this review, we have given a quantum control scenario for the enantiomer selective preparation of a model molecule, H_2POSH, in the low temperature limit. A one-dimensional potential model whose reaction path is given by rotation of S-H around the P-S axis was considered. The initial state was taken to be the torsional ground state, which has the same density as a racemate of two H_2POSH molecules, with both an L-form and an R-form. Pump and dump laser pulses with nearly constant relative phases control the selective preparation of exclusively the L- or R-forms of the enantiomers. Additional effects, such as multimode effects and temperature effects, should be taken into account in the near future.

ACKNOWLEDGMENTS

This work was partly supported by a German - Japan International Joint research project supported by Japan Society for the Promotion of Science and by DFG (Project Ma 515/17-1)., the Monbusho International Scientific Research Program (No. 10044054), by Grants-in-Aid for Scientific Research (No. 10640480), by Grant-in-Aid for Scientific Research on Priority Area(No. 11166205), by Development of High-Density Optical Pulse Generation and Advanced Material Control Techniques, and by Fonds der Chemischen Industrie.

REFERENCES

1. P. Brumer and M. Shapiro, Chem. Phys. Lett. **126**, 541 (1986).
2. S. M. Park, S. -P. Lu, and R. J. Gordon, J. Chem. Phys. **94**, 8622 (1991).
3. D. Neuhauser and H. Rabitz, Acc. Chem. Res. **26**, 496 (1993).
4. B. Kohler, J. L. Krause, F. Raksi, K. R. Wilson, V.V. Yakovlev, R. M. Whitnell and Y. J. Yan, Acc. Chem. Res. **28**, 133 (1995).
5. H. Kawashima, M. M. Wefers and K. A. Nelson, Annu. Rev. Phys. Chem. **46**, 627 (1995).

6. L-C.Zhu,V. D. Kleiman, X. Li, S, -P. Lu, K. Trentelman, and R. J. Gordon, Science **270**, 77 (1995).
7. D. J. Tannor and S. A. Rice, J. Chem. Phys. **83**, 5013 (1985).
8. S. Shi, A. Woody and H. Rabitz, J. Chem. Phys. **88**, 6870 (1988).
9. R. Kosloff, S. A. Rice, P. Gaspard, S. Tersigni and D. J. Tannor, Chem. Phys. **139**, 201 (1989).
10. W. Jakubetz, J. Manz and H.-J. Schreier, Chem. Phys. Lett. **165**, 100 (1990).
11. S. Shi and H. Rabitz, J. Chem. Phys. **92**, 364 (1990).
12. P. Gross, D. Neuhauser and H. Rabitz, J. Chem. Phys. **98**, 4557 (1993).
13. J. L. Krause, R. M. Whitnell, K. R . Wilson, Y. Yan and S. Mukamel, J. Chem. Phys. **99**, 6562 (1993).
14. Y. Yan, R. E. Gillian, R. M. Whitnell, K. R . Wilson, and S. Mukamel, J. Phys. Chem. **97**, 2320 (1993).
15. L. Shen, S. Shi and H. Rabitz, J. Phys. Chem. **97**, 8874 (1993).
16. L. Shen and H. Rabitz, J. Chem. Phys. **100**, 4811 (1994).
17. H. Shen, J.-P. Dussault and A. D. Bandrauk, Chem. Phys. Lett. **221**, 498 (1994).
18. M. Sugawara and Y. Fujimura, J. Chem. Phys. **100**, 5646 (1994).
19. M. Sugawara and Y. Fujimura, J. Chem. Phys. **101**, 6586 (1994).
20. M. Sugawara and Y. Fujimura, Chem. Phys. **196**, 113 (1995).
21. V. Dubov and H. Rabitz, Chem. Phys. Lett. **235**, 309 (1995).
22. J. Botina, H. Rabitz, and N. Rahman, J. Chem. Phys. **102**, 226 (1995).
23. J. Botina, H. Rabitz, and N. Rahman, Phys. Rev. A. **51**, 923 (1995).
24. V. Dubov and H. Rabitz, J. Chem. Phys. **103**, 8412 (1995).
25. J. Botina, H. Rabitz and N. Rahman, J. Chem. Phys. **104**, 4031 (1996).
26. H. Tang, R. Kosloff and S. A. Rice, J. Chem. Phys. **104**, 5457 (1996).
27. Y. Watanabe, H. Umeda, Y. Ohtsuki, H. Kono and Y. Fujimura, Chem. Phys. **217**, 317 (1997).
28. Y. Ohtsuki, H. Kono and Y. Fujimura, J. Chem. Phys. **109**, 9318 (1998).
29. Y.Fujimura, *Structure and Dynamics of Electronic Excited States* (Springer-Verlag, Berlin, 1999) p. 214.
30. H. Eyring, S. H. Lin and S. M. Lin, *Basic Chemical Kinetics* (John Wiley & Sons, New York, 1980).
31. W. L. Hase Editor, *Advances in Classical Trajectory Methods* Vol.1 (JAI Press Inc. Greenwich, 1992).
32. H. Umeda and Y. Fujimura, J. Chem. Phys. (in press).
33. M. Avalos, R. Babiano, P. Cintas, J. Jimenez, J. C. Palacios, and L. D. Barron, Chem. Revs. **98**, 2391 (1998).
34. J. A. Cina and R. A. Harris, J. Chem. Phys. **100**, 2531 (1994).
35. J. Shao and P. Hänggi, J. Chem. Phys. **107**, 9935 (1997).

36. Y. Fujimura, L. González, K. Hoki, J. Manz and Y. Ohtsuki, Chem. Phys. Lett. **306**, 1 (1999); ibid, **310**, 578 (1999).
37. M. Shapiro and P. Brumer, J. Chem. Phys. **95**, 8658 (1991).
38. C. M. Dion, S. Chelkowski, A. D. Bandrauk, H. Umeda, Y. Fujimura, J. Chem. Phys. **105**, 9083 (1996).
39. W. Jakubetz and B. L. Lan, Chem. Phys. **217**, 375 (1997).
40. H. Umeda, M. Sugawara, Y. Fujimura and S. Koseki, Chem. Phys. Lett. **229**, 233 (1994).
41. C.D. Schwieters and H. Rabitz, Phys. Rev. A. **48**, 2549 (1993).
42. Y. Ohtsuki, Y. Yahata, H. Kono, and Y. Fujimura, Chem. Phys. Lett. **287**, 627 (1998).

NONEXPONENTIAL DEPHASING AND CONTROL OF MOLECULAR QUANTUM SYSTEMS

MARTIN GRUEBELE AND VANCE WONG
Departments of Chemistry and Biophysics, and Beckman Institute for Advanced Science and Technology,
Urbana IL 61801, USA
E-mail: gruebele@scs.uiuc.edu

A variety of systems display power-law dephasing without a specific time scale at long times. Examples range from the classic spin-boson model at low temperature to highly laser-excited molecules undergoing vibrational dephasing. The specific mechanisms for this phenomenon as applied to molecular vibrations will be discussed, as well as general constraints that must be satified by the Hamiltonian in order to support slow decoherence. Experimental examples from the dephasing of the molecule $SCCl_2$ will be considered in detail. The data and calculations show anisotropic quantum diffusion of the vibrational wave packet in state space. The slow diffusion can be harnessed to control the vibrational motions of the molecule, leading to a 'static' coherent control scheme which achieves selective reactivity by combining coherent stabilization of vibrational wave packets with Franck-Condon control. This type of control scheme leads to a new criterion for degeneracy of interfering control paths: rather than $\Delta E=0$, which is usually required to achieve product control in the asymptotic time limit, ΔE in 'static' coherent control is inversely proportional to the dissociation time scale. To address the experimental data, analytical calculations use perturbation expansions on the quantum state space lattice; time-dependent simulations are carried out with quantum symplectic propagators. The localization conditions and hierarchically scaled anharmonic couplings required for slow decoherence exist for most realistic coupled molecular, spin, or environmental systems: power law decoherence on intermediate time scales could turn out to be a very general phenomenon.

1 Introduction

The dephasing and relaxation dynamics of atomic and molecular systems are important in a variety of applications, such as coherent control, NMR spectroscopy, reaction kinetics, and quantum computing. The simplest treatments of dephasing and relaxation generally yield Lorentzian lineshapes and an exponential decay of population or phase coherence. Examples range from the quantum-derivation of the Fokker-Planck equation to the Redfield treatment for spin relaxation in condensed phases, and to the application of the Golden Rule in molecular vibrational dephasing.

A number of approximations have to be invoked in order to arrive at these exponential decay laws. For example, a rapid and extensive cancellation of off-diagonal coherences is required for the Fokker-Planck equation, and the Golden Rule assumes that there is no correlation between the energy gap and coupling strength between bright and dark states. While these approximations are usually reasonable in the short time (classical) limit, their validity is less obvious for dynamics at long times, even in large systems where a lot of averaging can be expected to occur.

In molecules, vibrational dephasing from an initially prepared state is characterized by the survival probability $P(t) = |\langle\Psi(t)|\Psi(0)\rangle|^2$. Analytical calculations, detailed quantum dynamics simulations, and carefully constructed experiments have shown that P(t) can switch to a power-law dependence at intermediate times.[1-7] Recent work on backbone intramolecular vibrational energy redistribution (backbone IVR) reveals that even large organic molecules, excited to chemically interesting energies, switch from exponential to much slower power-law energy redistribution, after the initial state has explored only a few hundred state space cells, but long before the vibrations equilibrate. (Each state space cell occupies a volume h^N, where h is Planck's constant and N is the number of vibrational modes).[8-10]

Asymptotically slow IVR could be of great utility in the coherent control of molecular reactions.[11] Standard models of unimolecular dissociation such as the RRKM model predict that the *averaged* dissociation rate on the energy shell is independent of state preparation, as long as vibrational energy redistribution occurs on a time scale much faster than chemical reaction. More sophisticated analyses have shown that RRKM rates are a good approximation even when the exploration of state space is incomplete, as long as it is sufficiently representative, and have examined the conditions under which deviations may or may not occur.[12, 13] RRKM and similar theories have enjoyed great success in fitting average reaction rates on the energy shell, so molecules must at least approximately vibrationally randomize.

Approximately is not the same as completely, and the actual extent of randomization on the time scale of reaction is critical. Consider a hydrocarbon molecule with about 50 vibrational modes. At an energy corresponding to 600 THz, a femtosecond laser-prepared stretching overtone might have an IVR-induced bandwidth of 300 GHz, which contains over 10^6 eigenstates. In the statistical limit, the survival probability of the initially prepared overtone is therefore about 10^{-6}. Experiments can appear statistical even if not all 10^6 states are populated instantly. A $100h^N$ volume ($P(t) = 0.01$) may be sufficient to mimic statistical behavior fairly accurately at short times.[13]

From the point of view of coherent control,[14-19] there is a large difference between the two cases, reaching $P(t) = 10^{-6}$ 'instantly', or reaching only $P(t) = 0.01$ 'instantly' and $P(t) = 10^{-6}$ on a time scale longer than the reaction time. Each state in the molecular quantum state space has a simple structure and requires one phase and one amplitude in order to be controlled by an external electromagnetic field.[2, 11] In the first case, $2 \cdot 10^6$ phases and amplitudes need to be controlled to affect reactivity, in the second case only 200. The latter is easily within the reach of current technology.[17, 18] If the relaxation dynamics switch from an exponential to a power law on the time scale of chemical reactions, 'freezing' of an IVR wave packet would control reactivity (due to nonstatistical Franck-Condon overlaps of such a wave packet with product states).

Such a scheme will succeed even better well above the threshold of unimolecular reactions because reaction rates increase rapidly compared to IVR rates. As a result, less 'freezing' of the IVR wavepacket is required to catch up with the reaction rate. Of course, power-law IVR can only delay the onset of the RRKM regime, not prevent it: eventually the statistical baseline will be reached for most initial state preparations at high vibrational energy. However, if the time for completely statistical behavior is delayed by even 1-2 orders of magnitude, as observed, this could open up a variety of reactions to coherent control.

The recent demonstrations by experiment, theory, and simulation of power-law IVR dynamics show that while most IVR wave packets at high energy eventually cover the state space statistically, they reach this limit more slowly than predicted by Golden Rule theories.[2, 20] The IVR rate also increases much more slowly with energy because it depends on a local, not a total, density of states.[21] Most importantly, the nature of the vibrational Hamiltonian ensures that this behavior holds for large molecules also: the decays are very slow after an initial exponential 'deflationary' phase.[22, 23] Physically, this can be explained by the small fractal dimension of the IVR wave packet as it expands in the quantum state space. The dimension is small because even at high energy, vibrational resonances in large molecules are localized, allowing energy flow only in a few directions from each point in state space.[6] Vibrational resonances are localized because electron-pair bonding is largely a local property.

Here, we review some recent experimental data on energy flow, and briefly consider the models and quantum simulations which explain this data. In addition, we discuss new results on IVR lineshapes, matrix element distributions required of 'Golden-Rule-like' models to account for the experimental observations, and the minimal properties that have to be satisfied by local random matrix models to exhibit slow IVR at long times. The last topic is examined both with a molecular example and with the spin-boson model, which suggests extensions of slow long-time dynamics to the condensed phase in certain regimes. It turns out that the minimal requirements for a Hamiltonian matrix to produce retarded long-time dynamics are very simple and very general, and possibly satisfied by many real physical systems in addition to vibrationally excited molecules.

2 Power Law Dephasing in Molecules

In isolated gas phase molecules, energy redistribution is microcanonical, and IVR is therefore a dephasing process. Nonetheless, if we view the initially excited bond as the 'system' and the rest of the molecule as the 'bath,' IVR can also be described as a population relaxation process. Either way, calculations indicate that the system-bath Hamiltonian must be transformable to the following form to exhibit anomalously slow dephasing:[2]

1. Off-diagonal anharmonic couplings must decrease exponentially with the order of the coupling, and

2. the anharmonic coupling structure must be local in some representation if the system has many degrees of freedom.

Let us consider why these requirements are necessary for nonexponential dynamics, and when they may be fulfilled by a system-bath. Considering #1, if off-diagonal anharmonic couplings decrease too rapidly, the Hamiltonian becomes bilinear and the possibility of nonexponential processes is greatly restricted. If the couplings decrease too slowly, condition #2 cannot be satisfied, and fully chaotic behavior results. Too rapid or too slow a scaling therefore leads either to exponential dynamics or to Wigner coherence holes in a system with many degrees of freedom. Condition #2 is necessary for systems of many degrees of freedom: if the couplings are not local, probability density can hop anywhere in state space, the decay occurs within the full dimensionality of state space, leading to an exponential decay in many dimensions. (See $\delta \to \infty$ limit of eq. (1) on next page.)

The molecular vibrational Hamiltonian satisfies conditions #1 and #2. Although many detailed reasons can be given, in the end #1 is satisfied because electrons are much lighter than nuclei, which leads to the appropriate anharmonic scaling of potential constants $V^{(n)}$ of order n, $V^{(n)} \sim \omega^{n/2}$. ($\omega$ is a molecular vibrational frequency, and n is the overall difference in quantum numbers between two states.)[22] #2 is satisfied because the Born-Oppenheimer approximation leads to localized chemical bonds, which allows for the existence of a local representation for vibrational couplings. Vibrations localized in very different parts of large, nonsymmetrical molecules are not strongly coupled.

The latter reasoning applies to direct couplings, but what about indirect chains of couplings? If these chains are much more important than direct couplings, there could still be rapid transport of energy from one end of the molecule to another. If the signs of all coupling matrix elements were the same, this would indeed be the case. In reality, vibrational couplings have a ± 1 sign (they need not be complex). The many coupling chains connecting two states mostly cancel because of random phase, and the contribution of chains grows only as $k^{1/2}$ with the number k of chains. Numerical simulations have shown that for the long time IVR dynamics, chains and direct couplings are at best equally important; direct couplings usually dominate at long times.[23]

Macroscopic liquid and solid systems may also exhibit nonexponential behavior under certain circumstances, but the likelihood is much smaller. Solids are generally in the harmonic (phonon) limit, even when substantial energy is deposited. Liquids redistribute energy by long-range diffusion, which violates #2. As a result, the liquid system-bath Hamiltonian can be reduced to bilinear couplings to an excellent approximation; linear response theory is valid not because there are no anharmonic couplings in liquids (they are very strong), but because they are not localized. The onset of nonexponential scaling in the condensed phase may be too late to be practically useful, but this question remains to be investigated quantitatively.

A simple equation has been derived for the intermediate time power-law with which dephasing occurs:[6]

$$P(t) \approx (1-\sigma)(1-\frac{2t}{\delta\tau})^{\delta/2} + \sigma \tag{1}$$

where

$$\delta \approx \left.\frac{\ln \Sigma \mathcal{L}_{i0}}{\ln[n]-\ln[n-2]}\right|_0 , \quad \mathcal{L}_{i0} = \left[1+\left(\Delta E_{0i} / V_{i0}\right)^2\right]^{-1/2} . \tag{2}$$

In these equations, $P(t)$ is the survival probability of the initially prepared state (square of the time-autocorrelation function), σ the long time average survival probabilty $\langle P(t)\rangle$, τ the lifetime in the exponential limit $\delta\to\infty$. The logarithmic derivative in eq. (2) gives the dimension of the manifold of couplings of state "0" to nearby states "i", where 'nearby' corresponds to a quantum number difference of n between the state "0" and the states "i", and couplings are summed over a shell of width $\Delta n=2$ in state space. The coupling weights $\mathcal{L}_{i0}$ themselves are given by perturbation theory in terms of the coupling V_{i0} and the energy gap ΔE_{i0} between the state "0" and the set of states "i" in any local coupling representation of the Hamiltonian. $\delta<\mathcal{N}$ always, where $\mathcal{N}$ is the dimension of the vibrational state space, i.e. the number of vibrational modes. A large molecules could have a large δ: If $\delta \approx \mathcal{N}$, the decay in eq. (1) would be nearly exponential. In practice, δ is usually no greater than 2-5 even in large molecules because only a few resonances can transport energy away from the initial state due to localization. The power law then becomes evident.

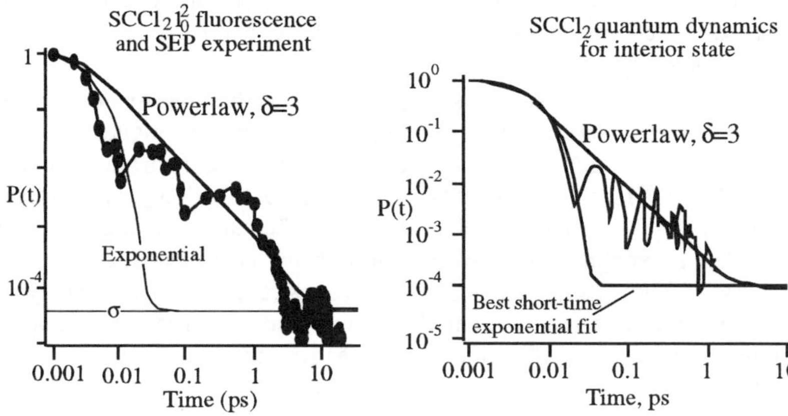

Figure 1. Experimentally measured decay of the Franck-Condon envelope of $SCCl_2$ excited from the 1^2 B vibronic level. The average behavior at short times can be approximated by an exponential decay, at long times better by eq. (1). A quantum dynamics calculation for an 'interior state' (with excitation in all 6 modes) shows a power law decay with a similar power law.

Fig. 1 illustrates an experimentally measured and a calculated IVR decay for the tetratomic molecule $SCCl_2$.[8] The calculation was carried out by standard quantum propagator methods[24] in 6-D using a spectroscopic Hamiltonian.[8] The observed power of ≈ 3 is only half the possible maximum, and this situation does not change for experiments or calculations of larger molecules.[10]

To use this behavior for control, the IVR wavepacket must be frozen in its initial shape by a laser pulse. This shape, unlike the dephased and highly complex shape of molecular eigenstates, can be very simple and therefore has very different Franck-Condon overlaps with different continuum states: control is achieved by freezing the initial wavepacket, rather than by moving it around, and by waiting for Franck-Condon overlaps to produce specific products.[11] This circumvents statistical behavior, but only for as long as the wavepacket can be frozen. Fig. 2 shows a calculation where an $SCCl_2$ vibrational wavepacket decays in about 100 fs, followed by some quantum beats. The calculation is repeated with the wave packet frozen by a shaped femtosecond pulse for up to 5-10 ps. This is made possible by the slow power-law dephasing at > 0.1 ps. If such freezing can be coupled with a unimolecular reaction having a sufficiently fast time scale τ_{rx}, 'static-coherent' control results. Because Franck-Condon effects take over on a time scale τ_{rx}, the final coherence can be created over an energy range $\Delta E \sim h/\tau_{rx}$. The multiple degenerate paths required by coherent control at asymptotic times (direct coherent product control) are not necessary.

Franck-Condon control has also been observed for bimolecular reactions.[25, 26] In the future the idea of wave packet freezing may be applied to bimolecular reactions also, at least in van der Waals clusters, where a controlled initiation of the reaction with short τ_{rx} is possible using femtosecond pulses.

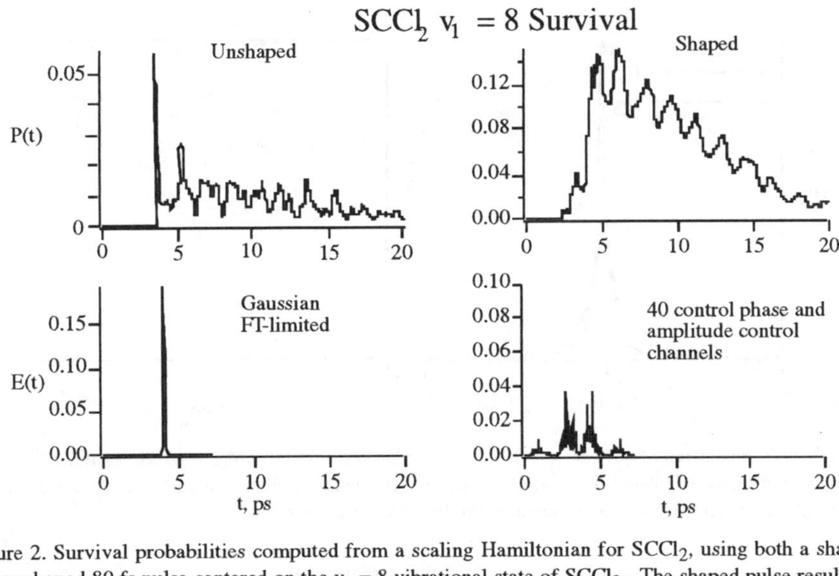

Figure 2. Survival probabilities computed from a scaling Hamiltonian for $SCCl_2$, using both a shaped and unshaped 80 fs pulse centered on the $v_1 = 8$ vibrational state of $SCCl_2$. The shaped pulse results in a wavepacket frozen for several picoseconds. Only 40 amplitude and phase control channels in a simulated liquid crystal pulse shaper are necessary to achieve the degree of control shown here.

3 Lineshapes and Other Examples

So far the dephasing has been discussed in the time domain. However, the power law behavior will also manifest itself in the frequency domain in terms of deviations from Lorentzian lineshapes. A convenient lineshape for discussion is the Fourier-transform of the real amplitude corresponding to the probability decay in eq. 1, named $L_{\delta/2}$. In the limit where the dimension coefficient δ goes to infinity, the lineshape approaches a Lorentzian as shown in fig. 3. The deviation from a Lorentzian can be large for small values of δ. In practice, as shown in Fig. 1, the early-time decay may be better represented by an exponential, in which case the reduction of the wings of the dephasing lineshape is not as obvious as in the example in fig. 3. If the power law scaling starts at $P(t) \approx 0.01$ (typical case in the molecular examples we have studied), deviations from a Lorentzian will amount to lesss than 1%, which requires vibrational lineshape measurements with excellent signal-to-noise ratio and very smooth vibrational envelopes. So far, only direct lineshape measurements in benzene have shown deviations from Lorentzian behavior;[27] the remaining data has come from time-resolved experiments.

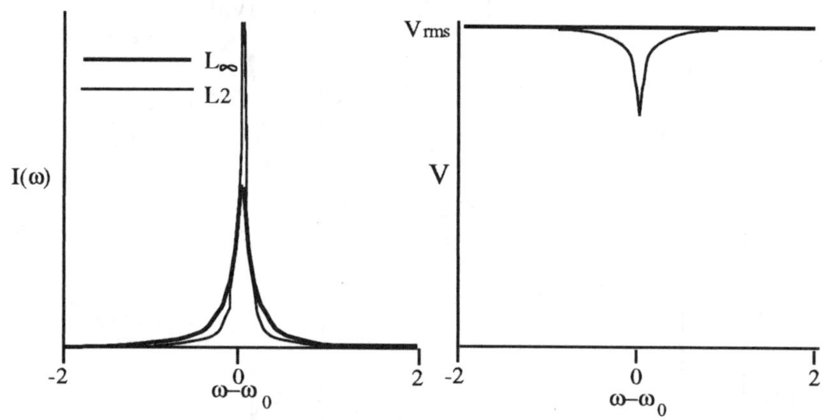

Figure 3. Comparison of power law line shapes for $\delta/2 = 2$ and $\delta/2 = \infty$ (Lorentzian). The corresponding matrix elements of a state "0" carrying oscillator strength to a dark manifold are shown on the right; they decrease in magnitude near the linecenter for $\delta < \infty$.

An interesting question is the following: What type of couplings from the initially excited state "0" dephasing to a 'dark' manifold of states "i" are needed to yield a power-law lineshape? Fortunately, this question can be answered rigorously. When applying the Golden Rule to obtain a Lorentzian lineshape, two criteria are usually applied. The 'dark' manifold is assumed to be uncoupled. Furthermore, the couplings from "0" to "i" are assumed to follow a random distribution. The latter implies that there should be no correlations of matrix element size with energy gap ΔE_{i0}. The Lawrance-Knight-Lehmann (LKL) algorithm allows a direct inversion from an N line spectrum (N-1 relative intensities and N-1 relative energies) to the Golden Rule representation (N-1 coupling matrix elements from the state "0" at $\Delta E = 0$ to states "i", and N-1 ΔE_{i0} energy gaps from "i" to "0"; see fig. 4).[28, 29] The result of such an inversion for the L_2 and L_∞ lineshapes is also shown in fig. 3. The ideal power law lineshape leads to a dip in the magnitude of coupling matrix elements for bath states near the bright state. Such downward fluctuations are a consequence of a local coupling structure, which has only a few states in resonance with state "0." Random couplings produce more uniform matrix element distributions. In practice, coupling fluctuations are of course not limited to a single dip near "0", but fluctuations further from state "0" will have a lesser effect on the line shape (see below).

This leads to the next question: What kind of properties must a Hamiltonian satisfy in order to be reducible to a 'Golden-Rule' format with a matrix element hole of the type shown in fig. 3? We have investigated a variety of coupling models, which have led us propose the anharmonic scaling and localization rules #1 and #2 in section II.[6, 22, 23] The simplest kind of model to satisfy these rules is a local random matrix model. A global random matrix ensemble such as the gaussian orthogonal ensemble (GOE) will yield coherence holes, not power laws. Such holes have been found in calculations only when the molecular coupling strength is increased beyond the normal average, at least for organic polyatomic molecules. Harmonic uncoupled ensembles will yield exponential decays.

We now turn to a simple HLRM (hierarchical local random matrix) model which retains some of the scaling and localization behavior found in real molecules, and examine the lineshapes obtained from such a model.[30] In the absence of couplings, the vibrational quanta in a molecule act like a Bose gas, i.e. like photons in a black-body cavity. At equilibrium, they are distributed according to the Bose-Einstein distribution. From this distribution of quanta one can calculate the probability distribution of n for pairs of states. The distribution of n yields the distribution of coupling constants via $V^{(n)} \sim \omega^{n/2}$, where ω is now a mean frequency of the vibrational modes. This provides the proper scaling #1. To introduce the minimal amount of localization, we impose a 'vibrational triangle rule' on the three matrix elements connecting triplets of states:

$$|n_{12} - n_{23}| \leq n_{13} \leq |n_{12} + n_{23}| \qquad (3)$$

This ensures that wavefunctions with similar difference in the numbers of nodes are connected by similar-sized matrix elements. In other words, if two states are strongly coupled, and one of them is coupled only weakly to a third state, the other of the pair will also be coupled only weakly to the third state. In practice, numerical calculations on *ab-initio* surfaces show this to be correct for 90% of matrix elements. Unlike the angular momentum triangle rule, the triangle rule for vibrations is only a propensity rule, albeit a good one. This model is called the Bose-statistics-triangle-rule model (BSTR).[30] One thing remains to be specified in the model: the diagonal state energies are chosen from a Poisson distribution, as expected for initially uncoupled oscillators. Fig. 4 below highlights a state in the energy, state space, and Bose-gas representation to illustrate the concepts.

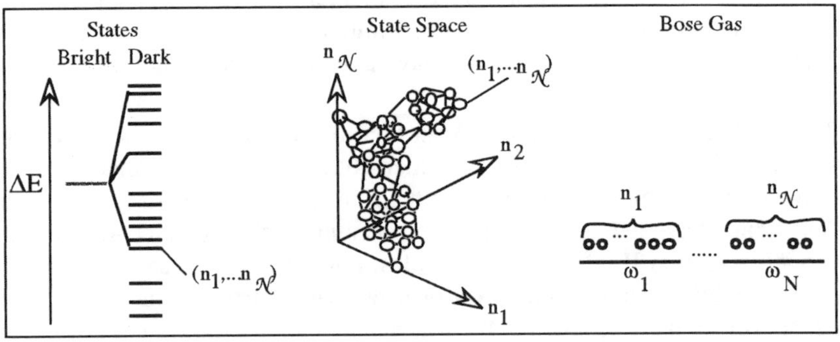

Figure 4. Molecular vibrational states in the 'Golden-Rule' energy representation, in state space, and in the Bose Gas representation. In state space, the coordinates are zero-order vibrational quantum numbers (classically: actions); in the Bose Gas representation, the quanta or vibrons which fill each molecular mode with different numbers. A particular state with quantum numbers n_i has been singled out for illustration.

To simulate the spectra and decays resulting from this model, we have used the shifted-update rotation (SUR) symplectic propagator[24] to compute the survival dynamics of an initial 'interior' state of a 10-D system with a frequency distribution similar to the CDBrClF molecule. An 'interior' state in state space is one which lies in the interior of the coordinate diagram in fig. 4 (middle panel).[7] It has nonzero excitation in all the modes, representative of the majority of states in state space at high energies. Fig. 5 on the next page shows the resulting decay and spectrum, as well as the LKL matrix elements in the 'Golden Rule' basis.

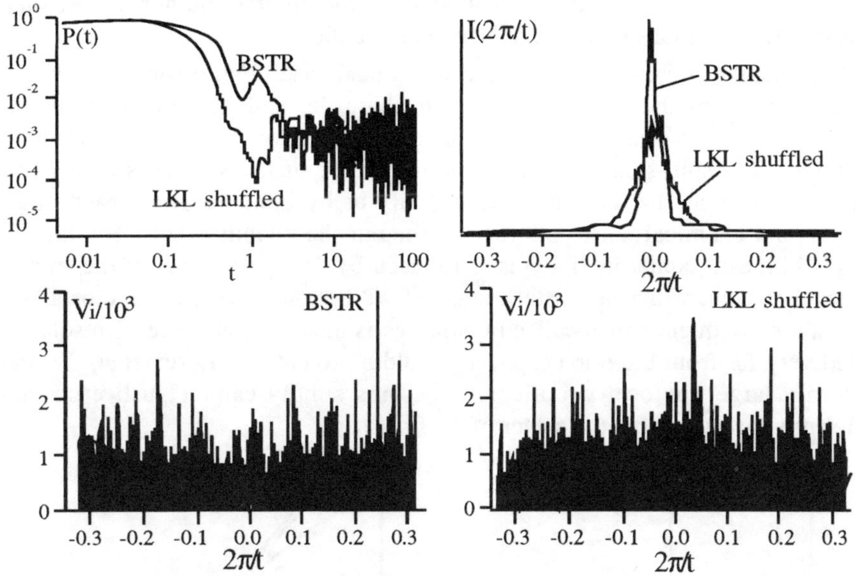

Figure 5. Survival probability, lineshape, and matrix element distribution of the initially excited state to all other states for a 10-D local random matrix calculation. Upper left: Survival probability of BSTR calculation and same calculation but with shuffled coupling matrix elements. Upper right: Lineshape for the BSTR and shuffled calculation; the latter is broader and more Lorentzian. Bottom: Matrix element distributions for the original and uncorrelated calculation; the original distribution shows stronger energy-correlated fluctuations than the randomized distribution, leading to the power law and line narrowing.

The HLRM model, like experiments and more accurate simulations, shows a nonexponential tail which decays to σ more slowly than an exponential. This tail can be made to disappear in several ways: if the scaling is eliminated by making all matrix elements equally large, a Wigner coherence hole appears; if the matrix elements are shuffled at random, the decay becomes exponential. Fig. 5 also shows that there are correlations among - and energy-dependent fluctuations in - the size of the LKL matrix elements. If these are shuffled to remove the correlation (bottom right), the decay also becomes exponential. Thus, energy dependent correlations have to be taken into account in any successful Golden Rule-like model. The use of randomly distributed couplings is physically unjustified.

The BSTR model is the simplest model we were able to find, which violates exponential decay dynamics. The only localization in the BSTR matrix results from the triangle rule. In this model, mixed matrix elements from distant localized vibrations are still as large as local anharmonic matrix elements. Realistic molecular Hamiltonians will tend to be even more local.

Localization occurs not only in the vibrational case. Fig. 6 shows the 1-D ^{1}H NMR spectrum of the 78 residue protein ubiquitin, and its computed survival probability. P(t) is again better represented by a power law up to the limit of resolution. The spin-spin coupling Hamiltonian for such a system is highly local because only spins that are adjacent spatially or by bonding affect each other's splitting and chemical shift patterns. Although the chemical shift Hamiltonian $J\mathbf{s}_1 \cdot \mathbf{s}_2$ is bilinear, 'anharmonicity' is introduced by the level cutoff of the spin 1/2 two level system, which is in stark contrast with an ∞-level harmonic oscillator. The main consequence of localized couplings is again the absence of resonances shifted very far from the line center, as would be found in a Lorentzian. Without arbitrarily large nonlocal couplings, two spins simply cannot push each other sufficiently far away from the center of the line shape.

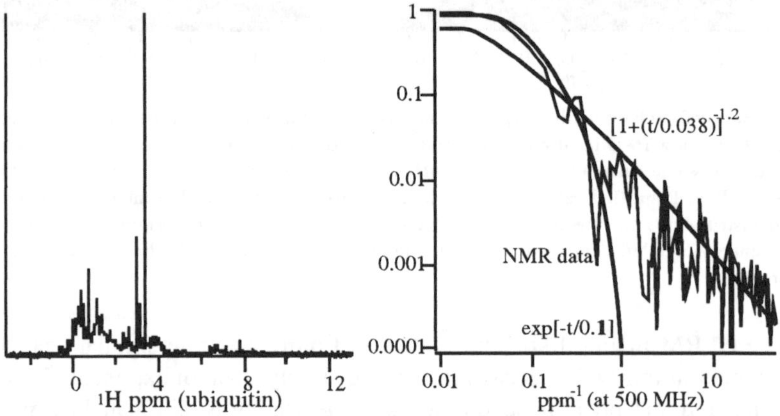

Figure 6. 1-D NMR spectrum of the protein ubiquitin and the corresponding survival probability. Like vibrational IVR features, the proton spectrum is more compact than expected for nonlocal couplings, resulting in a power law decay.

Let us consider explicit quantum dynamics simulations for the relaxation of a two-level system as a final example. Using the SUR algorithm we have computed the relaxation behavior of a single two level system coupled to one mode of a 6-D bath with randomly distributed frequencies at finite temperature. The calculation in fig. 7 shows the off-diagonal decays of the reduced system density matrix. Only at very small coupling strengths of the system to the bath can the decays be well-approximated by exponentials. At intermediate strengths (comparable to the intrabath anharmonicity and system splitting in the example in fig. 7), the coherence and population decay sub-exponentially.

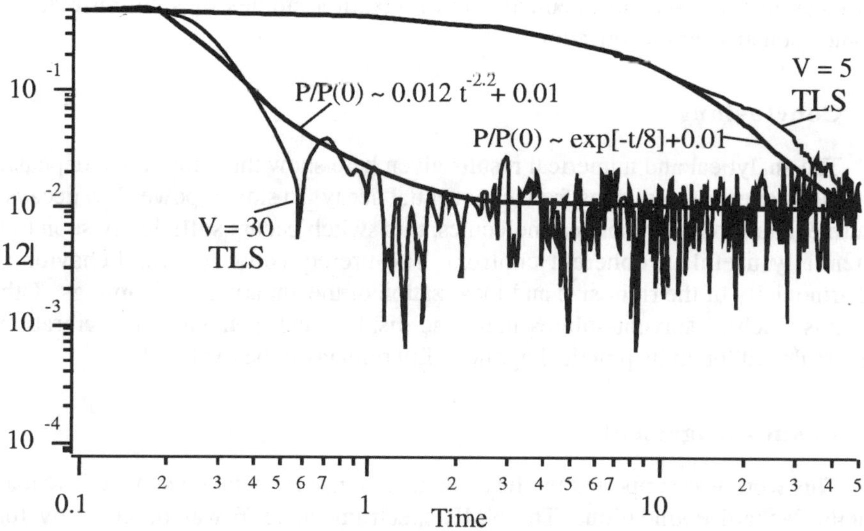

Figure 7. Decay of the off-diagonal coherence of a two level system coupled to a 6-D bath for V=5 and V=30. (The two-level system has ΔE = 100, time is in units of $2\pi/E$; the bath modes have randomly distributed mode energies ranging from 30 to 300.) At small couplings, the dynamics are exponential. At stronger couplings, the dynamics are better fitted by a power law after P(t) drops below 0.012.

The dimensionality of the above bath is of course very small. Chemical systems in solvents experience baths of much larger total dimensionality. However, the only directly coupled bath states are those in the solvent shell surrounding the system, and out of those, only a few may be actively coupled. At very short times, the effective dimensionality of a solvent bath is therefore much smaller than the maximum possible, and may very well be comparable to the D=6 bath shown in fig. 7. For how long a time this remains true depends on the diffusion constants in the solvation shell. Diffusion will rapidly increase the effective dimensionality of the solute in contact with the system, leading to the usual observed exponential dynamics. NMR line shape time scales (order of seconds in solution after dipolar averaging) are too long to ever expect anything but

Lorentzians for individual lines (as opposed to the overall spectrum shown in fig.6). Vibrational and reaction time scales are much faster, and it is conceivable that nonexponential deviations due to a reduced number of effective degrees of freedom in the solvent-solute interaction could play a role in condensed phase reactions. Recent advances in femtosecond chemistry and relaxation studies of solution reactions should soon answer this question.

4 Conclusions

The analytical and numerical results given here show that vibrational dephasing of molecular energy switches from exponential decays to slower power-law decays at intermediate times. In isolated molecules, this switch occurs sufficiently soon to be potentially useful in coherent control. The prerequisites for this behavior are anharmonicity of the right size and localization of the anharmonic couplings. Other systems, such as solvent-solutes may also display such behavior on a vibrational time scale, although its practical applicability remains to be explored.

5 Acknowledgements

This work was supported by the National Science Foundation and the David and Lucile Packard Foundation. The NMR spectrum in fig. 6 was obtained by John Ervin.

6 References

1. S. Schofield and P. G. Wolynes, *J. Chem. Phys.* **98**, 1123 (1993).
2. M. Gruebele, *Adv. Chem. Phys.* **114** (2000).
3. T. Mercouris and C. A. Nicolaides, *J. Phys. B: At. Mol. Opt. Phys.* **30**, 811 (1997).
4. J. F. Kauffman, M. J. Coté, P. G. Smith and J. D. McDonald, *J. Chem. Phys.* **90**, 2874 (1989).
5. D. Boyanovsky, C. Destri, H. J. D. Vega, R. Holman and J. Salgado, *Phys. Rev. D* **57**, 7388 (1998).
6. V. Wong and M. Gruebele, *J. Phys. Chem.* **103**, 10083 (1999).
7. R. Bigwood and M. Gruebele, *Chem. Phys. Lett.* **235**, 604 (1995).
8. R. Bigwood, B. Milam and M. Gruebele, *Chem. Phys. Lett.* **287**, 333 (1998).

9. P. G. Smith and J. D. McDonald, *J. Chem. Phys.* **92**, 1004 (1990).
10. M. Gruebele, *Proc. Nat. Acad. Sci. USA* **95**, 5965 (1998).
11. M. Gruebele and R. Bigwood, *Int. Rev. Phys. Chem.* **17**, 91 (1998).
12. D. V. Shalashilin and D. L. Thompson, *J. Chem. Phys.* **107**, 6204 (1997).
13. F. Remacle and R. D. Levine, *J. Phys. Chem.* **95**, 7124 (1991).
14. D. J. Tannor and S. A. Rice, *J. Chem. Phys.* **83**, 5013 (1985).
15. P. Brumer and M. Shapiro, *Acc. Chem. Res.* **22**, 407 (1989).
16. V. D. Kleiman, L. Zhu, X. Li and R. J. Gordon, *J. Chem. Phys.* **102**, 5863 (1995).
17. C. J. Bardeen, J. Che, K. R. Wilson, V. V. Yakovlev, V. A. Apkarian, C. C. Martens, R. Zadoyan, B. Kohler and M. Messina, *J. Chem. Phys.* **106**, 8486 (1997).
18. H. Kawashima, M. M. Wefers and K. A. Nelson, *Ann. Rev. Phys. Chem.* **46**, 627 (1995).
19. W. S. Warren, H. S. Rabitz and M. Dahleh, *Science* **259**, 1581 (1993).
20. S. A. Schofield and P. G. Wolynes, *J. Phys. Chem.* **99**, 2753 (1995).
21. R. Bigwood, M. Gruebele, D. M. Leitner and P. G. Wolynes, *Proc. Natl. Acad. Sci. USA* **95**, 5960 (1998).
22. D. Madsen, R. Pearman and M. Gruebele, *J. Chem. Phys.* **106**, 5874 (1997).
23. R. Pearman and M. Gruebele, *J. Chem. Phys.* **108**, 6561 (1998).
24. R. Bigwood and M. Gruebele, *Chem. Phys. Lett.* **233**, 383 (1995).
25. M. J. Bronikowski, W. R. Simpson, B. Girard and R. N. Zare, *J. Chem. Phys.* **95**, 8647 (1991).
26. A. Sinha, M. C. Hsiao and F. F. Crim, *J. Chem. Phys.* **92**, 6333 (1990).
27. A. Callegari, private communication .
28. W. D. Lawrance and A. E. W. Knight, *J. Phys. Chem.* **89**, 917 (1985).
29. K. K. Lehmann, *J. Phys. Chem.* **95**, 7556 (1991).
30. M. Gruebele, *J. Phys. Chem.* **100**, 12183 (1996).

WHAT ROLE CAN FOUR-WAVE MIXING TECHNIQUES PLAY IN COHERENT CONTROL?

VADIM V. LOZOVOY, EMILY J. BROWN, IGOR PASTIRK, BRUNA I. GRIMBERG AND MARCOS DANTUS

Department of Chemistry and Center for Fundamental Materials Research
Michigan State University, East Lansing MI 48824, dantus@cem.msu.edu

The role of four-wave mixing (FWM) techniques in coherent control is considered from the point of view of some of the most important developments in this field over the past years, namely multiphoton excitation, pump-dump methods, interference between coherent pulses, chirped laser pulses, and optimal control. FWM techniques provide a powerful platform for combining coherently multiple laser pulses. We explore the effectiveness of these techniques in controlling chemical reactions. The phase relationship between the pulses is maintained by detecting the signal in a phase-matching direction. The results presented show control over the observed dynamics from ground and excited state populations. The FWM signal results from the polarization of the sample following three different electric field interactions. The virtual echo sequence is achieved by the interactions of the sample with three consecutive electric fields characterized by $\exp[i(\mathbf{kx}-\omega t)]$, $\exp[-i(\mathbf{kx}-\omega t)]$ and $\exp[i(\mathbf{kx}-\omega t)]$. This sequence allows control over the observed ground or excited state dynamics. With the photon echo pulse sequence, characterized by interactions with $\exp[-i(\mathbf{kx}-\omega t)]$, $\exp[i(\mathbf{kx}-\omega t)]$, and $\exp[i(\mathbf{kx}-\omega t)]$, we find that control of ground and excited state populations is not achieved. Differences between these two pulse sequences are shown experimentally and illustrated using wave packet simulations. Data obtained using the 'mode suppression' technique, in which the timing between the first and third laser pulses is fixed while the second pulse is scanned are presented. We show that this technique does not suppress the observed vibrational coherence from the ground or excited state but it yields an additional component to the signal that is independent of the vibrational coherence of the sample. Spectrally dispersed FWM is shown to be an ideal tool for studying intramolecular dynamics and this idea is applied to understanding the role of chirp in controlling molecule-laser interactions. All coherent control methods are affected by the rate of decoherence of the sample. Here we show how these rates are measured with FWM techniques. The measurements presented here illustrate how photon echo measurements yield the homogeneous relaxation rate while the virtual echo measurements yield the sum between homogeneous and inhomogeneous relaxation rates.

1 Introduction

The road to laser control of chemical reactions has been one of many important turns.[1,2] The invention of the laser gave scientists the hope of controlling chemical reactions that went beyond photochemistry. One of the first paradigms was based on the monochromatic quality of the laser. This fostered the idea that chemical bonds could be selectively broken by multiple photon excitation MPE at a specific frequency determined by the vibrational frequency of the particular bond.[3-7] It was soon discovered that the energy being deposited in one vibrational mode was quickly dissipated to other intramolecular chemical bonds by a process termed intramolecular vibrational redistribution (IVR).[8-10] This realization meant that the MPE technique inevitably resulted in fission of the weakest bond. Therefore, the main use of MPE would be to deposit energy in selected molecules in a mixture, based on spectral absorption. Based on a thorough understanding of this method, bond selective fission would require excitation of a local mode in a molecular system with extremely slow IVR. This excitation would need to be followed by prompt bond fission achieved with UV excitation or a reactive collision. Vibrationally mediated chemistry was first proposed and demonstrated by Letokhov and coworkers.[11] Selective bond breaking of the OH or OD bonds in HOD was proposed by Imre and coworkers[12] and carried out experimentally by Crim and coworkers[13] and Valentini and coworkers.[14]

With the realization that IVR prevented the accumulation of energy in a given chemical bond for periods longer than a picosecond, two different approaches to control chemical reactions were developed independently. One approach was based on the continuous and coherent excitation of two competing reactive pathways. This method dubbed coherent control (CC) is based on the constructive or destructive interference among the different photochemical pathways.[15,16] This approach has yielded some very remarkable successes in controlling atomic ionization,[17] controlling the outcome of chemical reactions,[18] controlling the flow of current in semiconductors,[19] and controlling directionality of photoelectron emission in the gas phase.[20] Although it is still too early to have a complete outline of the limitations of this technique, we can identify coherence relaxation as an important parameter for the efficiency of this technique. In general, coherence relaxation is slower for gas-phase samples, especially samples cooled by molecular beam expansion. On the other hand, the inhomogeneous and homogeneous broadening that accompanies condensed phase measurements makes this the most challenging environment for CC. To achieve coherent control in condensed phase, one must ensure that the sample-laser interactions occur in a time scale that is faster than IVR and coherence relaxation.

The other approach to control chemical reactions circumventing IVR is founded on the use of ultrafast laser pulses. Conceptually, the laser-sample interactions take place in a time that is short compared to IVR. This concept was embodied in the theory of pump-dump.[21,22] On the experimental front, Zewail, who had been active in the determination of IVR rates,[23] realized that shorter pulses would be critical for achieving control of chemical reactions.[9] The development of lasers with femtosecond pulse duration made the pursuit of this work possible.[24-27] Zewail's group quickly incorporated these techniques and dedicated their work to the study the ultrafast dynamics of chemical reactions in the gas phase.[28-31] Ultrafast lasers have been used to make important contributions to the study of chemical reactions in gases, liquids, solids and surfaces.[29-32]

The experimental observation of vibrational wave packet dynamics caused by impulsive excitation of multiple vibrational levels using femtosecond pulses indicated that pump-dump control could be used to achieve mode selective chemistry using ultrafast lasers, for a comprehensive review of wavepacket dynamics see Reference 2. This is true provided the timing between the laser pulses was carefully controlled. Most of the femtosecond pump-probe experiments can be considered control experiments even though the coherence between the laser pulses does not play a role in the observed signals. The timing between the laser pulses controls the species that is excited with the probe laser. Limitations to these types of experiments relate primarily to wave packet spreading and the lack of a well-defined phase between pump and probe pulses. Even for isolated molecules, wave packet spreading results from the anharmonicity of the potential energy surface. This limitation can be addressed with chirped laser pulses, whereby the laser chirp can be used to cause a focusing of the vibrational wave packet at a particular point in time and space.[33-35] Single chirped laser pulses, in fact, have been used to control multiphoton excitation and chemical reactions.[36-41] The main limitation to the use of ultrafast, precisely-timed, chirped pulses has been our limited knowledge of the molecular Hamiltonian.

A new paradigm in control of chemical reactions was the proposition that an electric field exists such that a specific target can be optimally achieved.[42] This approach to control depends on the search for the optimal laser pulse. The search, initially carried out by computer algorithms (for reviews see References 42,43 and 1) has reached experimental realization in recent years.[44-50] Optimal control has the distinct advantage that the molecular Hamiltonian does not need to be known. The search is carried out by a number of statistical methods that operate on a trial and error basis. The solution is found by a number of iterations as the optimal field is found by convergence to a maximum attainment of the target. This iterative process without *a priori* knowledge implies that the technique can be applied to very complex systems. Interestingly enough, some of the theoretical work in the field of optimal control has sought that optimal solutions in many cases converge to combinations of a small number of pulses with a specific time delay and phase relationship; for example see References 51 and 52. This observation highlights the importance of every one of the methods that have been used to control chemical reactions: (a) multiphoton interactions, (b) short pulses, (c) pulse phase coherence, (d) timing between laser pulses, and (e) chirp.

The above combination of characteristics can be achieved by a setup combining collinear, phase-locked femtosecond laser pulses. Scherer *et al.* showed that phase locking provided control over the laser-molecule interaction in a collinear pump-probe arrangement.[53] Changing the phase of the laser pulses by π, the data from Scherer *et al.* showed control over the excited state wave packets of iodine.[53] The experimental realization of these measurements required active phase control over the two laser pulses in order to ensure a specific phase between the pulses in the interaction with the sample. This requirement makes the technique difficult. Bergmann and coworkers have developed a method for laser control of rovibrational population transfer based on stimulated Raman adiabatic passage (STIRAP). This technique uses two collinear pulses with different wavelengths that are partially overlapped in time such that the dump pulse precedes the pump pulse to achieve near 100% efficiency.[54] The combination of degenerate femtosecond laser pulses can be achieved using four-wave mixing methods. These techniques, developed over the last three decades, are nonlinear optical processes based on the coherent combination of laser pulses without requiring active phase stabilization.[55-59] Coherent Anti-Stokes Raman Scattering (CARS) measurements can be considered examples of control over excitation pathways, see for example the recent work of Schmitt *et al.*[60,61] In some cases the emission of signal can be controlled in FWM experiments, this is achieved through the interference of different nonlinear four-wave mixing signals.[62]

Our group has recognized nonlinear optical techniques, which use phase matching detection as a means to ensure coherence among a number of laser beams, to be an ideal platform for the coherent combination of laser pulses. Our work on FWM has shown that pulse sequences can take advantage of the vibrational dynamics of the sample molecules for achieving control of the electronic and vibrational process with three degenerate laser pulses.[63,64] Our work has yielded experimental observations as well as a theoretical understanding that shows that pulse sequences can be used to select among different Liouville pathways.[64,65] More recently, we have shown that coherently combined chirped femtosecond pulses can be used to control intramolecular dynamics of the system and developed a technique for the characterization of these dynamics based on spectrally dispersed FWM.[66,67] Realizing that coherent control can only be achieved while the system maintains its coherence, we have carried out coherence relaxation measurements that use some of the pulse sequences we have identified to provide information from ground or excited state coherence.

Gas-phase molecular iodine was chosen in this study for various reasons. (i) The visible B $^3\Pi_{0+u} \leftrightarrow X\ ^1\Sigma_{0+g}$ transition has been well characterized by frequency[68,69] and time domain spectroscopies.[60,64,66,67,70-74] (ii) The B-X transition is resonant with the fundamental wavelength of our ultrafast laser system. (iii) The vibrational periods of both X and B states are longer than the duration of our laser pulses, allowing us to impulsively excite wave packets in each state. (iv) The vibrational periods of the X and B states are quite different making assignment of the signal relatively easy.

In this paper, we demonstrate how pulse sequences can be used to control the observation of ground or excited state vibrational dynamics in three-pulse FWM experiments. One of the reasons for the success of this technique arises from the realization that laser-molecule interactions can take place with two components of the electric field $\exp[-i(\mathbf{kx}-\omega t)]$ or $\exp[i(\mathbf{kx}-\omega t)]$. Which one interacts first with the molecule changes the nonlinear optical processes that can be observed and controlled. This dual interaction is typically explained in terms of *bra* and *ket* interactions. Here we show how this selection is done in the laboratory and illustrate with different pulse sequences the effects each interaction with the electric field has on the sample. We show that the FWM signal can be spectrally dispersed and that this additional dimension is of great importance in the characterization of intramolecular dynamics. We illustrate the effects of chirped pulse with spectrally dispersed FWM. Finally, as with most laser control methods, FWM based techniques are limited by the rate of coherence loss. We present measurement on the rate of coherence loss using two methods in order to illustrate differences caused by inhomogeneous broadening.

2 Experimental

The experimental setup used for our measurements has been described earlier.[66,75] The experiments were performed with home-built amplified CPM dye laser producing 65 fs pulses (transform-limited) centered around 620 nm. The bandwidth at FWHM is typically about 9 nm. At the output of the laser amplifier, the femtosecond pulses were compensated for linear chirp in a double-pass prism arrangement and had 350 μJ of average energy. After the compression, the beam was split by two successive beam splitters into three beams of approximately equal intensity attenuated down to ~ 20 μJ each. The three beams were combined in the usual FWM forward box geometry and focused by 500 mm lens in a quartz cell containing the iodine vapor (See Figure 1). The amount of iodine in the cell was such that the optical density reaches a maximum of 0.4 OD and is independent of

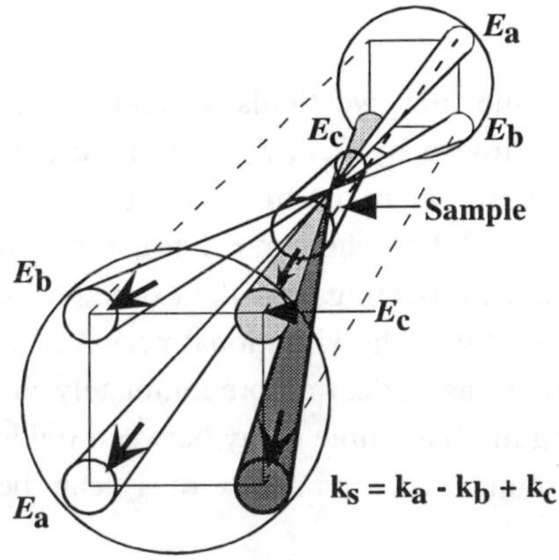

Figure 1 Schematic of the geometrical arrangement of the laser pulses. The lasers are incident from the corners of an imaginary box. The signal, at phase-matching angle, emerges in the corner indicated by the wave vector $\mathbf{k}_S$. The geometric location of each laser beam and the detection geometry determines the sign of each wave vector.

temperature for T > 80° C. The experiments presented here were carried out at 140° C. The coherence relaxation measurements were made in a different cell having excess iodine. The cell was heated evenly and the iodine vapor was in equilibrium with the temperature of the cell. Pulses in one of the beams (c or b, depending on the setup) were delayed with respect to the other two by a computer-controlled actuator (variable τ delay), while the fixed delay (τ_{nm}, where n, m = a, b, or c) was manipulated by a manual translator (micrometer) by delaying the pulses from the second beam with respect to the fixed one. In all cases beam b was opposite to the signal (see Figure 1) and beam a was fixed in time. The fixed time delays were calibrated by autocorrelation with a computer-controlled delay.

The spatially filtered three-pulse FWM signal arising in the phase-matching direction $\mathbf{k}_S = \mathbf{k}_a - \mathbf{k}_b + \mathbf{k}_c$ (see Figure 1) was collected by a spectrometer with wide spectral acceptance (2000 micron slits). For spectrally resolved transients, the slits were reduced to 250 microns. The transients were taken at 300-400 different time delays (about 200 shown) and averaged for 3-10 scans. At each time delay τ, the signal was collected for 10 laser shots. Laser pulses with intensity outside one standard deviation were rejected. The accumulated intensity for each given time delay was stored and the final transients were divided by this set of intensities.

3 Results and discussion

In Figure 2, we use a pulse sequence in which the first two fields, E_a and E_b, are delayed by 460 fs or 614 fs. Both pulses are followed by field E_c after a variable time delay τ. In the virtual photon echo (VE) measurement the first applied electric field is E_a; when τ_{ab} is 460 fs (Figure 2a) (460 fs = 3/2 τ_e where $\tau_e = 2\pi/\omega_e$ for I_2 in the B state), the dynamics show 307 fs oscillations, reflecting only the excited state contribution.[63,64] When τ_{ab} is 614 fs (Figure 2b) (twice the vibrational period of I_2 in the B state), the dynamics show 160 fs oscillations, reflecting predominately the ground state contribution.[63,64,66,67] By changing the fixed time delay between fields E_a and E_b, the observation of excited or ground state dynamics of I_2 can be controlled.

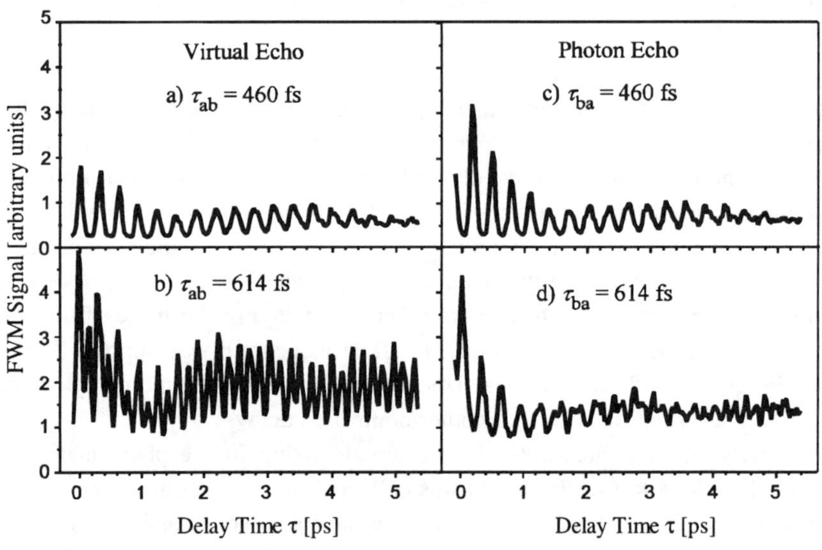

Figure 2. Three-pulse FWM data for which the time delay between the first two pulses is fixed at 460 or 614 fs. The transients shown in (a) and (b) are obtained for a virtual echo pulse sequence having pulse E_a preceding pulse E_b. The transients shown in (c) and (d) are obtained for a photon echo pulse sequence having pulse E_b preceding pulse E_a. In both cases the signal is plotted as a function of time delay between the second pulse and the laser pulse E_c.

In Figure 2c and 2d, transients are shown for a sequence in which the first interaction is with field E_b preceding field E_a. This setup is known as stimulated photon echo (PE).[59] In the photon echo configuration, when τ_{ba} is 460 fs, the dynamics reflect an excited state contribution with 307 fs oscillations; no ground state contribution is observed in this transient. When τ_{ba} is 614 fs, the 307 fs oscillations still dominate; however, after two picoseconds, some 160 fs oscillations can be seen. Fourier transforms of these two photon echo transients confirm that for τ_{ba} = 614 fs, there is ground state contribution absent when τ_{ba} = 460 fs (FFT not shown). The signal contribution due to rotational dynamics (slow 2 ps modulation) that is clearly observed in the virtual echo transients is much smaller in the PE transients. In both cases, VE and PE, the value of the time delay τ_{ab}/τ_{ba} selects the Liouville pathways that yield to the dynamics of the ground or excited state. Note that this selection is much more efficient for the virtual echo setup.

In Figure 3, enlargements of the virtual echo and photon echo transients for $\tau_{ab} = \tau_{ba} = 460$ fs are shown. It is clear that the two transients are exactly out-of-phase with each other; when one is at a maximum, the other is at a minimum. In both cases, field E_c with $\exp[i(\mathbf{k}_c\mathbf{x}\text{-}\omega t)]$ must interact with the wave packet formed in the excited state by field E_b with $\exp[-i(\mathbf{k}_b\mathbf{x}\text{-}\omega t)]$ producing the de-excitation of the wave packet $\Psi^{(1)}{}_b$. In Figure 3 (right), schematics based on calculated wave packets of the excitation process after the first two electric fields are shown. The wave packet resulting from excitation by field E_a is shown as a solid line; the wave packet resulting from excitation by field E_b is shown as a dashed line. In the virtual echo case, field E_b interacts with the system secondly with τ_{ab} = 460 fs. At $\tau = 0$ fs (time delay for field E_c with respect to the second pulse), the dashed wave packet is at the inner turning point of the potential where the transition probability is maximized. Therefore the interaction results in a maximum signal. However, in the photon echo, field E_b interacts first with the system (still with τ_{ba} = 460 fs) and at $\tau = 0$ fs, the wave packet is at the outer turning point of the potential with negligible transition probability upon interaction with field E_c. After half a vibrational period, the wave packet returns to the inner turning point, where the transition probability is maximized and the interaction with E_c produces a maximum signal. Therefore, the signals for photon echo and virtual echo are exactly out-of-phase. This is true in the particular case chosen where both PE and VE signals contain only excited state contributions. Pshenichnikov *et al.* have observed that an optical setup that interferes the electric fields resulting from the VE and PE signals leads, in some cases, to the complete loss of signal due to destructive interference.[62] Here we show that the excited state vibrational coherence is exactly out of phase.

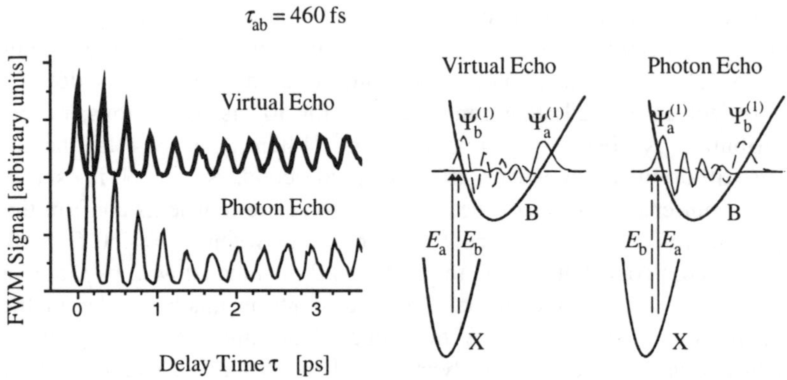

Figure 3. (Left) Experimental data for virtual echo (thick line) and photon echo (thin line) measurements with $\tau_{ab} = \tau_{ba} = 460$ fs. Notice that the signals, corresponding to excited state dynamics, are exactly out-of-phase. The signal intensity has not been re-scaled, and it is interesting to note that the photon echo signal peaks at 154 fs. The 180° phase difference between these transients can be understood by looking at the dynamics of the wave packet in the B state of I_2 resulting from the interaction with pulse E_b. (Right) Simulation of the wave packet motion on the B state of I_2 under the conditions of virtual echo and photon echo when $\tau_{ab} = \tau_{ba} = 460$ fs. In both cases signal formation depends on de-excitation of $\Psi_b^{(1)}$ (dashed curve). At time delay $\tau = 0$, which corresponds to the time delay between the second and third pulses, $\Psi_b^{(1)}$ is in the Franck-Condon region for the VE case, maximizing the transition probability when the third pulse is applied there. However, for the PE case $\Psi_b^{(1)}$ is at the outer turning point of the excited state potential when $\tau = 0$, minimizing the transition probability when the third pulse is applied in the Franck-Condon region. In the PE case, a delay of half a vibrational period is required to achieve the maximum observation of signal when the wave packet returns to the inner turning point.

Shank and coworkers introduced a technique aimed at suppressing the contribution of vibrational coherence to photon echo signals.[76,77] In this technique the time delay between the first and third pulse is fixed and the second beam is scanned. The goal of this method was to achieve accurate relaxation rate measurements in liquids by minimizing the observation of vibrational dynamics. The extent of apparent suppression has been shown to be proportional to the extent of inhomogeneous broadening in the sample.[78] Here we present this type of mode suppression measurement on a gas phase sample. Mode suppression should take place when the time delay between pulses E_b and E_a are in phase with the vibrational motion in the sample, here 614 fs, and mode suppression should not take place when the time delay τ_{ba} is out of phase, here 460 fs.[76,77] Beam E_c is scanned between time zero, where pulse E_b is fixed, and positive times, see Figure 4. The data for 460 fs, mode suppression is OFF, shows pronounced 307 fs vibrations corresponding to the excited state. Notice that the data shows no background or ground state contributions. The data for 614 fs, mode suppression is ON, shows a considerable background signal as well as ground and excited state vibrational dynamics. Our data indicates that the background of the 'mode suppressed' signal

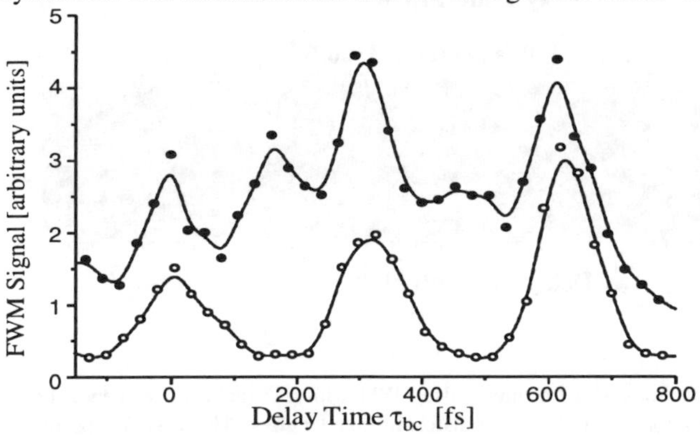

Figure 4. Photon echo transients obtained for a sequence consistent with the mode suppression technique. The delay between the first and the third pulses is fixed with τ_{ba} = 614 fs, mode suppression is ON (solid circles), or τ_{ba} = 460 fs, mode suppression is OFF (open circles). The two transients are plotted as obtained without being shifted or normalized to each other. When mode suppression is ON the signal contains excited state oscillations (307 and 614 fs features), ground state oscillations (see 160 and 480 fs features), and a background that seems to be independent of the vibrational dynamics. When mode suppression is OFF, the transient shows strong excited state modulation with no background and no ground state vibrational features.

corresponds to a contribution that is independent of vibrational motion. However, excited and ground state vibrations are not suppressed in the gas phase where the contributions of inhomogeneous broadening are small. Note that with mode suppression we do not achieve control of ground or excited state excitation as we do using the virtual echo setup,[63,64,79] compare the results presented in Figures 2 and 4.

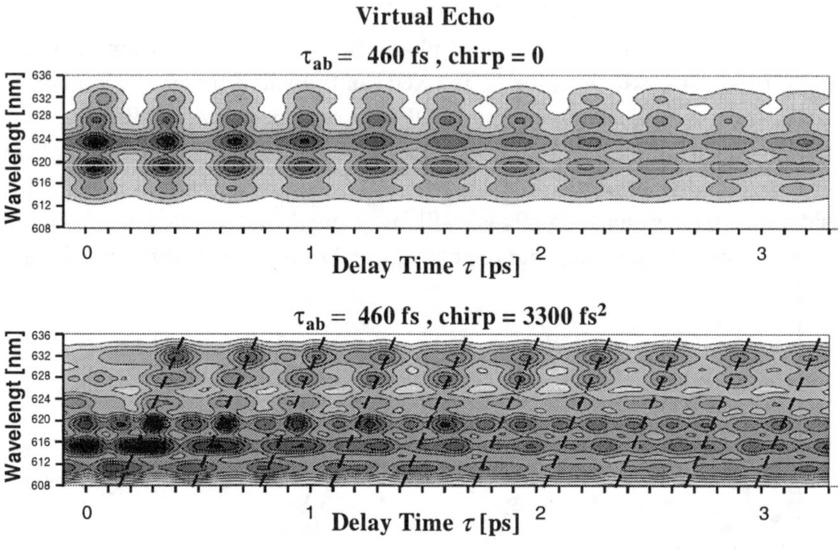

Figure 5. Spectrally dispersed data obtained for three-pulse FWM with a virtual echo sequence. (Top) Experimental data obtained with τ_{ab} = 460 fs and transform-limited pulses. The first 3.3 ps of the spectrally dispersed data are presented as a contour plot. The data show primarily excited state vibrational dynamics (307 fs period). (Bottom) Same as above but the laser pulses were chirped with ϕ'' = 3300 fs^2. The data show a mixture of ground and excited state vibrational coherence. Dashed lines corresponding to the magnitude of the linear chirp are drawn to indicate that the observed vibrational dynamics are chirped. The observation of the different phases as a function of wavelength would be lost in spectrally integrated FWM measurements.

All the transients presented so far are plots of the spectrally integrated signal as a function of time. The FWM emission contains valuable spectral information.[66,67,80] Figure 5 shows spectrally dispersed data obtained as a function of τ for a VE set-up. When transform-limited pulses were applied with (a) τ_{ab} = 460 fs, the spectrally dispersed signal can be assigned to the dynamics of the excited state with an oscillation period of τ_e = 307 fs corresponding to vibrational levels v′ = 6-11. No evidence of ground state dynamics is evident in this transient. In order to explore the role of the pulse chirp in the control of the molecular dynamics, experimental data for VE setup with τ_{ab} = 460 fs was obtained when beams E_a and E_b are equally chirped, ϕ''=+3300 fs^2 (see Figure 5b). From these data, it is clear that chirp pulses add a new contribution from the ground state dynamics. Near the 620 nm region of the spectrum, the mixing of both states dynamics is evident, whereas for longer wavelength the excited state dynamics prevails. Notice that the vibrational dynamics observed at redder wavelengths (632 nm) are advanced in time compared to those at bluer wavelengths (611 nm). This time delay is caused by the positive linear chirp in the laser pulses, which corresponds to ~300 fs in the 28 nm spectral range. Dashed lines with the calculated chirp rate are drawn in Figure 5 and are shown to be in very good agreement with the experimental observation. The clear observation of a linearly chirped vibrational wave packet, as shown in Figure 5, cannot be obtained from spectrally integrated transients. Future experiments are planned with oppositely chirped pulses and with different sign and magnitude of chirp for each pulse. For the measurements presented here the three beams were equally chirped in magnitude and sign.

Coherent control by definition depends on the rate of coherence loss in the sample. Therefore, measurements of the rate of coherence loss are of great importance. Since the first observation of photon echoes in molecules,[55] the technique has been heralded for the fact that inhomogeneous contributions to the relaxation rate cancel. This property has been utilized in numerous measurements primarily in condensed-phase samples.[81] Here we illustrate photon echo and virtual echo measurements for gas-phase iodine. The results, shown in Figure 6, demonstrate the temperature dependence of both methods. The reduced temperature dependence of the virtual echo measurements results from the overwhelming inhomogeneous contribution to the observed relaxation rates.

Measurements with the virtual echo pulse sequence are dominated by inhomogeneous broadening. Contributions to inhomogeneous broadening in this sample arise primarily from Doppler effects as well as from the Boltzman distribution of initial rotational and vibrational states that are excited by the broadband ultrafast laser pulses. The photon echo data show clear temperature dependence and yields the homogeneous rate of decoherence in the sample. It is interesting to note that one can extrapolate to room temperature where the coherence in the sample lasts for several nanoseconds. These long coherence times may be useful for performing nonlinear optical operations related to quantum computing.

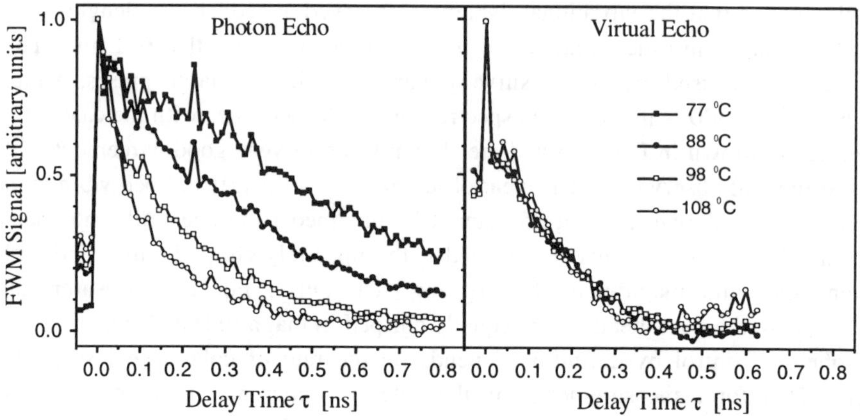

Figure 6. Measurements of coherence relaxation between ground $(X)^1\Sigma_g+$ and excited $(B)^3\Pi_0+_u$ electronic states of I_2. (Left) Experimental data obtained using a photon echo pulse sequence, E_b precedes beams E_c and E_a which are overlapped in time. The intensity of the FWM signal shows a temperature dependent exponential decay. The values obtained as a function of temperature are: 644 ± 48 ps (77° C), 390 ± 17 ps (88° C), 204 ± 9 ps (98° C), and 168 ± 10 ps (108° C). (Right) Experimental data obtained using a virtual echo pulse sequence, E_c precedes beams E_a and E_b which are overlapped in time. The intensity of the FWM signal, in this case, does not show clear temperature dependence and the decay is not a simple exponential. The value obtained when the transients are fit to a single exponential is 150 ± 10 ps. The difference between the photon echo and virtual echo measurements stems from the cancellation of inhomogeneous broadening achieved by the photon echo process (see text).

4 Conclusions

In this paper, we explore the role of FWM methods in studying laser control of chemical reactions. The number of ultrafast pulses, the timing between them (sequence), their phase, and chirp are shown to be effective control parameters on the molecular system. All of these parameters play a role in the search for optimal laser fields for control. Here we demonstrate that FWM techniques are very useful in coherently combining a discrete number of laser pulses. This allows us to learn about the role of one or more control parameters in controlling a process. From our experiments, we obtain a set of conditions that achieves a particular target in terms of parameters that can be related to the Hamiltonian of the system being controlled. Theoretical as well as some experimental studies in optimal control have found that the optimal fields in many cases correspond to two or three pulses with a particular phase relationship and chirp.

One of the clearest applications of FWM methods to coherent control is their usefulness in measuring coherence relaxation. In recent years, ultrafast FWM methods have focused on measuring the rate of coherence loss in condensed phase. Here we show how the methods work for gas-phase samples and make a comparison between photon echo and virtual echo type measurements. The main difference is that for photon echo measurements the inhomogeneous broadening cancels giving a more accurate homogeneous coherence relaxation time. Our data can be used to extrapolate a homogeneous coherence time for iodine in the nanosecond scale, giving the possibility of performing a very large number of coherent interactions with this gas-phase sample at room temperature.

Degenerate FWM methods are shown here to be very useful in the characterization of intramolecular dynamics. Our measurements show the dynamics of the wave packet prepared by the first two laser pulses. When the pulses are transform limited and the delay between them is 460 fs, only excited state dynamics are observed. However, when the laser pulses are chirped, the observed dynamics are quite different. The results obtained from spectrally dispersed FWM show how the phase of different components of the wave packet are advanced or delayed. Lozovoy *et al.*[66] have shown that with two pulses that are equally chirped (sign and magnitude) the coherence of the wave packet can be controlled but not the population transfer. To affect the population transfer between two different states, the two pulses should have the same chirp magnitude but opposite signs.[66]

In this paper, we illustrate differences between photon echo and virtual echo phenomena. We show that the signals have opposite phase. The distinction between these phenomena is of great importance for non-collinear excitation. Our results show control of the molecular system within the framework of phase-matching detection. This implies that only the subset of molecules forming a transient grating in the sample is controlled and yields the coherent signal observed. To achieve control over all molecules in the laser path, one must combine the lasers collinearly using phase-locking techniques.[53] Measuring changes in the overall population must then be done with a phase incoherent method. We show that with the virtual echo setup controlling the time between the first two pulses allows to control the observation of ground or excited state dynamics. This type of control is not available with the photon echo setup. Shank's and Wiersma's group have used the timing between the first and third pulses to suppress the contribution of vibrational coherence in the measurement of coherence relaxation of molecules in a condensed phase.[76-78] Mode suppression methods are shown here to add a component to the signal that is independent of vibrational coherence to the signal. However, in our data we do not observe the suppression of ground or excited state dynamics. Our finding is consistent with the observation of Wiersma indicating that mode suppression depends on large inhomogeneous broadening. Presumably, under large inhomogeneous broadening the vibrational coherence is lost leaving only the signal that is independent of the coherence.

In summary, we have shown that degenerate FWM methods can be used to measure coherence life times and to characterize the intramolecular dynamics of the system that is being controlled. We have shown the effect of chirp in controlling the vibrational coherence of the system. We have shown that the sequence in which pulses are coherently combined can lead to control of the dynamics that are observed. With these observations, we conclude that FWM methods can play an important role in coherent control of chemical reactions.

Acknowledgements

This research was partially funded by a grant from the National Science Foundation (CHE-9812584). MD is a Lucille and David Packard Science and Engineering Fellow, a Camille Dreyfus Teacher-Scholar, and an Alfred P. Sloan Fellow. EJB is supported by a National Science Foundation Graduate Fellowship. IP gratefully acknowledges a James L. Dye Endowment Fellowship.

References

1 R. J. Gordon and S. A. Rice, *Annu. Rev. Phys. Chem.* **48**, 601 (1997).
2 J. Manz, in *Femtochemistry and Femtobiology*, edited by V. Sundström (World Scientific, Singapore, 1998).
3 J. Stone, M. F. Goodman, and D. A. Dows, *Chem. Phys. Lett.* **44**, 411 (1976).
4 S. Mukamel and J. Jortner, *Chem. Phys. Lett.* **40**, 150 (1976).
5 V. S. Letokhov and A. A. Makarov, *Appl. Phys.* **16**, 47 (1978).
6 M. Quack, *J. Chem. Phys.* **69**, 1282 (1978).
7 M. Quack, in *Laser Induced Processes in Molecules*, Vol. 6, edited by K. Kompa and S. D. Smith (Springer Verlag, Berlin, 1979).
8 P. J. Robinson and K. A. Holbrook, *Unimolecular Reactions* (Wiley, New York, 1972).
9 A. H. Zewail, *Physics Today* **33**, 27 (1980).
10 J. Manz and C. S. Parmenter (Eds.), *Chem. Phys.* **139**, 1 (1989) and articles in this special issue.
11 V. S. Letokhov, *Science* **180**, 451 (1973).
12 D. G. Imre and J. Zhang, *Chem. Phys.* **139**, 89 (1989).
13 R. L. Vander Wal, J. L. Scott, and F. F. Crim, *J. Chem. Phys.* **92**, 803 (1990).
14 I. Bar, Y. Cohen, D. David, S. Rosenwaks, and J. J. Valentini, *J. Chem. Phys.* **93**, 2146 (1990).
15 M. Shapiro and P. Brumer, *Chem. Phys. Lett.* **208**, 193 (1993).
16 M. Shapiro and P. Brumer, *Int. Rev. Phys. Chem.* **13**, 187 (1994).
17 C. Chen, Y. Y. Yin, and D. S. Elliott, *Phys. Rev. Lett.* **64**, 507 (1990).
18 L. C. Zhu, V. Kleiman, X. N. Li, S. P. Lu, K. Trentelman, and R. J. Gordon, *Science* **270**, 77 (1995).
19 N. B. Baranova, A. N. Chudinov, and B. Y. Zeldovich, *Opt. Commun.* **79**, 116 (1990).
20 H. G. Muller, P. H. Bucksbaum, D. W. Schumacher, and A. Zavriyev, *J. Phys. B-At. Mol. Opt. Phys.* **23**, 2761 (1990).
21 D. J. Tannor and S. A. Rice, *J. Chem. Phys.* **83**, 5013 (1985).
22 D. J. Tannor, R. Kosloff, and S. A. Rice, *J. Chem. Phys.* **85**, 5805 (1986).
23 P. M. Felker and A. H. Zewail, *Adv. Chem. Phys.* **70**, 265 (1988) and references therein.
24 R. L. Fork, B. I. Greene, and C. V. Shank, *Appl. Phys. Lett.* **38**, 671 (1981).
25 R. L. Fork, C. V. Shank, and R. T. Yen, *Appl. Phys. Lett.* **41**, 223 (1982).
26 J. A. Valdmanis, R. L. Fork, and J. P. Gordon, *Opt. Lett.* **10**, 131 (1985).
27 C. V. Shank, *Science* **233**, 1276 (1986) and references therein.
28 *Chemical Reactions and Their Control on the Femtosecond Time Scale, XXth Solvay Conference on Chemistry*, Vol. 101, edited by P. Gaspard, I. Burghardt, I. Prigogine, and S. A. Stuart (Wiley, New York, 1997).
29 *Femtosecond Chemistry*, Vol. I and II, edited by J. Manz and L. Wöste (VCH, Weinheim, 1995).
30 *Femtochemistry and Femtobiology*, edited by V. Sundström (World Scientific, Singapore, 1998).
31 *Femtochemistry: Ultrafast Chemical and Physical Processes in Molecular Systems*, edited by M. Chergui (World Scientific, Singapore, 1996).

32 *Femtochemistry*, Vol. I and II, edited by A. H. Zewail (World Scientific, Singapore, 1994).
33 V. V. Lozovoy, S. A. Antipin, F. E. Gostev, A. A. Titov, D. G. Tovbin, O. M. Sarkisov, A. S. Vetchinkin, and S. Y. Umanskii, *Chem. Phys. Lett.* **284**, 221 (1998).
34 R. Uberna, Z. Amitay, R. A. Loomis, and S. R. Leone, *Faraday Discuss.*, 385 (1999).
35 J. L. Krause, R. M. Whitnell, K. R. Wilson, Y. Yan, and S. Mukamel, *J. Chem. Phys.* **99**, 6562 (1993).
36 I. Pastirk, E. J. Brown, Q. Zhang, and M. Dantus, *J. Chem. Phys.* **108**, 4375 (1998).
37 C. J. Bardeen, J. Che, K. R. Wilson, V. V. Yakovlev, P. Cong, B. Kohler, J. L. Krause, and M. Messina, *J. Phys. Chem. A* **101**, 3815 (1997).
38 V. Yakovlev, C. J. Bardeen, J. Che, J. Cao, and K. R. Wilson, *J. Chem. Phys.* **108**, 2309 (1998).
39 C. J. Bardeen, V. Yakovlev, J. A. Squier, and K. R. Wilson, *J. Am Chem. Soc.* **120**, 13023 (1998).
40 A. H. Buist, M. Muller, R. I. Ghauharali, G. J. Brakenhoff, J. A. Squier, C. J. Bardeen, V. V. Yakovlev, and K. R. Wilson, *Opt. Lett.* **24**, 244 (1999).
41 R. Zadoyan, N. Schwenter, and V. A. Apkarian, *Chem. Phys.* **233**, 353 (1998).
42 D. Neuhauser and H. Rabitz, *Acc. Chem. Res.* **26**, 496 (1993).
43 S. A. Rice, J. Manz, B. Kohler, S. Mukamel, P. W. Brumer, M. Quack, M. Shapiro, R. A. Marcus, K. Schaffner, H. Rabitz, G. R. Flaming, D. J. Tannor, and G. Gerber, in *Chemical Reactions and Their Control On the Femtosecond Time Scale Xxth Solvay Conference On Chemistry*, Vol. 101 (1997), pp. 213 and references therein.
44 J. L. Krause, R. M. Whitnell, K. R. Wilson, Y. J. Yan, and S. Mukamel, *J. Chem. Phys.* **99**, 6562 (1993).
45 B. Kohler, J. L. Krause, F. Raksi, C. Rosepetruck, R. M. Whitnell, K. R. Wilson, V. V. Yakovlev, Y. J. Yan, and S. Mukamel, *J. Phys. Chem.* **97**, 12602 (1993).
46 Y. J. Yan, R. E. Gillilan, R. M. Whitnell, K. R. Wilson, and S. Mukamel, *J. Phys. Chem.* **97**, 2320 (1993).
47 Y. J. Yan, *J. Chem. Phys.* **100**, 1094 (1994).
48 J. L. Krause, R. M. Whitnell, K. R. Wilson, and Y. J. Yan, in *Femtosecond Chemistry*, edited by J. Manz and L. Wöste (VCH Verlagsgesellschaft, Weinheim, 1995).
49 A. M. Weiner, D. E. Leaird, G. P. Wiederrecht, and K. A. Nelson, *Science* **247**, 1317 (1990).
50 A. Assion, T. Baumert, M. Bergt, T. Brixner, B. Kiefer, V. Seyfried, M. Strehle, and G. Gerber, *Science* **282**, 919 (1998).
51 S. P. Shah and S. A. Rice, *Faraday Discuss.*, 319 (1999).
52 Y. Fujimura, L. Gonzalez, K. Hoki, J. Manz, and Y. Ohtsuki, *Chem. Phys. Lett.* **306**, 1 (1999).
53 N. F. Scherer, A. J. Ruggiero, M. Du, and G. R. Fleming, *J. Chem. Phys.* **93**, 856 (1990).
54 K. Bergmann and B. W. Shore, in *Molecular dynamics and spectroscopy by stimulated emission pumping*, Vol. 4, edited by H.-L. Dai and R. W. Field (World Scientific, Singapore, 1995).

55 N. A. Kurnit, I. D. Abella, and S. R. Hartmann, *Phys. Rev. Lett.* **13**, 567 (1964).
56 C. K. N. Patel and R. E. Slusher, *Phys. Rev. Lett.* **20**, 1087 (1968).
57 Y. R. Shen, *The Principle of Nonlinear Optics* (Wiley, New York, 1984).
58 R. W. Boyd, *Nonlinear Optics* (Academic Press, San Diego, 1992).
59 S. Mukamel, *Principles of Nonlinear Optical Spectroscopy* (Oxford University Press, New York, 1995).
60 M. Schmitt, G. Knopp, A. Materny, and W. Kiefer, *Chem. Phys. Lett.* **270**, 9 (1997).
61 M. Schmitt, G. Knopp, A. Materny, and W. Kiefer, *J. Phys. Chem. A* **102**, 4059 (1998).
62 M. S. Pshenichnikov, W. P. d. Boeij, and D. A. Wiersma, *Phys. Rev. Lett.* **76**, 4701 (1996).
63 E. J. Brown, I. Pastirk, B. I. Grimberg, V. V. Lozovoy, and M. Dantus, *J. Chem. Phys.* **111**, 3779 (1999).
64 I. Pastirk, E. J. Brown, B. I. Grimberg, V. V. Lozovoy, and M. Dantus, *Faraday Discuss.* **113**, 401 (1999).
65 B. I. Grimberg, V. V. Lozovoy, M. Dantus, and S. Mukamel, *(manuscript in preparation).*
66 V. V. Lozovoy, B. I. Grimberg, E. J. Brown, I. Pastirk, and M. Dantus, *J. Raman Spectrosc.*, (in press).
67 I. Pastirk, V. V. Lozovoy, B. I. Grimberg, E. J. Brown, and M. Dantus, *J. Phys. Chem. A* **103**, 10226 (1999).
68 R. S. Mulliken, *J. Chem. Phys.* **55**, 288 (1971).
69 J. Tellinghuisen, *J. Quant. Spectrosc. Radiat. Transfer* **19**, 149 (1978).
70 R. M. Bowman, M. Dantus, and A. H. Zewail, *Chem. Phys. Lett.* **161**, 297 (1989).
71 M. Dantus, R. M. Bowman, and A. H. Zewail, *Nature* **343**, 737 (1990).
72 M. Gruebele, G. Roberts, M. Dantus, R. M. Bowman, and A. H. Zewail, *Chem. Phys. Lett.* **166**, 459 (1990).
73 M. Schmitt, G. Knopp, A. Materny, and W. Kiefer, *Chem. Phys. Lett.* **280**, 339 (1997).
74 V. V. Lozovoy, A. A. Titov, F. E. Gostev, D. G. Tovbin, S. A. Antipin, S. Y. Umanskii, and O. M. Sarkisov, *Chem. Phys. Reports* **17**, 1267 (1998).
75 E. J. Brown, Q. Zhang, and M. Dantus, *J. Chem. Phys.* **110**, 5772 (1999).
76 R. W. Schoenlein, D. M. Mittleman, J. J. Shiang, A. P. Alivisatos, and C. V. Shank, *Phys. Rev. Lett.* **70**, 1014 (1993).
77 C. J. Bardeen and C. V. Shank, *Chem. Phys. Lett.* **203**, 535 (1993).
78 W. P. de Boeij, M. S. Pshenichnikov, and D. A. Wiersma, *Annu. Rev. Phys. Chem.* **49**, 99 (1998).
79 V. V. Lozovoy, I. Pastirk, E. J. Brown, B. I. Grimberg, and M. Dantus, *J. Chem. Phys.*, (submitted).
80 L. D. Book and N. F. Scherer, *J. Chem. Phys.* **111**, 792 (1999).
81 G. R. Fleming, *Chemical Applications of Ultrafast Spectroscopy* (Oxford University Press, NY, 1986).

USING OPTICAL COLLISION TO CONTROL THE TRANSITION-REGION DYNAMICS

KENJI OHMORI
Research Institute for Scientific Measurements, Tohoku University, Sendai 980-8577, Japan
E-mail: ohmori@rism.tohoku.ac.jp

We have developed the "optical collision (OC) " approach to reach and control the transition regions of atomic and molecular interactions. The approach is based on the photoabsorption by a collisional quasimolecule during a single binary collision. Here I present two different OC approaches; one is the time-domain approach based on the fs pump-probe technique, and the other is the frequency-domain approach based on what we call the "second-order optical collision (SOOC)". The time-domain approach has been applied to the Hg–CO van der Waals (vdW) complex, which is regarded as the bound quasimolecule, to realize the first observation of wave packet motion associated with an intermolecular energy flow in the vdW interaction. Possibilities of chirped pump-pulse and double pump-pulse control of the relevant energy flow are discussed. For the frequency-domain approach, we have applied the "wing-wing double resonance" technique to the Hg–Ar collisional quasimolecule to observe an ultrafast SOOC, where two-color, two-photon absorption takes place during a single binary collision, with a pair of nanosecond laser pulses. This experiment serves as the optical switching of an ultrafast thermal atomic collision; a collision pathway is controlled by changing the frequencies of those laser pulses. It is also shown that an ultrafast pump/probe measurement can be carried out with sub-picosecond time resolution using nanosecond pulsed lasers.

1 Introduction

An "optical collision" is the process where a collisional quasimolecule absorbs photons during a single atomic and molecular collision. It triggers the second-half collision on an excited-potential surface, making a direct access to the transition region. Here I use the term "transition region" as all intermediate configurations between reagents and products, including what we call the "transition state" at the saddle point of a potential-energy surface. The optical collision serves as the basis of our strategy to explore and control the transition-region dynamics of atomic and molecular interactions. The strategy, which we call the "optical collision" approach, is subdivided into two categories; one is the far wing approach, and the other is the van der Waals (vdW) approach. As is now well understood, a far-wing spectrum of a pressure-broadened atomic line is ascribed to an optical transition of an ultrashort-lived quasimolecule formed during a thermal binary-collision[1] and has been used by several groups to extract the potential curve and the transition-region dynamics.[2] We have developed our own far-wing approach.[3–10] It is based on the laser pump/probe method applied to the far wings of a pressure-broadened atomic line. A half collision is thus triggered by excitation of a thermal quasimolecule in this approach. The vdW approach, on the other hand, is based on laser excitation of a vdW complex, which is generated in a supersonic jet expansion and regarded as a bound quasimolecule, and a half collision is triggered by excitation of a rotationally and vibrationally cold quasimolecule. These two approaches are combined complementarily to be useful in understanding a nonadiabatic coupling induced by nuclear rotation and vibration and in determining an interaction potential for a wide range of nuclear configuration.

It is also possible to classify our optical collision approach from a different point of view; one is the frequency-domain approach, and the other is the time-domain approach. In the frequency-domain approach, a pair of nanosecond laser pulses are used as pump and probe pulses; the pump wavelength is scanned to yield an excitation spectrum of a quasimolecule for a specific scattering product of a half collision detected by the probe pulse.[3–10] In the time-domain approach, on the other hand, we use a pair of femtosecond laser pulses and scan the pump-probe delay-time to clock a half collision in real time.[9,10] The frequency- and time-domain approaches are complementary in our optical-collision studies.[9,10] Both realize direct observation and quantum control of the transition regions in atomic and molecular interactions.

In what follows, I present the selected number of topics associated with our optical-collision approach. In Sec. II, the time-domain approach is applied to the Hg–CO vdW complex to realize the first observation of wave packet dynamics involved in the electronic to rovibrational intermolecular energy flow in the vdW interaction; the process is closely connected with the initial stage of the solute-solvent energy dissipation in solution. Furthermore, possibilities of the quantum control of the relevant energy-flow will be discussed. Recently we have developed a new technique in the frequency domain, which we call the "wing-wing double resonance (WWDR)" technique. In Sec. III, it is applied to the Hg–Ar collisional quasimolecule. A pair of nanosecond laser pulses are simultaneously applied to far wings of the Hg atomic lines broadened by Ar to make it possible to take an ultrafast pump/probe measurement with subpicosecond time-resolution. This experiment serves as the optical switching of an ultrafast thermal atomic collision; a collision pathway is controlled by changing the frequencies of those laser pulses.

2 Intermolecular Energy Flow in the Binary vdW Interaction

One of the important pathways of the Hg(6 $^3P_1$)–molecule interaction is the fine-structure transition yielding the Hg(6 $^3P_0$) atom

$$\mathrm{Hg}(6^3P_1) + \mathrm{X} \rightarrow \mathrm{Hg}(6^3P_0) + \mathrm{X} \tag{1}$$

where the electronic energy in the Hg atom is converted into the rovibrational energy in X.[11–16] Few experimental investigations have been made on the transition regions of this process, which are the $\widetilde{A}^3 0^+ \rightarrow \widetilde{a}^3 0^-$ and $\widetilde{B}^3 1 \rightarrow \widetilde{a}^3 0^-$ nonadiabatic transitions of the Hg–X quasimolecule.[4,17,18] The $\widetilde{A}$ and $\widetilde{B}$ states correlate adiabatically to the asymptote Hg(6^3P_1)+X, and the $\widetilde{a}$ state correlates to Hg(6^3P_0)+X. Fuke *et. al*[17] and Yamanouchi *et al.* [18] have carried out such experiments by means of laser excitation of the Hg–X vdW complex, scanning the excitation wavelength to measure the Hg(6^3P_0) action spectrum. In some cases, these vdW studies in the frequency domain have provided detailed information on the excited states of the complex correlated directly to the early stage of process (1). For the Hg–N_2 interaction, for example, the action spectrum yields a number of important information on the transition-region dynamics.[17,18] For the Hg–CO interaction, on the other hand, this type of frequency domain approach does not work well since the life time of the excited state is short, so that the spectrum is highly diffuse and complicated.[9,10,17] Here I report the application of the time domain approach to this system. The fs pump-probe technique is employed to clock the relevant nonadiabatic decay of the excited complex in real time and to observe wave packet motion involved in that decay. Although some groups have successfully applied such an ultrafast approach to the weakly bound vdW complexes,[19] the present study serves as the first one to visualize wave packet motion associated with an intermolecular energy flow in the binary vdW interaction.

2.1 Experiment

The Hg-CO vdW complex was produced by the supersonic jet expansion of heated Hg vapor (470 K) seeded in high-pressure CO/He mixture (0.6 atm CO and 1.4 atm He) into a vacuum chamber through a pulsed valve of 0.3mm in diameter.

The output of a mode-locked Ti:Sapphire laser pumped by Ar^+ laser was amplified by a double-stage regenerative amplifier pumped by two 10Hz Nd:YAG lasers to give the final output around 760 nm. It was frequency tripled to generate the pump and probe pulses around 253.7 nm. Portions of these pulses were reflected by a quartz disk and introduced into a nitric oxide ionization cell where two-photon ionization signal was monitored as a function of the pump-probe delay to measure their cross correlation. The pulse width was about 300 fs and the pulse energy was about 10 μJ for both the pump and probe pulses. They were introduced collinearly into the vacuum chamber to intersect the jet perpendicularly at 7 mm downstream from the valve.

Figure 1 shows the present pump-probe scheme. The Hg–CO complex was excited by the pump pulse to the coherent superposition of the bound wave functions on the $\tilde{B}$-state potential to generate a wave packet moving on the excited-state surface. Every time it reaches a particular point of a nonadiabatic transition, a fraction hops on the $\tilde{a}$-state surface and moves towards the asymptote Hg(6^3P_0) + CO; the $\tilde{A}$ state may possibly seves as an intermediate in the present $\tilde{B} \rightarrow \tilde{a}$ hopping. The nascent Hg(6^3P_0) atom was excited by the probe pulse to the 7^3D_1 state. The fluorescence from this state was sampled through the viewing port perpendicular to both of the jet and the laser axis and then focused onto the entrance slit of a 20cm monochromator equipped

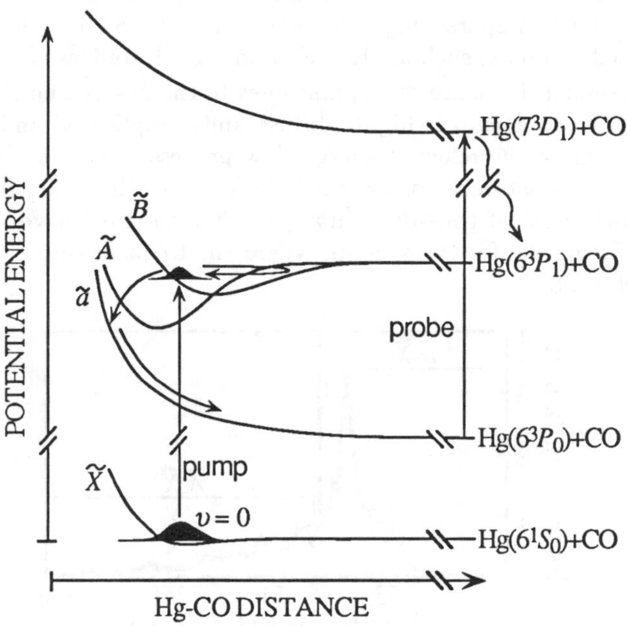

Figure 1. Pump-probe scheme for the real-time clocking of the $\tilde{B} \rightarrow \tilde{a}$ nonadiabatic transition of the Hg-CO vdW complex.

with a 1200 grooves/mm grating blazed at 400 nm. The fluorescence due to the $6^3P_1 - 7^3D_1$ transition (265.4 nm) was detected with a bandwidth of ~ 2.3 nm by a photomultiplier attached to the monochromator. The signals were amplified and then fed into a fast gated integrator. The output of the gated integrator was digitized and then stored in a microcomputer. A mercury-vapor cell was inserted in the laser beam path to remove a wavelength portion at 253.7 nm. Without this vapor cell, the true signal was smeared out by the huge background attributed to the Hg $6^1S_0 \leftarrow 6^3P_1$ atomic fluorescence. By scanning the pump-probe delay, the $\widetilde{B} \rightarrow \widetilde{a}$ nonadiabatic decay was clocked in real time.

2.2 Results and Discussion

Nonadiabatic Transition Dynamics

For convenience, we show in Fig. 2 our previous frequency-domain measurement of the excitation spectrum of the Hg–CO vdW complex in the $\widetilde{A} - \widetilde{X}$ and $\widetilde{B} - \widetilde{X}$ bands for the formation of the Hg(6^3P_0) atom.[9,10] The highly dense structure is seen in the shorter wavelength side of the Hg $6^1S_0 - 6^3P_1$ resonance line, and the relatively sparse one in the other side. By analogy with the other mercury related complex, such as Hg – N_2 and Hg – Rare Gas (RG),[20] we have tentatively assigned the dense and sparse ones to the $\widetilde{B} - \widetilde{X}$ and $\widetilde{A} - \widetilde{X}$ bands, respectively. The spectrum is highly diffuse and complicated and yields very little information on the relevant energy flow process. The diffuseness should arise from the ultrafast decay of the excited states. I will demonstrate in what follows that the ultrafast time-domain approach is quite effective in extracting dynamical information for the system, where the frequency-domain approach does not work well.

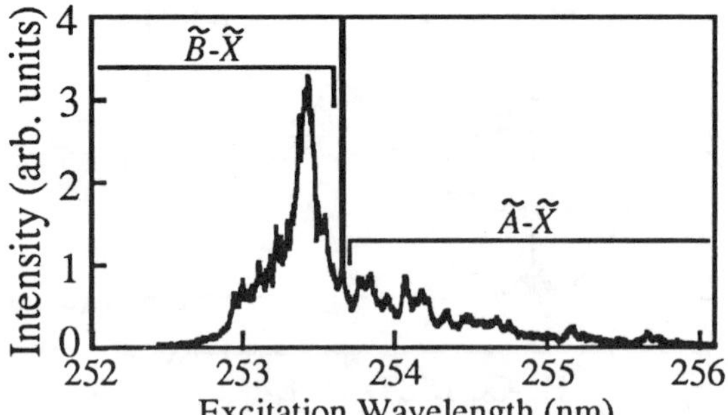

Figure 2. $\widetilde{A}$–$\widetilde{X}$ and $\widetilde{B}$–$\widetilde{X}$ excitation spectra of the Hg-CO vdW complex for the formation of the Hg(6^3P_0) atom.

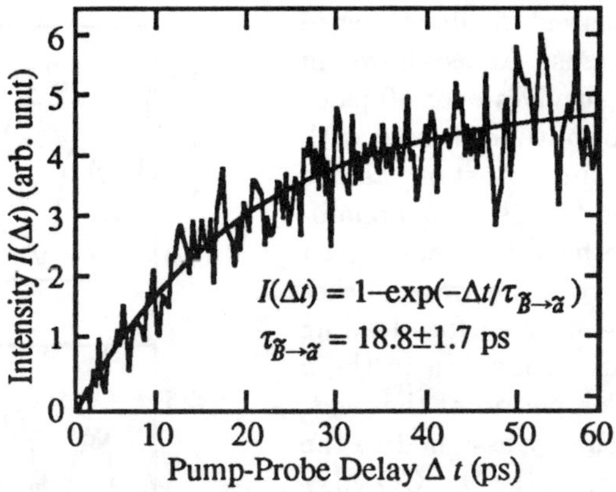

Figure 3. Temporal evolution of the fluorescence intensity with the pump wavelength tuned to the $\widetilde{B}$-$\widetilde{X}$ band of the complex and with the probe wavelength tuned to the Hg 7^3D_1-6^3P_0 atomic line, respectively.

Figure 3 shows the result of the time-domain measurement: the fluorescence intensity I plotted against the pump-probe delay Δt. For this measurement, the center wavelength of the pump and probe pulses was 253.5 nm, tuned to the $\widetilde{B}-\widetilde{X}$ band of the Hg–CO complex and to the Hg $6^3P_0 - 7^3D_1$ atomic transtition. It is thus concluded that Fig. 3 represents the real-time clocking of the $\widetilde{B} \rightarrow \widetilde{a}$ nonadiabatic transition. The observed time dependence, $I(\Delta t)$, is well fitted by a single-exponential function

$$I(\Delta t) = I(\Delta t = \infty)(1 - \exp \frac{\Delta t}{\tau_{\widetilde{B}\rightarrow\widetilde{a}}}), \tag{2}$$

with $I(\Delta t = \infty)$ and $\tau_{\widetilde{B}\rightarrow\widetilde{a}}$ being adjustable parameters, and $\tau_{\widetilde{B}\rightarrow\widetilde{a}}$ is the time constant for the $\widetilde{B} \rightarrow \widetilde{a}$ nonadiabatic decay of the excited Hg–CO complex. The least-square fit is shown as a solid curve in Fig. 3. We have thus found for the first time that the $\widetilde{B} \rightarrow \widetilde{a}$ nonadiabatic transition proceeds with a time constant of 18.8 ± 1.7 ps, where the error represents one standard deviation.

Then we changed the pump and probe wavelengths by - 0.1 nm, making the pump pulse still tuned to the $\widetilde{B} - \widetilde{X}$ band and the probe pulse detuned from the Hg $6^3P_0 - 7^3D_1$ atomic line. This detuning should allow the probe pulse to detect a wave packet at a specific Hg–CO distance on the $\widetilde{a}$ - state surface before it reaches to the asymptote Hg(6^3P_0) + X. In this measurement, we have observed the fluorescence signal beating against Δt as shown in Fig. 4. The beat survives over 30 ps or longer, and its amplitude decays with almost the same time constant as has been obtained for the $\tilde{B} \rightarrow \tilde{a}$ nonadiabatic decay. We have thus concluded that the beat mirrors the wave packet motion on the $\tilde{B}$ - state surface leading to the nonadiabatic transition to the $\tilde{a}$ - state surface. The period of the beat, T, is 1.2 ps, reasonably assigned to the stretching vibration of the Hg–CO vdW mode of the $\tilde{B}$ state by the analogy with similar systems, such as the Hg–RG and Hg–N_2 vdW complexes;[20] the bending vibration and the overall rotaion on the $\tilde{B}$ - state surface should have longer periods by an order and two orders of magnitudes, respectively.[20] It is thus elucidated that the radial motion between Hg and CO plays an important role in this nonadiabatic coupling between the $\tilde{B}$ and $\tilde{a}$ states, and the nonadiabatic transition is localized at a specific Hg–CO distance. Moreover, it is seen in Fig. 4 that the first peak appears at $\Delta t = 1.2$ ps coinciding with the period of the beat mentioned above. This elucidates that the nonadiabatic transition occurs near the point where the complex is initially excited by the pump pulse from the $\tilde{X}$ state. The point should be the inner turning point in the Hg–CO stretching coordinate of the $\tilde{B}$ - state potential by the analogy with other Hg–X systems, such as Hg–N_2 and Hg–RG.[20] Hence the observed initial phase of the beat suggests that the $\tilde{B} \rightarrow \tilde{a}$ nonadiabatic transition occurs near the point where Hg and CO

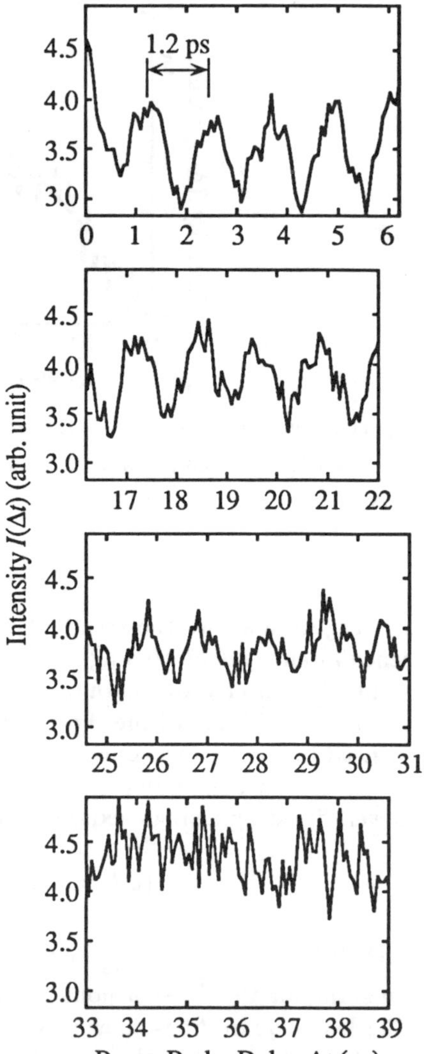

Figure 4. Temporal evolution of the fluorescence intensity with the pump wavelength tuned to the $\tilde{B}$-$\tilde{X}$ band of the complex and with the probe wavelength detuned from the Hg 7^3D_1-6^3P_0 atomic line.

come nearest to each other on the $\widetilde{B}$-state surface. The period of the beat, T, is devided by the time constant, $\tau_{\widetilde{B}\to\widetilde{a}}$, to give the $\widetilde{B} \to \widetilde{a}$ transition probability, $P_{\widetilde{B}\to\widetilde{a}}$, for a single approach of Hg and CO:

$$P_{\widetilde{B}\to\widetilde{a}} = \frac{T}{\tau_{\widetilde{B}\to\widetilde{a}}}. \tag{3}$$

$P_{\widetilde{B}\to\widetilde{a}}$ is thus determined to be 0.07, meaning that the transition occurs one time in every 14 that the wave packet enters the region of the nonadiabatic transition. This is consistent with our previous far-wing study on the same system, where $P_{\widetilde{B}\to\widetilde{a}}$ has been obtained for the free $\widetilde{B}$ state to be 0.1–0.2. [4] Figure 5 shows the Fourier transform (FT) of the beat observed for Δt = 0–6 ps. A strong dominant peak is seen around $\omega = 27$ cm^{-1}. This, of course, corresponds to the main component of the beat, discussed above, with a period of 1.2 ps and is assigned to the stretching vibration of the Hg–CO vdW mode. The progression ascribed to this stretching vibration should be observed in the $\widetilde{B} - \widetilde{X}$ band of the complicated excitation spectrum shown in Fig. 1. Figure 6 shows one of the plausible assignments of the relevant progression for the vibrational levels v_s = 0–5. This assignment gives the Morse potential parameters of the $\widetilde{B}$-state to be $\omega_e = 31.4 \pm 0.8$ cm^{-1}, $\omega_e x_e = 2.0 \pm 0.1$ cm^{-1}, and $D_e = 123 \pm 12$ cm^{-1}. D_e was calculated from the Morse relation $D_e = \omega_e^2/(4\omega_e x_e)$. The Morse potential thus obtained is plotted against the Hg–CO distance R in Fig. 6, where R_e is arbitrary. According to the present assignment, the observed beat is attributable to the coherent superposition of $v_s = 0$ and 1.

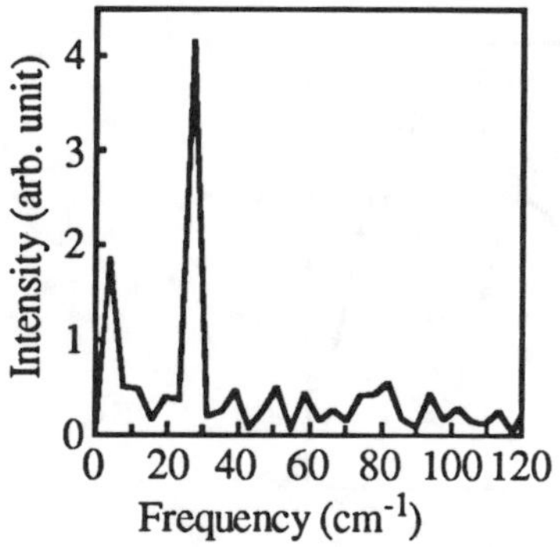

Figure 5. Fourier transform of the beat shown in Fig. 4.

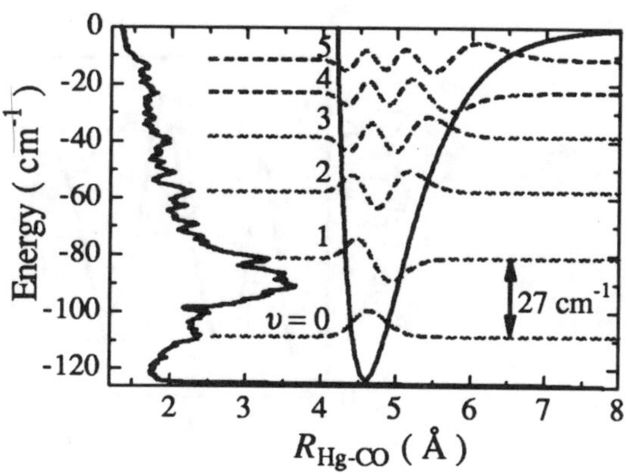

Figure 6. Vibrational assignment of the $\widetilde{B}$-$\widetilde{X}$ band shown in Fig. 2 and the Morse potential curve of the $\widetilde{B}$ state. R_e is arbitrary.

Possibility of Quantum Control

Using this $\widetilde{B}$-state potential, we have examined theoretically the possibilities to control the relevant nonadiabatic process. Since it is found in this study that the nonadiabatic transition occurs near the inner turning point of the $\widetilde{B}$-state potential, we try to control the probability P_{target} of finding the excited complex in the target region around that turning point. The present target region lies between $R = 3.97\text{Å}(= R_1)$ and $R = 4.82\text{Å}(= R_2)$ with R_e being 4.7Å where the $\widetilde{B}$-state potential obtained by *ab-initio* calculations takes its minimum value for the T-shape geometry, which is most stable for the *ab-initio* $\widetilde{X}$-state potential.[21] We solve the one-dimentional time-dependent Schrödinger equation

$$i\frac{\partial}{\partial t}\Psi(R,t) = [H(R) - \mu\varepsilon(t)]\Psi(R,t) \tag{4}$$

to obtain the time dependence of P_{target}, which is defined as

$$P_{\text{target}}(t) = \int_{R_1}^{R_2} |\Psi(R,t)|^2 dR, \tag{5}$$

for several different light fields $\varepsilon(t)$. $H(R)$ is the total Hamiltonian of the complex, the sum of a nuclear kinetic energy operator and an electronic part, and μ is an electronic dipole-moment operator. The initial state is $v = 0$ on the $\widetilde{X}$-state potential obtained for the most stable T-shape geometry by *ab-initio* calculations.[21] The nonadiabatic decay is not included in the present calculation.

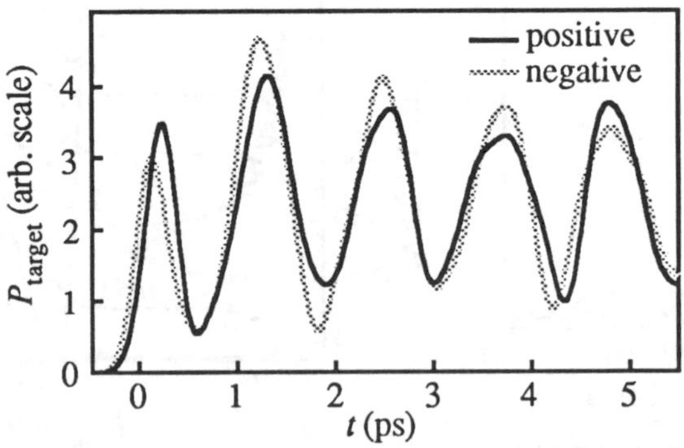

Figure 7. Temporal evolution of the probability P_{target} calculated for the linearly chirped pulse excitation (see text). The frequency-sweeping rate $\zeta = +80\ \text{cm}^{-1}/\text{ps}$ (positive chirp) and $-80\ \text{cm}^{-1}/\text{ps}$ (negative chirp). $t = 0$ ps denotes the arrival of the chirped pulse.

First we have investigated the effect of chirpes of the pump pulse. Equation (4) is solved for a linearly chirped pulse:

$$\varepsilon(t) = \varepsilon_{\mathbf{max}} \exp(-t^2/2\delta^2) \cos[(\omega_{t=0} + \zeta t)t], \tag{6}$$

where δ is the temporal width, $\omega_{t=0}$ is the center frquency at $t = 0$, and ζ is the frequency-sweeping rate. Figure 7 shows the results for δ = 200 fs, $\omega_{t=0}$ = 39463 cm^{-1}, and ζ = +80 and -80 cm^{-1}/ps. These pulses create coherent superpositions of v=0 and 1. It is seen in Fig. 7 that no big difference appears in $P_{\mathrm{target}}(t)$ between the positive and negative chirp excitations; the averaged values for $P_{\mathrm{target}}(t)$ differ by only 5% between those two conditions. Chirped pump pulse is not effective very much for the present purpose.

Next we have examined the possibility of a double pump-pulse control, where the Hg–CO complex is irradiated sequentially by two identical laser pulses with their particular relative phase. Equation (4) is solved for the double-pulse excitation:

$$\varepsilon(t) = \varepsilon_{\mathbf{max}} \exp[-t^2/2\delta^2] \cos \omega t + \varepsilon_{\mathbf{max}} \exp[-(t - \Delta t)^2/2\delta^2] \cos \omega(t - \Delta t), \tag{7}$$

where the second pulse is delayed from the first one by Δt. $P_{\mathrm{target}}(t)$ is thus calculated for $\delta = 200$ fs and $\omega = 39463$ cm^{-1}, for which 0.4 fs in Δt corresponds to π in $\omega\Delta t$. Figure 8 shows the results for Δt around 1.2 ps, the period of the wave packet recurrence. Under this condition, the second pulse arrives when the wave packet created by the first pulse comes back to the Franck-Condon region, that is, the inner turning point. It is seen in Fig. 8 that $P_{\mathrm{target}}(t)$ completely changes in changing Δt by only 0.4 fs. One is constructive and the other is destructive interferences. Figure 9, on the other hand, shows the results for Δt around 1.8 ps. This condition gives the situation that the second pulse arrives when the wave packet reaches the outer turning point, that is, the furthermost point from the Franck-Condon region. As is seen in Fig. 9, $P_{\mathrm{target}}(t)$ drastically changes again in changing Δt in steps of only 0.2 fs. This type of double pump-pulse scheme has been successfully applied to Rydberg-electron dynamics in a number of studies.[22] V. Blanchet *et al.* and N. F. Scherer *et al.* have applied a similar technique to nuclear motion,[23] and it will be promising also to control the wave packet dynamics under the present situation. The double pump-pulse experiment mentioned above is now in preparation in our group.

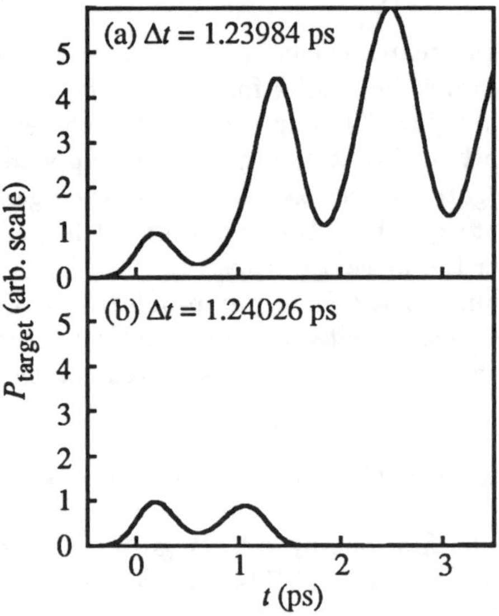

Figure 8. Temporal evolution of the probability P_{target} calculated for the double pulse excitation (see text). The second pulse is delayed by (a) 1.23984 ps and (b) 1.24026 ps from the first pulse. $t = 0$ ps denotes the arrival of the first pulse.

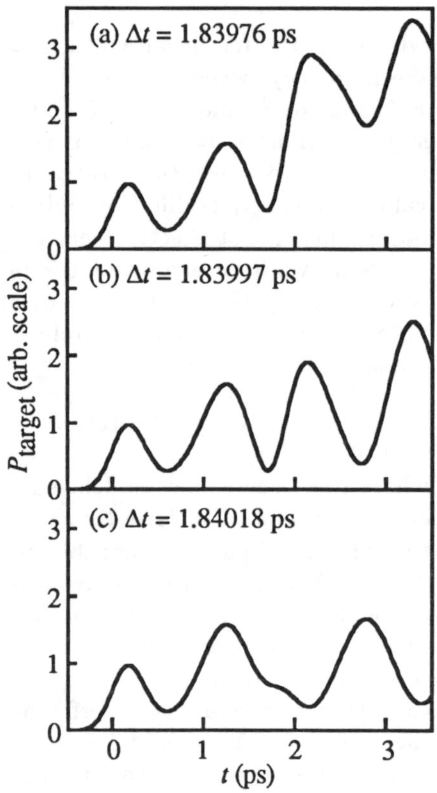

Figure 9. Temporal evolution of the probability P_{target} calculated for the double pulse excitation (see text). The second pulse is delayed by (a) 1.83976 ps, (b) 1.83997 ps, and (c) 1.84018 ps from the first pulse. $t = 0$ ps denotes the arrival of the first pulse.

3 Optical Switching of Ultrafast Thermal Atomic Collisions

In the "second-order optical collision (SOOC)", two-color, two-photon absorption takes place during a single binary collision. For the thermal collision energy, those two photons need to be absorbed sequentially in a few picoseconds since a single collision is completed in that time scale. It is, therefore, quite difficult to observe the SOOC under the thermal condition, and there has been only one previous measurement of that ultrafast event.[24] We have recently developed the "wing-wing double resonance (WWDR)" technique to observe that ultrafast SOOC using nanosecond pulsed lasers. In this section, it is applied to the Hg–Ar thermal collision to demonstrate that the WWDR technique realizes ultrafast pump/probe measurements with sub-picosecond time resolution using a pair of nanosecond laser pulses. Furthermore, a collision pathway can be controlled by changing the frequencies of those laser pulses. This experiment serves as the optical switching of an ultrafast thermal atomic collision.

3.1 Experiment

The third harmonic of a pulsed Nd:YAG laser was used to pump two dye lasers simultaneously. One of their outputs was frequency doubled to generate a pump pulse around 253.7 nm. The other dye-laser output around 435.8 nm was used as a probe pulse. The pulse duration was about 7 ns for both the pump and probe pulses. These pump and probe pulses were simultaneously introduced into the heat-pipe oven where the Hg vapor was generated and Ar was admitted to give a total pressure of 5.5 Torr. Figure 10 shows the present pump-probe scheme. The pump wavelength was fixed in the vicinity of the Hg $6^1S_0 - 6^3P_1$ resonance line at 253.7 nm. For the present measurements, the detuning Δ_{pump} of the pump laser from the resonance line was -25cm^{-1} or $+25\text{cm}^{-1}$, giving the $A-X$ or $B-X$ excitation, respectively, of the Hg–Ar collisional quasimolecule at the Condon point R_{pump}. The probe wavelength was scanned in the vicinity of the Hg $6^3P_1 - 7^3S_1$ resonance line at 435.8 nm. The quasimolecule travelling on the $A-$ or B-state potential absorbed this probe pulse at the Condon point R_{probe} on the way to the asymptote Hg(6^3P_1) + Ar to be further excited to the E state. And then the quasimolecule dissociates adiabatically on this E-state potential to the asymptote Hg(7^3S_1) + Ar. The fluorescence from this excited Hg(7^3S_1) atom was detected.

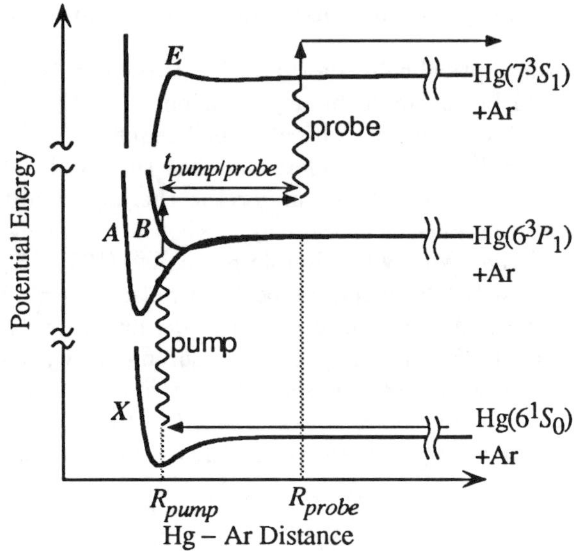

Figure 10. Schematic representation of the wing-wing double resonance (WWDR) technique applied to the Hg–Ar system. The quasimolecule travels from R_{pump} to R_{probe} in time $t_{pump/probe}$ on the *A* or *B* state potential.

3.2 Results and Discussion

We have measured the $X \to A \to E$ and $X \to B \to E$ SOOC spectra with Δ_{pump} being -25cm^{-1} and $+25\text{cm}^{-1}$, respectively. They are plotted in Fig. 11 against the detuning Δ_{probe} of the probe laser from the Hg $6^3P_1 - 7^3S_1$ resonance line. The resonances seen at $|\Delta_{probe}| = 25$ cm^{-1} are ascribed to nonresonant direct two-photon transitions between the E and X states, and they are not discussed here. As is seen in Fig. 11, the spectrum changes dramatically according to Δ_{pump}, indicating that the collision pathway has been switched from $X \to A \to E$ to $X \to B \to E$ by changing Δ_{pump} from -25cm^{-1} to $+25\text{cm}^{-1}$. This is the optical switching of an ultrafast thermal atomic collision.

Suppose that the quasimolecule travels on the A- or B-state surface from R_{pump} to R_{probe} in time $t_{pump/probe}$. Then $t_{pump/probe}$ is a function of the orbital angular momentum ℓ and expressed as

$$t_{pump/probe}(\ell, R_{probe}) = \sqrt{\frac{\mu_{\rm Hg-Ar}}{2}} \int_{R_{pump}}^{R_{probe}} \frac{dR}{\sqrt{E_{pump}(\ell) - V_{A/B}(R)}}, \quad (8)$$

where $\mu_{\rm Hg-Ar}$ is the reduced mass, and $E_{pump}(\ell)$ is the most probable translational energy, corresponding to the peak of thc Maxwellian distribution, at the moment of excitation by the pump pulse. E_{pump} is measured from the dissociation limit $\rm Hg(6^3P_1)$ + Ar. And $V_{A/B}(R)$ is the A- or B-state potential. $t_{pump/probe}$ is thus calculated for $\ell = 0$ and $\Delta_{pump} = \pm 25 \rm cm^{-1}$ and plotted against Δ_{probe} in Fig. 12. It is seen from this figure that one can carry out an ultrafast pump/probe measurment with sub-picosecond time-resolution using the present nanosecond-pulsed lasers. We have applied this methodology to the real-time probing of the alignment of $\rm Hg(6^3P_1)$ during a collision with Ar. The analyses of the data are now in progress.

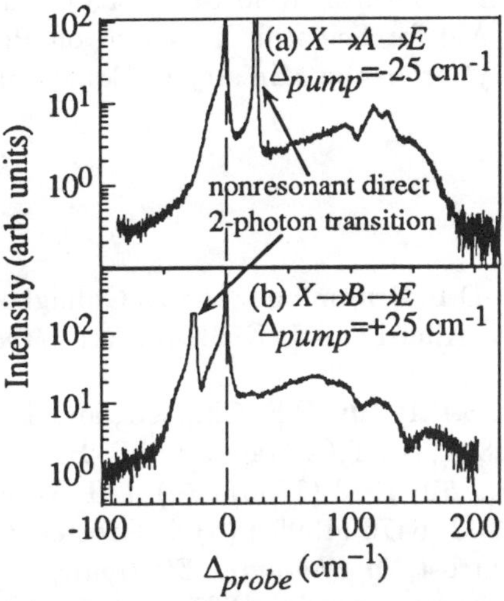

Figure 11. Second-order optical collision (SOOC) spectra of the Hg–Ar collisional quasimolecule. Δ_{pump} is the detuning of the pump laser from the Hg 6^1S_0–6^3P_1 resonance line, and Δ_{probe} is the detuning of the probe laser from the Hg 6^3P_1–7^3S_1 resonance line. The temperature is 373K.

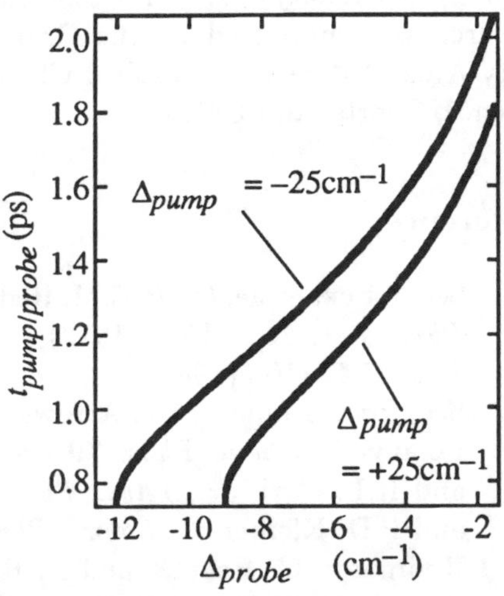

Figure 12. Correlation between the pump-probe delay $t_{pump/probe}$ and Δ_{probe} for the $X \to A \to E$ (Δ_{pump}=-25cm^{-1}) and $X \to B \to E$ (Δ_{pump}=+25cm^{-1}) SOOC's (see text) of Hg–Ar obtained for the orbital angular momentum ℓ =0 and the temperature T=373K.

Since $t_{pump/probe}$ depends on ℓ, a better time resolution is achieved by using the rotationally-cold vdW complex instead of the thermal collisional quasimolecule as the initial state to be excited by the pump pulse. In this measurement, for example, the Hg–Ar vdW complex is excited by the pump pulse to the repulsive wall of the B state to trigger a half collision, where the Maxwellian ditribution of the translational energy is no longer need to be considered. This project is now in progress in our group.

Acknowledgment

The author thanks the members of the laboratory for UV spectroscopy at RISM (Tohoku University) : Mr. K. Amano, Mr. H. Chiba, Mr. M. Shoji, Dr. M. Okunishi, and Professor Y. Sato. He also thanks Dr. T. Kurosawa (AIT), Mr. T. Takahashi (Tokyo Electron co.,ltd.), Professor Evgueni Nikitin (Technion - Israel Institute of Technology), Professor Alex Devdariani (St. Petersburg University), and Professor Satoshi Yabushita (Keio University). This research was supported by the Grant-in Aid for Scientific Research on Priority Areas "Molecular Physical Chemistry" from the Ministry of Education, Science, Sports, and Culture.

References

1. See, for example, (a) R.E.M. Hedges, D.L. Drummond, and A. Gallagher, *Phys. Rev.* **A 6**, 1519 (1972) ; (b) N. Allard and J. Kielkopf, *Rev. Mod. Phys.* **54**, 1103 (1982).
2. See, for example, (a) P. Arrowsmith, S.H.P. Bly, P.E. Charters, and J.C. Polanyi, *J. Chem. Phys.* **79**, 283 (1983) ; (b) T.C. Maguire, P.R. Brooks, and R.F. Curl, Jr., *Phys. Rev. Lett.* **50**, 1918 (1983) ; (c) T.H. Wong and P.D. Kleiber, *J. Chem. Phys.* **102**, 6476 (1995); (d) J. Grosser, O. Hoffmann, C. Rakete, and F. Rebentrost, in *Photonic, Electronic, and Atomic Collisions*, XX ICPEAC, Vienna, Austria, 1997, edited by F. Aumayr and HP. Winter (World Scientific, 1997), pp. 571.
3. (a) K. Ohmori, T. Takahashi, H. Chiba, K. Saito, T. Nakamura, M. Okunishi, K. Ueda, and Y. Sato, *J. Chem. Phys.* **105**, 7464 (1996) ; (b) K. Ohmori, T. Takahashi, H. Chiba, K. Saito, T. Nakamura, M. Okunishi, K. Ueda, and Y. Sato, *ibid.* **105**, 7474 (1996).
4. (a) K. Ohmori, T. Kurosawa, H. Chiba, M. Okunishi, K. Ueda, Y. Sato, and E.E. Nikitin, *J. Chem. Phys.* **102**, 7341 (1995) ; (b) K. Ohmori, T. Kurosawa, H. Chiba, M. Okunishi, and Y. Sato, *ibid.* **100**, 5381 (1994).

5. T. Kurosawa, K. Ohmori, H. Chiba, M. Okunishi, K. Ueda, Y. Sato, E.E. Nikitin, and A.Z. Devdariani, *J. Chem. Phys.* **108**, 8101 (1998).
6. K. Amano, K. Ohmori, T. Kurosawa, H. Chiba, M. Okunishi, K. Ueda, Y. Sato, E.E. Nikitin, and A.Z. Devdariani, *J. Chem. Phys.* **108**, 8110 (1998).
7. (a) K. Ohmori, K. Amano, M. Okunishi, H. Chiba, K. Ueda, and Y. Sato, *Chem. Lett.* 765 (1996) ; (b) Y. Sato, T. Nakamura, M. Okunishi, K. Ohmori, H. Chiba, and K. Ueda, *Phys. Rev.* **A 53**, 867 (1996) ; (c) T. Maeyama, H. Ito, H. Chiba, K. Ohmori, K. Ueda, and Y. Sato, *J. Chem. Phys.* **97**, 9492 (1992).
8. K. Ohmori and Y. Sato, in *Photonic, Electronic, and Atomic Collisions*, XX ICPEAC, Vienna, Austria, 1997, edited by F. Aumayr and HP. Winter (World Scientific, 1997), pp. 593–602.
9. K. Ohmori, in *Spectral Line Shapes*, Vol. 10, edited by R. Herman (AIP Press, 1999), pp. 352–363.
10. K. Ohmori, in *LASERS'98*, edited by V.J.Corcoran and T.A.Goldman (STS Press, McLean VA, 1999), pp. 106–115.
11. J. Pitre, K. Hammond, and L. Krause, *Phys. Rev.* **A 6**, 2101 (1972).
12. F. J. Van Itallie, L. J. Doemeny, and R. M. Martin, *J. Chem. Phys.* **56**, 3689 (1972).
13. H. F. Krause, S. Datz and S. G. Johnson, *J. Chem. Phys.* **58**, 367 (1973).
14. H. Horiguchi and S. Tsuchiya, *Bull. Chem. Soc. Jpn.* **44**, 1213 (1971); **47**, 2768 (1974); **50**, 1657, 1661 (1977).
15. H. Horiguchi and S. Tsuchiya, *J. Chem. Soc., Faraday Trans.* 2, **71**, 1164 (1975).
16. B. B. Brady, G. B. Spector, L. Chia, and G. W. Flynn, *J. Chem. Phys.* **86**, 3245 (1987).
17. K. Fuke, T. Saito, and K. Kaya, *J. Chem. Phys.* **81**, 2591 (1984).
18. K. Yamanouchi, S. Isogai, and S. Tsuchiya, M. -C. Duval, C. Jouvet, O. B. d'Azy, and B. Soep, *J. Chem. Phys.* **89**, 2975 (1988).
19. See, for example, (a) L. Krim, P. Qiu, C. Jouvet, C. Lardeux-Dedonder, J. G. McCaffrey, B. Soep, D. Solgadi, O. Benoist d'Azy, P. Ceraolo, N. Dai Hung, M. Martin, Y. Meyer, and J. P. Visticot, *Chem. Phys. Lett.* **89**, 200 (1992); (b) L. Krim, B. Soep, and J. P. Visticot, *J. Chem. Phys.* **103**, 9589 (1995); (c) C. Dedonder-Lardeux, M. Berdah, C. Jouvet, S. Martrenchard-Barra, J. M. Mestdagh, D. Sogadi, and J. P. Visticot, *J. Chem. Phys.* **104**, 2740 (1996).
20. M. -C. Duval, B. Soep, and W. H. Breckenridge, *J. Phys. Chem.* **95**, 7145 (1991), *and references cited therein.*

21. K. Amano, K. Ohmori, H. Chiba, M. Okunishi, Y. Sato, and S. Yabushita, *to be submitted.*
22. See, for example, (a) J. A. Ramswell, V. G. Stavros, J. Lei, Q. Hong, and H.H. Fielding, *Phys. Rev.* **A 59**, 2186 (1999); (b) D. W. Schumacher, J. H. Hoogenraad, D. Pinkos, and P. H. Bucksbaum, *Phys. Rev.* **A 52**, 4719 (1995); (c) R. R. Jones, C. S. Raman, D. W. Schumacher, and P. H. Bucksbaum, *Phys. Rev. Lett.* **71**, 2575 (1993); (d) J. F. Christian, B. Broers, J. H. Hoogenraad, W. J. van der Zande, and L. D. Noordam, *Opt. Commun.* **103**, 79 (1993); (e) L. D. Noordam, D. I. Duncan, and T. F. Gallagher, *Phys. Rev.* **A 45**, 4734 (1992).
23. (a) V. Blanchet, C. Nicole, M. -A. Bouchene, and B. Girard, *Phys. Rev. Lett.* **78**, 2716 (1997); (b) N. F. Scherer, R. J. Carlson, A. Matro, M. Du, A. J. Ruggiero, V. Romero-Rochin, J. A. Cina, G. R. Fleming, and S. A. Rice, *J. Chem. Phys.* **95**, 1487 (1991).
24. D. A. Olsgaard, M. D. Harvey, A. Sieradzan, and R. A. Lasell, *Phys. Rev. Lett.* **69**, 1745 (1992).

Part II. Coherent Phase Control

WHAT HAVE WE LEARNED FROM THE PHASE LAG IN COHERENT CONTROL EXPERIMENTS?

ROBERT J. GORDON, JEANETTE A. FISS, ANI KHACHATRIAN, KASPARS TRUHINS, AND LANGCHI ZHU

Department of Chemistry (m/c 111), University of Illinois at Chicago, 845 West Taylor Street, Chicago, IL 60607-7061

TAMAR SEIDEMAN

Steacie Institute for Molecular Sciences, National Research Council of Canada, Ottawa K1A-0R6, Canada

In coherent control experiments the product signal intensity is modulated by interference between two excitation paths. This modulation is produced by varying the relative phase of the electromagnetic fields used to excited the target. It is observed that the modulated signals for different channels may be out of phase with respect to each other. The phase lag between different channels is energy dependent and contains information about the dynamics of the system. This paper explores different mechanisms that produce such phase lags and assesses what may be learned from them.

1 Introduction

Interference between competing paths plays a central role in the formulation of quantum mechanics[1] and is responsible for a wide variety of physical phenomena. Examples of processes that depend on quantum mechanical interference are asymmetric spectroscopic lineshapes,[2] rainbow scattering,[3] photofragment and photoelectron angular distributions,[4,5] and atomic and molecular interferometry.[6,7] The key ingredient in all of these phenomena is the relative phases of the wave functions for various paths connecting the initial and final states of a system. The values of these phases are important because they carry the information about the scattering dynamics of the system. This fact is explicit in the semiclassical description of a wave function[8] and in inversion schemes used to extract potential energy functions from spectroscopic data.[9]

In the examples cited above, interference is produced by the superposition of two waves having a definite phase relation. The resultant signal is either enhanced or diminished, depending on whether the waves composing it are in or out of phase. As was demonstrated by Brumer and Shapiro over a decade ago,[10] this fact can be exploited to control the yields and branching ratios of chemical reactions.[11,12] If the two excitation paths consist of absorption of

m and n photons, then the total (angle-integrated) probability of obtaining product S is given by[10]

$$p^S(E) \;=\; p_n^S(E) \;+\; p_m^S(E) \;+\; 2|p_{mn}^S(E)|\cos(\phi + \delta_{mn}^S), \qquad (1)$$

where E is the total energy, and p_m^S and p_n^S are the transition probabilities for the absorption of n and m photons, respectively.[a] The cross term has an amplitude, $2|p_{mn}^S(E)|$, and a phase, $\phi + \delta_{13}^S$, where $\phi = n\phi_m - m\phi_n$ is the relative phase of the electromagnetic fields, and δ_{mn}^S is given by

$$|p_{mn}^S|e^{i\delta_{mn}^S} \;=\; e^{-i\phi}\int d\hat{k}\; < g|D^{(m)}|ES\hat{k}^- >< ES\hat{k}^-|D^{(n)}|g >, \qquad (2)$$

where $|ES\hat{k}^- >$ is a continuum eigenstate, $|g>$ is the ground state, and $D^{(j)}$ is the j- photon dipole operator. The quantity δ_{13}^S, which we shall refer to as the *channel phase*, depends solely on the properties of the material target.

In a typical coherent control experiment, the laser phase, ϕ, is varied continuously, producing a sinusoidal modulation of the product signals. Because δ_{13}^S depends on the product channel, the modulated signals for any pair of channels, A and B, differ in phase by a quantity,

$$\Delta\delta(A,B) \;=\; \delta_{mn}^A - \delta_{mn}^B, \qquad (3)$$

which we refer to as a *phase lag*. An example is shown in the bottom two panels of Fig. 1 for the ionization and dissociation of HI.

The phase lag is important in control experiments because it provides a tool for altering the branching ratio of a reaction. In recent years it was realized that the value (and energy dependence) of the phase lag is inherently interesting because of fundamental information that it provides about the system. In the present paper we present a catalogue of interference effects that produce a phase lag, summarizing the main theoretical conclusions and giving, when possible, experimental examples. In addition, we present new data on a "molecular interferometer" that allows one to extract the absolute phase of a wave function.

2 Experimental Method

All the experiments discussed here were performed using one- and three-photon excitation, with laser frequencies $\omega_3 = 3\omega_1$. The details of the experiment have been presented elsewhere[13] and need not be repeated here.

[a] In keeping with the convention used in nonlinear optics, we will use subscripts m and n to denote, respectively, the $n-$ and $m-$photon paths, such that $n\omega_m = m\omega_n$. But it will also be convenient to use the parenthetical superscript (j) to denote a $j-$photon transition; e.g., $D^{(3)}$ denotes the three-photon dipole operator.

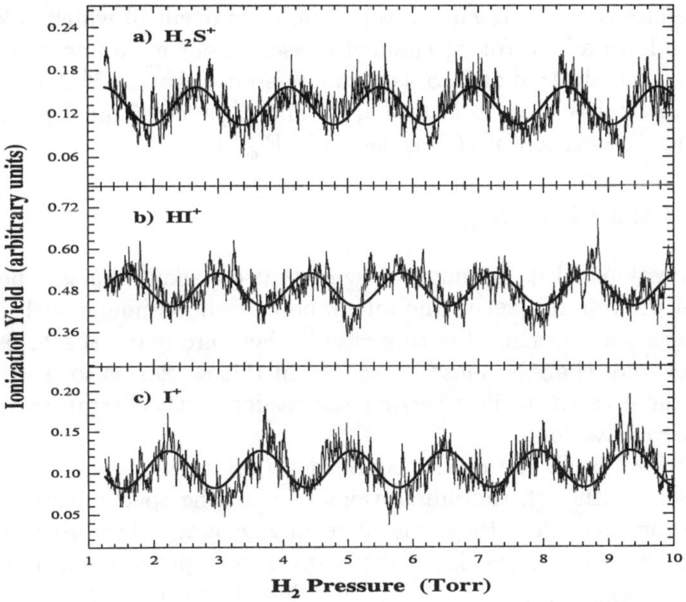

Figure 1. Modulation curves for H_2S^+, HI^+ and I^+ for a mixture of H_2S and HI. The wavelength of the ω_1 field is 353.80 nm. Taken by permission from Ref. 14.

Suffice it to say that ω_3 was produced from ω_1 by third harmonic generation in a rare gas, thereby establishing a definite phase relation between the two fields. The relative phase of the fields was controlled by passing the laser beams through a gas that has a different index of refraction at the two frequencies. Products from the various reaction channels were measured with a time-of-flight mass spectrometer.

One detail which needs to be emphasized is the method used to isolate individual channel phases. Although the observed phase lag is the difference between two channel phases, by selecting one of the channels, say channel A, to have a real transition amplitude (i.e., $\delta_{13}^B = 0$ or π), it is possible to determine δ_{13}^A absolutely (modulo $\pm\pi$). A convenient way of selecting such a reference channel is to use a gas mixture, one component of which is known empirically to have a zero (or π) channel phase. In our experiments to date we have used the ionization of H_2S as such a reference channel.[14] An example of how this channel is used to isolate $\delta_{13}^{HI^+}$ and δ_{13}^{I} (the channel phases for ionization and dissociation of HI) is shown in Fig. 1.

3 Origin of the Phase Lag

A casual inspection of Eq. (2) might suggest that δ_{13}^{S} should be zero because the phases of the bra and ket in the integrand cancel. Although such cancelation does occur in a certain limiting case,[15] there are many scenarios that produce a non-zero channel phase. Our goal in this section is to provide an overview of these cases, while referring the reader to the literature for the details of their derivation.

Figure 2 is a catalogue of the conditions that might lead to non-zero channel phases. Although the illustrations refer to the specific case of one- vs. three-photon excitation, they may be readily generalized to other pairs of excitation paths. Broadly speaking, there are three types of conditions that lead to non-zero channel phases. First, coupling of the three-photon continuum state to some other continuum introduces a channel phase which has been referred to in some of the literature as a "molecular phase." Second, the existence of resonances at the three-photon level may also produce a channel phase, sometimes referred to as a "resonance phase." Third, resonances at intermediate energies in the three-photon path contribute a Breit-Wigner phase to the overall channel phase.

Figure 2a depicts the limiting case in which there is no channel phase. If there are no resonances present, and if the continuum is not coupled to any other continua, then cancelation occurs and the channel phase vanishes. The continuum in this case produces only elastic scattering. The phase shifts of the component partial waves of the scattered state are independent of coordinates, resulting in a term-by-term cancelation of the phases in a partial wave expansion. An example of this case is seen in $\Delta\delta(HI^+, H_2S^+)$ shown in Fig. 3, where the peaks near 355.0 and 356.0 nm ride on top of a baseline of zero phase lag.[16]

If the three-photon continuum is coupled to some other continuum (i.e., if it is inelastic), then term-by-term cancelation of the phases does not occur. Another way of stating this result is to express $|ES\hat{k}^- >$ in Eq. (2) as a linear combination of zero-order continuum states. The integrand then contains cross terms with different bras and kets, so that the phases do not cancel. An example of this case is seen in the dissociation channel of HI in Fig. 3 (i.e.,

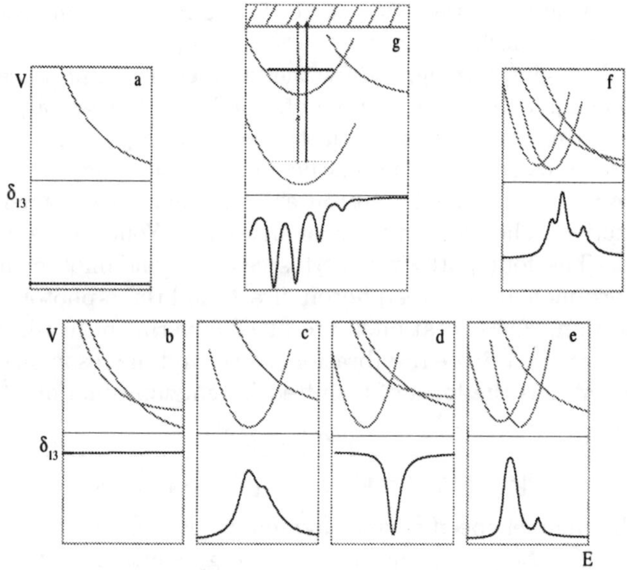

Figure 2. Schematic drawing showing different sources for the channel phase, δ_{13}^S. (a) In the case that the continuum potential induces only elastic scattering, δ_{13}^S vanishes. (b) If the continuum is coupled to a second continuum, δ_{13}^S is a smooth, nonzero function of energy. (c) An isolated resonance that interacts with a purely elastic continuum gives rise to a maximum at the resonance energy. The relative phase arises from the interference of the direct and resonance-mediated routes and is maximized where interference is constructive. (d) In the case that an isolated resonance interacts with a coupled continuum, the phase shift is nonzero off-resonance and falls to a minimum on-resonance. (e) Although an isolated resonance with no direct route to the continuum does not produce a phase, coupled resonances give rise to a nonzero δ_{13}^S regardless of the availability of a direct route. (f) Coupled resonances interacting with a coupled (i.e., inelastic) continuum produce a structured maximum in δ_{13}^S superimposed on a nonzero, slowly varying background. (g) A qualitatively different source of a relative phase is a resonance located at an intermediate energy in the three-photon path. In the case that the continuum at the three-photon level is elastic, δ_{13}^S is the Breit-Wigner phase of the intermediate resonance.

$\Delta\delta(I^+, H_2S^+)$), where the weak structure rides on top of a molecular phase baseline of approx. -120°.

Panels (c) through (f) of Fig. 1 show the channel phases produced by resonances located energetically at the three-photon level. In case (c), an isolated resonance interacting with a purely elastic continuum gives rise to a maximum at the resonance energy. The relative phase arises from the interference of the direct and resonance-mediated routes and is maximized where interference is constructive. The mechanism is analogous to Young-type interference from four slits. The four pathways in this case are one-photon direct, one-photon resonance-mediated, three-photon direct, and three-photon resonance-mediated excitation. For a rotationally resolved experiment in which the parent state is prepared in a single ro-vibrational level and the resonance vibration is resolved, tan δ_{13}^S has the form of a shifted Lorentzian function,[17]

$$tan\ \delta_{13}^S = \frac{2(q^{(1)} - q^{(3)})}{\left[\epsilon - \frac{1}{2}(q^{(1)} + q^{(3)})\right]^2 + \left[4 - \frac{1}{4}(q^{(1)} - q^{(3)})^2\right]}, \qquad (4)$$

where $q^{(j)}$ is the channel-specific Fano asymmetry parameter[2] for a $j-$ photon transition, $\epsilon = (E - E_0)/2\Gamma$ is the reduced energy shift, E_0 is the resonance energy, and Γ is the width of the resonance. The reader is referred to Ref. 17 for the general case of many rotational levels. An example of a resonance phase where many rotational levels are present is given by the 356.0 nm peak in Fig. 3, which is produced by the $5d(\pi, \delta)$ resonance of HI.[16] (The smaller peak at 355.0 nm is probably caused by another resonance that has only been tentatively assigned.) Another example is the $5s\sigma$ ionization resonance of HI and DI shown in Ref. 18.

If an isolated resonance is present and the direct process is negligible compared to the resonance-mediated one (case (d)), the phase shift vanishes at the resonance energy. In this limit, the excitation and decay processes decouple, and all memory of the excitation scheme is lost once the molecule reacts. An experimental signature of this case is a *minimum* in $|\delta_{13}^S|$ on resonance, with a non-zero value of the phase lag produced off-resonance by a molecular phase. The reader is referred to Ref. 15 for a derivation of this result. An example of such a "window" is the $5s\sigma$ dissociation resonance for DI shown in Fig. 4.[18] The strong dip in $\Delta\delta(HI^+, I)$ is also caused by this mechanism; however, the additional structure near 353.7 nm is as yet unexplained.

A more complex pattern is produced in case (f), where coupled resonances interacting with a coupled (i.e., inelastic) continuum produce a structured maximum in δ_{13}^S superimposed on a nonzero, slowly varying background.[17] There are as yet no reported examples of this case. At first it was believed that the $5d(\pi, \delta)$ feature in Fig. 3 is produced by such a coupled resonance;

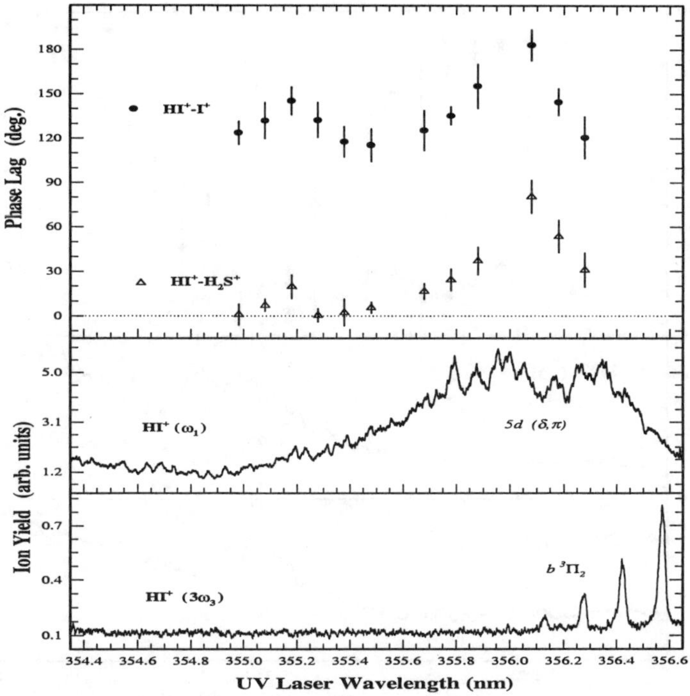

Figure 3. Phase lag spectrum (top panel) for the photodissociation and photoionization of HI (circles) and for the photoionization of a mixture of HI and H_2S (triangles) in the vicinity of the $5d(\pi, \delta)$ resonance of HI. The bottom panel is the one-photon ionization spectrum of HI. Taken with permission from Ref. 16.

however, in this instance the three-photon resonant-mediated transition is much weaker than the other paths. There results a "three-slit" mechanism, with one-photon direct, one-photon resonance mediated, and three-photon direct paths.

4 A Molecular Interferometer

A qualitatively different mechanism is depicted in Fig. 1g. In this case a resonance is present at the two-photon level, making $D^{(3)}$ a complex operator.

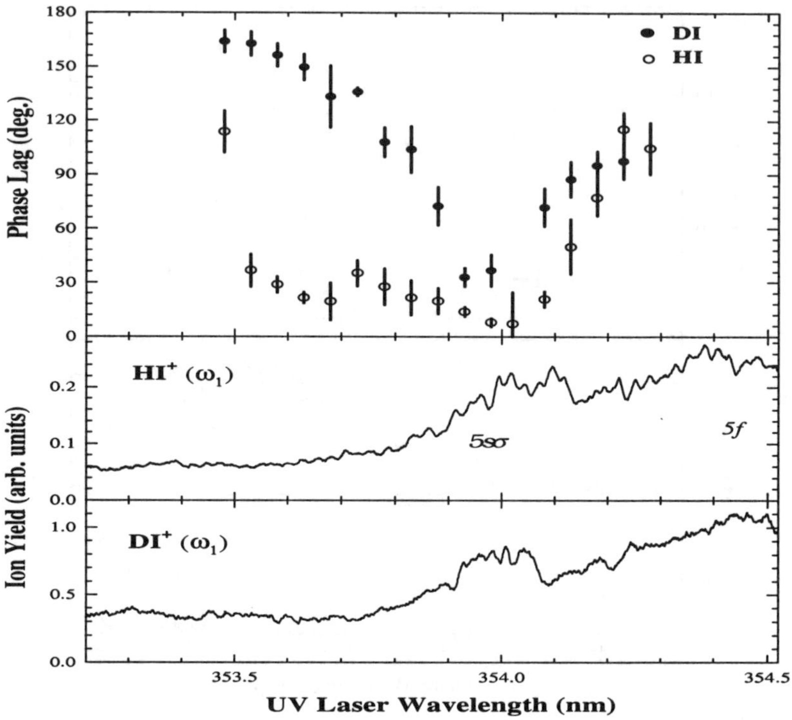

Figure 4. Phase lag spectrum (top panel) for the photodissociation and photoionization of HI (open circles) and DI (closed circles) in the vicinity of the $5s\sigma$ resonance. The bottom two panels are the one-photon ionization spectrum of HI and DI. Taken with permission from Ref. 18.

In the event that there are no molecular or resonance phases present at the three-photon level, the channel phase is identical to the phase of the two-photon state. The system behaves as a "molecular interferometer," with δ_{13}^{S} providing a measure of the phases of states present in only one arm.

The quantity of interest in this case is the Breit-Wigner phase of a quasi-bound state. The formalism required for computing the relative phase arising from complex intermediates is developed in Ref. 17 and only briefly summarized here. Integration over scattering angles and thermal averaging transform the relative phase in Eq. (2) to the form

$$\delta_{13}^S = \arg \sum_{J_g} W_{J_g} \sum_J \mathcal{T}^{(1)}(J_g|ESJ)\mathcal{T}^{(3)^*}(J_g|ESJ), \quad (5)$$

where $\mathcal{T}^{(j)}$ is an angle-averaged $j-$photon transition dipole matrix element, defined in the body-fixed frame, J_g and J are total angular momenta in the initial and continuum states, and W_{J_g} are Boltzmann weights, determined by the rotational temperature of the molecular beam. Complex intermediates in the three-photon process give rise to a complex energy denominator in $\mathcal{T}^{(3)}$, which translates into an observable phase, δ_{13}^S.

In the case that the rotational levels are well-separated, $B_e \gg \Gamma$ where B_e is the rotational constant, δ_{13}^S traces in the vicinity of each line position the Breit-Wigner phase of that resonance. That is,

$$\delta_{13}^S = \arg\{\mathcal{T}^{(1)}(J_g|ESJ)\mathcal{T}^{(3)^*}(J_g|ESJ)\} = \delta_\infty - \delta_{\rm res}(E), \quad (6)$$

where δ_∞ is an asymptotic phase that that reduces to 0 or $\pm\pi$ in the absence of the coupling mechanisms discussed in the previous section, and

$$\delta_{\rm res} = -\arg(E - E_{J_2}) = -\tan^{-1}[(\Gamma/2)/(E - E_{J_2}^R)] \quad (7)$$

is the Breit-Wigner phase, E_{J_2} being the complex energy eigenvalue of the J_2 rotational resonance and $E_{J_2}^R$ its real part (the line position). The Breit-Wigner phase has a value of $-\pi$ to the red of the resonance position, reaches $-\pi/2$ on resonance, and approaches 0 above resonance. The energy dependence of $\delta_{\rm res}$ for an isolated rotational level is plotted in Fig. 5a.

For a thermally averaged parent state, Eq. (6) generalizes as[17]

$$\tan\,\delta_{13}^S = \frac{\sum_{J_gJ_2} A_{J_g,J_2}\frac{1}{2}\Gamma_{J_2}[(E_{J_g}+2\omega-E_{J_2}^R)^2+\Gamma_{J_2}^2/4]^{-1}}{\sum_{J_gJ_2} A_{J_gJ_2}(E_{J_g}+2\omega-E_{J_2}^R)[(E_{J_g}+2\omega-E_{J_2}^R)^2+\Gamma_{J_2}^2/4]^{-1}}, \quad (8)$$

where A_{J_g,J_2} is a multiple sum over products of dynamical, geometric, and Boltzmann weighting factors, given explicitly in 17, and δ_∞ is set equal to zero. The exact form of Eq. (8) depends (through the A_{J_g,J_2}) on the ground, intermediate, and continuum potential energy surfaces and is thus useful only in cases where this information is available. A fully analytical expression, which does not rely on knowledge of the electronic structure of the molecule and also provides better insight into the origin and structure of δ_{13}^S, can be derived by introducing a single (and common[19]) approximation in Eq. (8). Namely, we neglect the dependence of the body-fixed eigenfunctions and the corresponding eigenvalues on the total angular momentum. Within this approximation, it is readily shown[17] that all dynamical factors in the $A_{J_gJ_2}$

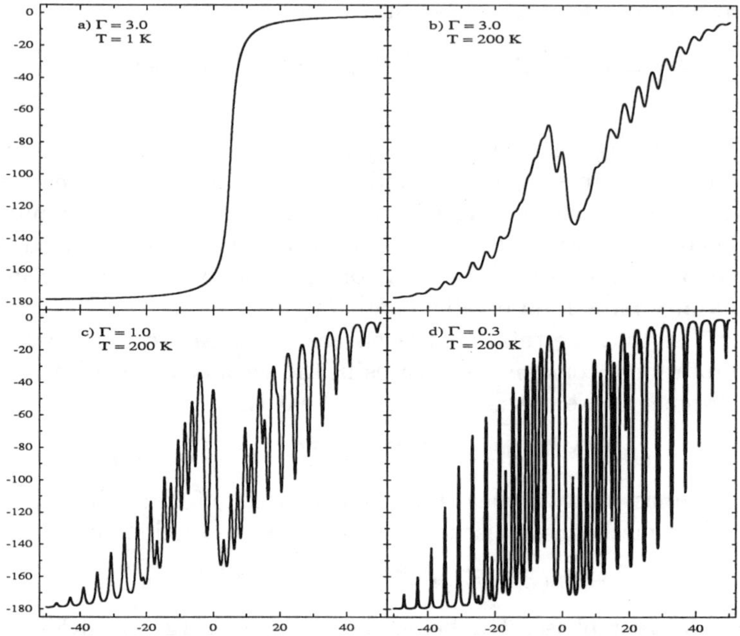

Figure 5. Model calculation of the Breit-Wigner phase for a Boltzmann ensemble of molecules for various values of the temperature, T, and resonance width, Γ. The abscissa is the displacement of the energy from the $0-0$ transition divided by the rotational constant of the excited state. The rotational constant was taken to be 6.427 cm^{-1} for both the ground and excited states, and a $\Sigma \rightarrow \Sigma \rightarrow \Pi \rightarrow \Pi$ three-photon transition was assumed. (a) T=1 K, $\Gamma = 3B_e$. In this case only the $S(0)$ transition occurs. (b) T=200K, $\Gamma = 3B_e$. (c) T=200K, $\Gamma = 1B_e$. (d) T=200K, $\Gamma = 0.3B_e$.

cancel out between the numerator and denominator of Eq. (8), obtaining an analytical expression that depends only on the resonance width and the beam temperature.

Model calculations of δ_{res} for a thermal mixture at 200 K are presented in Fig. 5b-d for a range of Γ values. Throughout the range the overall S shape of the phase is evident. For the broadest case considered ($\Gamma = 3B_e$ in Fig. 5b), the $O-$ and $S-$type rotational branches appear as weak oscillations far from resonance, whereas the $P-$, $Q-$, and $R-$branches strongly overlap. For the sharpest case considered ($\Gamma = 0.3B_e$ in Fig. 5d), all the rotational branches are well resolved.

In an experimental study we measured the Breit-Wigner phases of complex intermediate states of hydrogen iodide molecules.[21] These states are members of the rotational manifold of the $b^3\Pi_1$ Rydberg state that are located energetically at two-thirds of the ionization energy, i.e., at $E_{J_g} + 2\omega_3$. A partial level scheme of HI showing the relevant levels is given in Fig. 6. The inset shows the rotational levels of the quasi-bound $b^3\Pi_1$ state, which is predissociated by the $A^1\Pi$ continuum state. The lasers are tuned away from any three-photon resonances, so that only a direct transition to the ionization continuum occurs. The ionization of HI and H_2S at the three-photon energies considered here do not contribute to the phase lag, and hence $\Delta\delta(HI^+, H_2S^+)$ is a direct measure of the absolute phase of the two-photon resonant states of HI.

The experimental results are shown in Fig. 7. In panel (a), δ_{13}^S for the ionization of HI (obtained as the phase lag between the HI^+ and H_2S^+ yield curves) is plotted as a function of the three-photon wavelength. The three-photon spectrum of HI^+ is given in panel (b), with the peaks assigned by the O-branch rotational resonances at the two-photon level.[20] In panels (c) and (d) are shown the one-photon ionization spectra of HI and H_2S. The lack of structure in the one-photon spectra demonstrates the absence of resonances at the three-photon level.

The observed δ_{13}^S (Fig. 7a) is seen to follow the shape predicted by Eq. (8), providing a direct measure of the Breit-Wigner phase of the quasi-bound $b^3\Pi_1$ rotational manifold. The solid curve in Fig. 7a shows a least squares fit of the data to the analytical form of Eq. (8). Allowing up to a linear dependence of the resonance width on J_2, we find a rotational temperature of 236 K and $\Gamma_{J_2} = (5.5 + 0.61J_2)$ cm^{-1}. (A quadratic expansion of Γ_{J_2} gives a similar quality fit.) The Breit-Wigner phases of individual rotational resonances [see Eq. (7)] obtained from the fit are shown by the dot-dashed curves in Fig. 7a. The increase of Γ_{J_2} with J_2 is indicative of a rotational perturbation coupling the $b^3\Pi_1$ state to the continuum and is consistent with our spectroscopic observation that the branching ratio of predissociation vs. ionization increases with J_2.[22]

5 Future Directions

The theoretical and experimental results summarized in this paper demonstrate how the phase lag obtained in coherent control experiments can be a

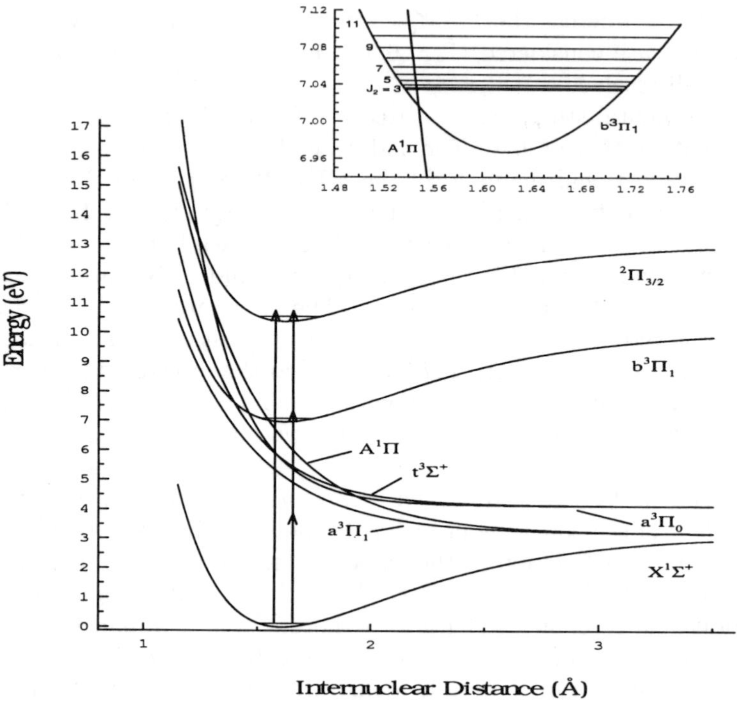

Figure 6. Excitation scheme of HI. One and three photons are used to produce ground state $HI^+(^2\Pi_{3/2})$ ions. The $b^3\Pi_1$ state, located near the two-photon level, is predissociated by several continuum states. The inset shows the rotational levels of the $b^3\Pi_1(v=0)$ state, which are predissociated by the $A^1\Pi$ continuum state.

powerful tool for studying the properties of the continuum. The availability of both the phase and modulus of the dipole transition matrix element opens up various exciting possibilities for future research. One of these is the direct inversion of the phase lag spectrum to determine the cross correlation function that underlies the intermediate state dynamics. Another direction that we are exploring is use of the phase lag to detect spectroscopic transitions that are too weak to observe by conventional absorption spectroscopy. Finally, with a more complete understanding of its physical origin, we are better equipped to exploit the phase lag as a tool for coherent control.

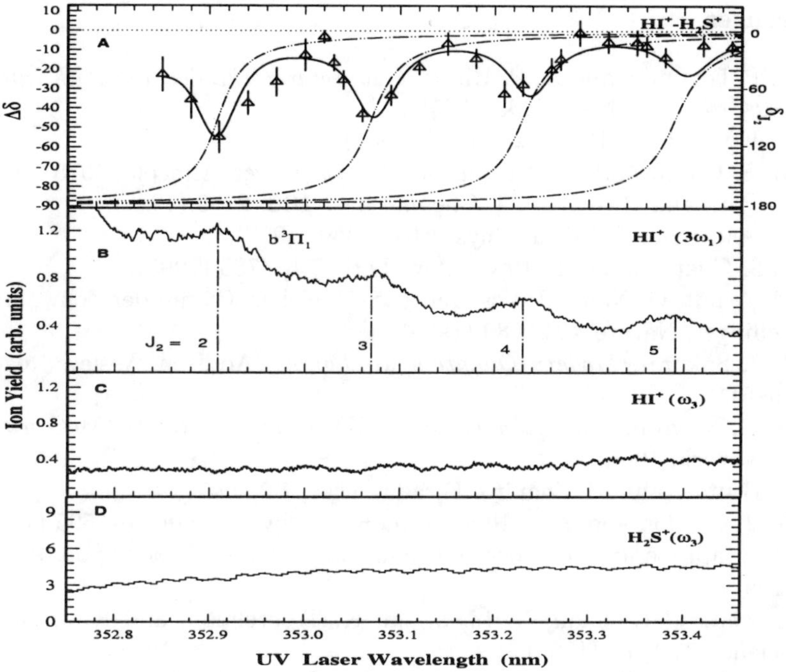

Figure 7. Phase lag and ionization spectra of HI and H_2S. Panel (a) shows the observed phase lag between the HI^+ and H_2S^+ signals. The solid curve is a least squares fit of Eq. (8) to the data. The dot-dashed curves are the Breit-Wigner phases of individual rotational resonances, deduced from the data using Eq. (8). Panel (b) is the three-photon ionization spectrum of HI, showing the two-photon, O-type rotational branch of the $b^3\Pi_1(v_2 = 0, J_2) \leftarrow X^1\Sigma^+(v_g = 0, J_g)$ transition. Panels (c) and (d) show the one-photon ionization spectra of HI and H_2S, respectively.

We wish to thank the National Science Foundation for its generous support.

References

1. R. P. Feynman and A. R. Hibbs, *Quantum Mechanics and Path Integrals* (McGraw Hill, New York, 1965).
2. U. Fano, Phys. Rev. **124**, 1866 (1961).
3. M. S. Child, *Molecular Collision Theory* (Dover, Mineola, 1975)
4. H. Park and R.N. Zare, J. Chem. Phys. **106**, 2239 (1997).
5. T. Seideman, J. Chem. Phys. **107**, 7859 (1997).
6. M.S. Chapman et al., Phys. Rev. Lett. **74**, 4783 (1995).
7. M. Arndt, O. Nairz, J. Vos-Andreae, C. Keller, G. van der Zouw, and A. Zellinger, Nature **401**, 680 (1999).
8. J. J. Sakurai, *Modern Quantum Mechanics* (Addison Wesley, Reading, 1995).
9. R. G. Newton, *Scattering Theory of Waves and Particles*, (McGraw-Hill, New York, 1966).
10. P. Brumer and M. Shapiro, Faraday Disc. **82**, 177 (1986).
11. R. J. Gordon and S. A. Rice, Ann. Rev. Phys. Chem. **48**, 595 (1997).
12. M. Shapiro and P. Brumer, J. Chem. Soc., Trans. Faraday Soc. **93**, 1263 (1997).
13. L. Zhu, V. Kleiman, X. Li, S. Lu, K. Trentelman, and R. J. Gordon, Science **270**, 77 (1995).
14. J. A. Fiss, L. Zhu, R. J. Gordon, and T. Seideman, Phys. Rev. Lett. **81**, 65 (1998).
15. T. Seideman, J. Chem. Phys. **108**, 1915 (1998).
16. J. A. Fiss, A. Khachatrian, L. Zhu, R. J. Gordon, and T. Seideman, Disc Faraday Soc. **113**, 61 (1999).
17. T. Seideman, J. Chem. Phys. **111**, 9168 (1999).
18. L. Zhu, K. Suto, J. A. Fiss, R. Wada, T. Seideman, and R. J. Gordon Phys. Rev. Lett. **79**, 4108 (1997).
19. W. M. McClain and R. A. Harris, in *Excited States*, E. C. Lim, ed., (Academic, New York, 1977), Vol. 3, pp 1-56.
20. J. A. Fiss, L. Zhu, K. Suto, G. He, and R. J. Gordon, Chem. Phys. **233**, 335 (1998).
21. J.A. Fiss, A. Khachatrian, K. Truhins, L. Zhu, R. J. Gordon, and T. Seideman (submitted).
22. A. Khachatrian, K. Suto, L. Zhu, and R. J. Gordon, unpublished results.

PHASE-SENSITIVE PHOTOIONIZATION OF ATOMS AND ITS APPLICATION

TAKASHI NAKAJIMA

Institute of Advanced Energy, Kyoto University
Gokasho, Uji, Kyoto 611-0011, Japan

We propose a new method for the *direct* determination of the quantum phase of continua utilizing the phase of linearly/circularly polarized lasers. With this method the phase difference of continua with opposite as well as same parities can be directly determined from the *phase lag* of the angle-resolved photoelectron signal with respect to the relative phase of lasers. For illustration, the proposed method is specifically applied to the Na atom to determine the phase differences between s, p, and d waves.

1 Introduction

The control of photoabsorption processes of atoms and molecules through the relative phase of lasers represents one of the schemes to manipulate light-matter interactions, which has been under extensive study in the last several years.[1–12] Starting from the control of transition strength between two bound states,[1–5] the study has been extended to bound-free[6–12] and bound-autoionizing transitions,[8–10,12] where not only the control of photoionization/-dissociation yields themselves but also the control of branching ratio into different channels has been investigated . The product yield modulates as a function of relative phase ϕ of two lasers, and in general, the phase of modulation differs for each product,[11,12] since the phase originated from intra-atomic/-molecular interactions differs from channel to channel. Thus it might come to mind that the determination of such a phase difference might be possible utilizing phase-controlled lasers. In the case of an atomic system, on which we focus in this work, such a phase represents nothing but a phase shift due to electrostatic interaction between an electron and a core. If, in particular, we consider a single-valence-electron atom, the total phase shift δ_l is given by

$$\delta_l(W) = \delta_l^C(W) + \pi\mu_l(W) \ , \tag{1}$$

where $\delta_l^C(W)$ and $\mu_l(W)$ are a Coulomb phase shift and a quantum defect, respectively, for the l symmetry at energy W (in Rydberg units) above the ionization threshold. δ_l^C has been defined, as usual, as

$$\delta_l^C(W) = arg\left(\Gamma(l+1-i/\sqrt{W})\right) - \frac{\pi}{2}l\,. \tag{2}$$

In general, a precise value of the quantum defect for a continuum state is not known, theoretically or experimentally, even for a single-valence electron system, since it reflects short-range and therefore many-body interactions of the valence electron with other electrons in the core. Thus the determination of the phase shift is not a simple issue.

It is well-known that a photoelectron angular distribution (PAD) contains a lot of information on transition amplitudes and the difference of the phase shifts of wavefunctions, and in principle, these quantities could be *indirectly* determined by measuring PAD's [13,14] and fitting theoretical curves to them. It should be recognized, however, that the experimental determination of these quantities is not trivial even for alkali-metal atoms, since there are several parameters, such as ratios of transition amplitudes and the difference of phase shifts, to be fitted.

In Fig. 1 we plot the PAD of Na($3s$) by non-resonant two-photon ionization to the final state energy of W=0.0278 Ryd above ionization threshold.

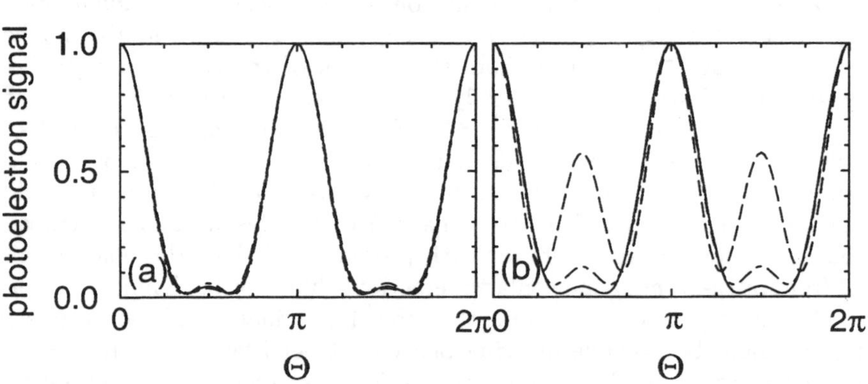

Figure 1. Variation of the PAD of Na with slightly modified values of radial matrix elements and phase shifts under single-color, non-resonant two-photon ionization. (a) PAD with the ratio of radial matrix elements $R^{(2)}_{ks}/R^{(2)}_{kd}$ kept at its original value obtained from single-channel quantum defect theory (solid), and changed by $\pm 15\%$ (dashed and dot-dashed) from its original value. The phase difference $\delta_s - \delta_d$ is kept at its original value. (b) PAD with phase difference kept at its original value (solid), and changed by $\pm 15\%$ (dashed and dot-dashed) from its original value. The ratio of radial matrix elements is kept at its original value for both.

Parameters used for calculations are obtained using single-channel quantum defect theory. In Fig. 1(a), we show the variation of the PAD with slightly modified ratios (± 15%) of the radial matrix elements to the two continua *ks* and *kd*, where *k* stands for the kinetic energy of an outgoing photoelectron. Similarly, in Fig. 1(b), the variation of the PAD with slightly modified values (±15%) of phase differences are plotted. Again it can be seen that the determination of phase difference is a rather delicate issue.

The potential use of the phase of lasers for the determination of the phase difference of continua has been briefly mentioned some time ago,[6] but a straightforward application of phase-controlled two-color lasers instead of a single-color laser does not improve the situation for that purpose. Under the simultaneous action of two laser fields, it is quite essential to know the precise values of transition strengths into each channel, which are products of transition matrix elements and laser intensities, if the standard PAD technique mentioned above is to be applied to extract the values of phase difference of continua. The use of two laser fields simply introduces additional complexity, leading to the even lower reliability of the fitted quantities.

In this work, we propose that the phase difference of continua can be directly and precisely determined with the use of phase-controlled lasers with frequencies ω and 2ω. It should be noted that the proposed method allows us to determine the phase difference of continua with *different parities* as well as the same parity, for example, *s* and *p* continua, which cannot be measured otherwise. Such a determination would become possible by a careful choice of laser polarization and also by *isolating* a particular pair of partial waves of interest through angle-resolved photoelectron detection. In contrast to the conventional PAD technique where the phase difference is determined by fitting a theoretical curve of the PAD with several free parameters to that of an experimental one, the phase difference is observed directly as a *phase lag*[11,12] with this proposed method with respect to the relative phase of two lasers. It is interesting to point out that, when this method is applied to light alkali-metal atoms such as Li, Na, or K, the ratios of radial matrix elements do *not* play a role in the determination of phase difference. Although the determination of phase difference by the conventional PAD technique relies on those ratios whose accuracy is always questionable, our method is not affected by those quantities. For heavier alkali-metal atoms such as Rb or Cs, however, the spin-orbit interaction plays an important role, and more careful consideration is necessary. Nevertheless, we find that the direct determination of the phase difference is possible, in principle, even under the presence of strong spin-orbit interaction.

2 Photoelectron Angular Distribution

We start with the description of a continuum of an alkali-metal atom. For a given wave vector $\vec{k}$ and magnetic quantum number m_s of the spin of the outgoing photoelectron, the continuum state is expanded using partial waves:

$$|\vec{k}\,; m_s\rangle = \sum_{l,m_l} a_{lm_l} |\vec{k}\,; l\, 1/2\, m_l\, m_s\rangle\ , \tag{3}$$

where

$$a_{lm_l} = 4\pi i^{\,l} e^{-i\delta_l}\, Y^*_{lm_l}(\Theta, \Phi) \tag{4}$$

and

$$|\vec{k}\,; l\, 1/2\, m_l\, m_s\rangle = |l\, m_l\rangle |1/2\, m_s\rangle\ . \tag{5}$$

For an alkali-metal atom, the total phase shift δ_l of the l partial wave is a sum of the Coulomb phase shift and a shift due to the quantum defect as given by Eqs. (1) and (2). Y_{lm} are the spherical harmonics, and $(\Theta, \Phi)(\equiv \Omega)$ defines the direction of the outgoing photoelectron with respect to the quantization axis. As usual, the quantization axis is taken along the polarization direction for linearly polarized light, and the propagation direction for circularly polarized light, respectively. Under the presence of two laser fields with amplitudes ε_ω and $\varepsilon_{2\omega}$ for frequencies ω and 2ω, and relative phase ϕ, the total laser field is written as

$$E(t) = (\varepsilon_\omega\, e^{i\omega t} + \varepsilon_{2\omega}\, e^{i(2\omega t+\phi)}) + c.c. \tag{6}$$

For a given initial state defined by $|(l'\ 1/2)\ j'\ m_{j'}\rangle$, the angular distribution of photoelectron with spin m_s is given by

$$\left.\frac{d\sigma}{d\Omega}\right|_{m_{j'},m_s} = \left|\sum_{lm_l} a^*_{lm_l}\, (-1)^{l-1/2+m_j} \sqrt{2j+1} \begin{pmatrix} l & s & j \\ m_l & m_s & -m_j \end{pmatrix} \right.$$
$$\left. \times\ \langle \vec{k}\,; (l\ 1/2)\ j\ m_j | e^{i\phi} D_q + D_q^{(2)} | (l'\ 1/2)\ j'\ m'_j\rangle \right|^2\ , \tag{7}$$

where D_q and $D_q^{(2)}$ are the single- and two-photon dipole operators, respectively, which are nothing but the qth spherical components of the dipole moments r_q and $r_q^{(2)}$ multiplied by some appropriate factors, i.e., $D_q = C_{2\omega} r_q$ and

$D_q^{(2)} = C_\omega r_q^{(2)}$, with $C_{2\omega} = 0.767\sqrt{I_{2\omega}}$ and $C_\omega = 2.05 \times 10^{-9} I_\omega$, respectively.[9] $I_{2\omega}$ and I_ω are the laser intensities (in W/cm^2) with frequencies 2ω and ω. If the initial state is unpolarized and the spin of photoelectrons is not resolved upon detection, the angular distribution is obtained by summing over m_j and m_s, which reads,

$$\frac{d\sigma}{d\Omega} = \sum_{m_{j'}, m_s} \left. \frac{d\sigma}{d\Omega} \right|_{m_{j'}, m_s} . \tag{8}$$

The matrix element can be computed by using the following relation:

$$\begin{aligned} \langle (l\ 1/2)\ j\ m_j | r_q | (l'\ 1/2)\ j'\ m_{j'} \rangle = &\ (-1)^{j-m_j+1/2+1+l+j'} (-1)^{l+l_>} R \\ & \times \sqrt{(2j+1)(2j'+1)} \begin{Bmatrix} l & j & 1/2 \\ j' & l' & 1 \end{Bmatrix} \begin{pmatrix} j & 1 & j' \\ -m_j & q & m_j' \end{pmatrix} , \end{aligned} \tag{9}$$

where R and $l_>$ denote the radial matrix element from the initial to the final states and the larger of l and l', respectively.

Now we further assume that the initial state is the ground state, i.e., $s_{1/2}$ state. In such a case, if both fields are linearly polarized, Eq. (7) becomes

$$\begin{aligned} \frac{d\sigma}{d\Omega} \propto & \left| -i\frac{\sqrt{3}}{3} C_{2\omega}\, Y_{10}\, (R_1 + 2R_2)\, e^{i(\delta_p+\phi)} + C_\omega \left\{ \frac{1}{3} Y_{00}\, (R_1^{(2)} + 2R_2^{(2)})\, e^{i\delta_s} \right.\right. \\ & \left.\left. - \frac{2\sqrt{5}}{75} Y_{20} (5R_3^{(2)} + R_4^{(2)} + 9R_5^{(2)})\, e^{i\delta_d} \right\} \right|^2 \\ & + \left| i\frac{2\sqrt{3}}{3} C_{2\omega}\, Y_{11} (R_1 - R_2)\, e^{i(\delta_p+\phi)} \right. \\ & \left. - \frac{\sqrt{30}}{75} C_\omega\, Y_{21}\, (-5R_3^{(2)} - R_4^{(2)} + 6R_5^{(2)})\, e^{i\delta_d} \right|^2 , \end{aligned} \tag{10}$$

where R_1, R_2 are single-photon radial matrix elements from $s_{1/2}$ to $kp_{1/2}$ and $kp_{3/2}$, respectively, and $R_n^{(2)}$ $(n = 1, 2, .., 5)$ are effective two-photon radial matrix elements defined as, respectively, $s_{1/2} \to p_{1/2} \to ks_{1/2}$, $s_{1/2} \to p_{3/2} \to ks_{1/2}$, $s_{1/2} \to p_{1/2} \to kd_{3/2}$, $s_{1/2} \to p_{3/2} \to kd_{3/2}$, and $s_{1/2} \to p_{3/2} \to kd_{5/2}$. If, on the other hand, both fields are right circularly polarized, Eq. (7) becomes

$$\frac{d\sigma}{d\Omega} \propto \left| -i\sqrt{3}C_{2\omega}\,Y_{11}\,R_2\,e^{i(\delta_p+\phi)} - \frac{\sqrt{30}}{5}\,C_\omega\,Y_{22}\,R_5^{(2)}\,e^{i\delta_d}\right|^2$$
$$+ \left| -i\,\frac{\sqrt{3}}{3}C_{2\omega}Y_{11}(R_1+2R_2)e^{i(\delta_p+\phi)}\right.$$
$$\left. -\frac{\sqrt{30}}{75}C_\omega Y_{22}(10R_3^{(2)}+2R_4^{(2)}+3R_5^{(2)})\,e^{i\delta_d}\right|^2$$
$$+ \left| -i\frac{\sqrt{6}}{3}C_{2\omega}Y_{10}(R_1-R_2)e^{i(\delta_p+\phi)}\right.$$
$$\left. +\frac{\sqrt{30}}{75}C_\omega Y_{21}(5R_3^{(2)}+R_4^{(2)}-6R_5^{(2)})e^{i\delta_d}\right|^2 . \qquad (11)$$

It should be noted that in Eqs. (10) and (11), the j−dependence of the phase shifts δ_l has been ignored, since, even for Cs, such a dependence is known to be very small compared with that of the radial matrix elements.[15] Eqs. (10) and (11) are the basic equations we use in the work. In the following subsections, we consider two cases depending on the strength of spin-orbit interaction.

2.1 Weak Spin-Orbit Interaction

First, consider the case in which the spin-orbit interaction is sufficiently weak and the radial matrix elements have approximately no j-dependence, i.e., $R_1 \simeq R_2$, $R_1^{(2)} \simeq R_2^{(2)}$, and $R_3^{(2)} \simeq R_4^{(2)} \simeq R_5^{(2)}$. Such an assumption is valid for Li, Na, and K. Then, the last terms in both Eqs. (10) and (11) disappear. It can be seen from Eq. (10) that the angle-resolved photoelectron signal contains information on the phase shifts of δ_s, δ_p, and δ_d. From an experimental point of view, however, it is not as easy as it appears to determine these quantities, since other free parameters for fittings, i.e., laser intensities and radial matrix elements, are not precisely known, in general. It is interesting to point out that the situation becomes much simpler if we focus on the photoelectron signal emitted to some particular direction. For example, in this particular case, the photoelectron signal emitted to the *magic angles* $\Theta = \cos^{-1}(1/\sqrt{3})$ and $\pi - \cos^{-1}(1/\sqrt{3})$ contains a phase-dependent factor given by $\sin(\delta_p - \delta_s + \phi)$, since Y_{20} becomes zero at this angle. This means that, by measuring the phase lag of the photoelectron signal with respect to the relative phase of lasers ϕ at this Θ (magic angle), the phase difference $\delta_p - \delta_s$ could be directly and precisely determined as a phase lag of photoelectron signal with respect to ϕ. It should be emphasized that this phase difference can be experimentally determined *independent of the values of radial matrix elements and laser intensities I_ω and $I_{2\omega}$*. Thus, the key to the direct determination of the phase difference is to focus on the photoelectron signal emitted to the magic angle where only a single pair of the partial waves contributes to the signal, and to measure the phase lag as a function of the relative phase ϕ of the lasers. With this particular measurement, however, the phase difference $\delta_p - \delta_d$ or $\delta_s - \delta_d$ cannot be determined. Now we consider Eq. (11). The advantage of utilizing circularly polarized light is that we can isolate the other pair of partial waves, which are p and d waves, respectively, for the single- and two-photon ionization from the $s_{1/2}$ ground state. Thus, by using circularly polarized light, the phase difference $\delta_p - \delta_d$ can be determined. Obviously the use of linearly and circularly polarized light is complementary.

2.2 Strong Spin-Orbit Interaction

As a second case, assume that the spin-orbit interaction is not negligible. Heavy alkali-metal atoms such as Rb and Cs fall into this category. In this case, the last terms in Eqs. (10) and (11) cannot be ignored any more. For the case of circularly polarized light, this does not cause a serious problem, as can be seen from Eq. (11). For linearly polarized light, however, this is not the case, since the contribution of the d partial wave cannot be ignored even at the magic angle for determining $\delta_p - \delta_s$. In order to correlate the strength of spin-orbit interaction with the additional phase lag due to the non-vanishing d wave, we define generalized perturbation functions x_1 and x_2 for single- and two-photon processes, respectively, as

$$x_1 = \frac{2R_2 + R_1}{R_2 - R_1}, \tag{12}$$

$$x_2 = \frac{5R_3^{(2)} + R_4^{(2)} + 9R_5^{(2)}}{5R_3^{(2)} + R_4^{(2)} - 6R_5^{(2)}}, \tag{13}$$

and

$$w = \frac{5(R_1^{(2)} + 2R_2^{(2)})}{5R_3^{(2)} + R_4^{(2)} + 9R_5^{(2)}}. \tag{14}$$

Note that all of these parameters, x_1, x_2, and w, are functions of photoelectron energy, where w is a kind of ratio of "averaged" two-photon radial matrix elements for s and d waves. Using these quantities and Eq. (10), the ϕ-dependent modulation M of the photoelectron signal emitted to the magic angle $\Theta = \cos^{-1}(1/\sqrt{3})$ is obtained, after some algebra, as,

$$M \propto \sin(\delta_p - \delta_s + \phi) + \frac{2\sqrt{2}}{wx_1x_2}\sin(\delta_p - \delta_d + \phi)\ . \tag{15}$$

For weak spin-orbit interaction, x_1 and x_2 tend to take large values, and the second term of the above equation may be dropped out as in the case of light alkali-metal atoms. In general, however, in order to use Eq. (10), the quantities w, x_1, x_2, and $\delta_p - \delta_d$ must be known in advance. Although the phase difference $\delta_p - \delta_s$ can be extracted from the measurement with circularly polarized light, the determination of other quantities, w, x_1, and x_2, would be an experimental as well as theoretical challenge itself.

One way out of this difficulty is to use *polarized* atoms in the $m_j = 1/2(-1/2)$ ground state and conduct a spin-resolved photoelectron detection. Then, for spin-up (-down) photoelectrons, only the first (second) term of Eq. (10) contributes to the photoelectron signal, and by detecting photoelectrons at the magic angle (any angle), the phase difference $\delta_p - \delta_s$ ($\delta_p - \delta_d$) can be unambiguously determined *even under the presence of strong spin-orbit interaction.* In other words, under the presence of strong spin-orbit interaction, not only $\delta_p - \delta_s$ but also $\delta_p - \delta_d$ can be directly determined using linearly polarized light only, at the expense of more complicated experimental setup.

Now we consider the way to calibrate the relative phase of lasers, which is quite essential to employ this technique. One way is to make use of the PAD of hydrogen, where the phase involved is nothing but a Coulomb phase shift, and to use it as a reference to determine the relative phase of the lasers. This method would be particularly useful for circularly polarized light, since the second harmonic generation technique which has been demonstrated to be useful for the calibration of the relative phase of linearly polarized lasers[16] cannot be immediately applied for circularly polarized lasers. An excited state, for example $2s_{1/2}$ state of hydrogen, might be used for calibration, for the photon energies of the lasers might not be sufficient to ionize a hydrogen atom in the ground state. For heavier alkali-metal atoms (Rb and Cs), it might be sufficient to calibrate the laser phase only for linear polarization, because both $\delta_p - \delta_s$ and $\delta_p - \delta_d$ can be determined by linearly polarized light, in principle, as discussed above.

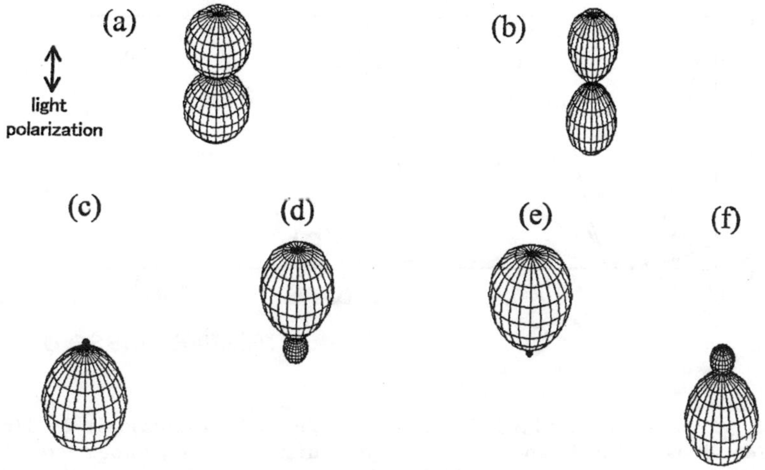

Figure 2. 3D plots of the PAD of Na by linearly polarized light under (a) single-photon process only with frequency 2ω, (b) two-photon process only with frequency ω, and (c)-(f) both fields with well-defined phase. In (c)-(f), ϕ is varied from 0 to $3\pi/2$ by $\pi/2$ step.

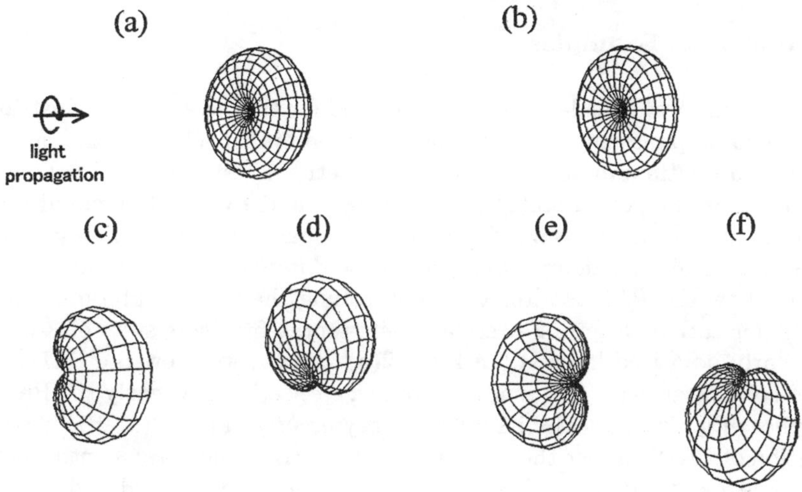

Figure 3. Same with Fig. 2 except that circularly polarized light is employed instead of linearly polarized light.

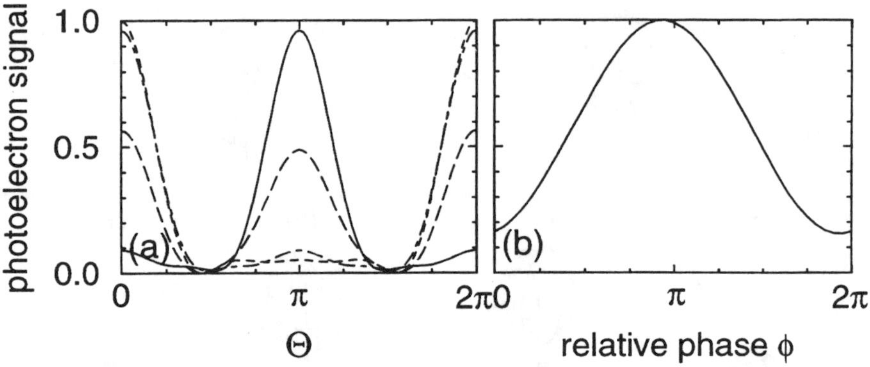

Figure 4. (a) Variation of the PAD of Na at $\phi = 0$ (solid), $\pi/3$ (long-dashed), $2\pi/3$ (dotted), and π (dot-dashed) under two-color fields. (b) Variation of the photoelectron signal at $\Theta = \cos^{-1}(1/\sqrt{3})$ under two-color fields as a function of the relative phase ϕ. Both lasers are *linearly* polarized. The final state energy is 0.0278 Ryd above ionization threshold for both (a) and (b).

3 Numerical Examples

Specific numerical calculations are performed for the Na atom for illustration. The Coulomb phase shifts are computed based on Eq. (1), and the quantum defect of a continuum state for each symmetry, is estimated by extrapolating the value for bound states of Na. Single- and two-photon radial matrix elements are obtained using single-channel quantum defect theory. Before presenting results for determining the phase difference, we first show in Figs. 2 and 3 how the PAD is altered by controlling the relative phase ϕ and the laser polarization. Figure 2 presents some representative results of 3-D PAD by linearly polarized laser(s). In Figs. 2(a) and (b) we show the PAD's by a single-color laser, with frequency 2ω and ω, respectively. Clearly the PAD via two-photon ionization exhibits stronger asymmetry. Figs. 2(c)-2(f) show the variation of PAD's under the presence of phase-controlled lasers with frequencies ω and 2ω at different values of the relative phase ϕ. As is already known,[6] the direction of photoelectron emission can be controlled. In Fig. 3, we show similar results for circularly polarized laser(s). Although the shapes of PAD's themselves are very different from those produced by linearly polarized lasers, the direction of photoelectron emission is again found to be quite controllable. Figure 4(a) shows a theoretical PAD of Na by linearly polarized lasers based on the single-channel quantum defect theory at the final state energy of W=0.0278 Ryd. A significant variation of the PAD is seen as ϕ is varied from 0 to π. Figure 4(b) shows the variation of the photoelectron signal emitted at the magic angle $\Theta = \cos^{-1}(1/\sqrt{3})$, where the phase-sensitive modulation comes from the factor $\sin(\delta_p - \delta_s + \phi)$ as explained before. This way, the phase difference of continua can be directly determined from the phase lag, which is insensitive to the ratio of radial matrix elements or the laser intensities employed. A similar calculation has been performed for circularly polarized lasers as shown in Fig. 5. Under the single-color photoionization by circularly polarized light, the PAD, of course, does not depend on the angle Φ which defines the azimuthal angle around the propagation axis of the laser. Under the simultaneous action of two lasers with well-defined phase, however, the photoelelctron signal does depend on the angle Φ as well as Θ, as shown in Fig. 5(a). The variation of the photoelectron signal to the particular direction defined by $(\Theta, \Phi) = (\pi/2, 0)$ is shown in Fig. 5(b). Again, this result implies that the phase difference $\delta_p - \delta_d$ can be directly determined with the use of circularly polarized lasers.

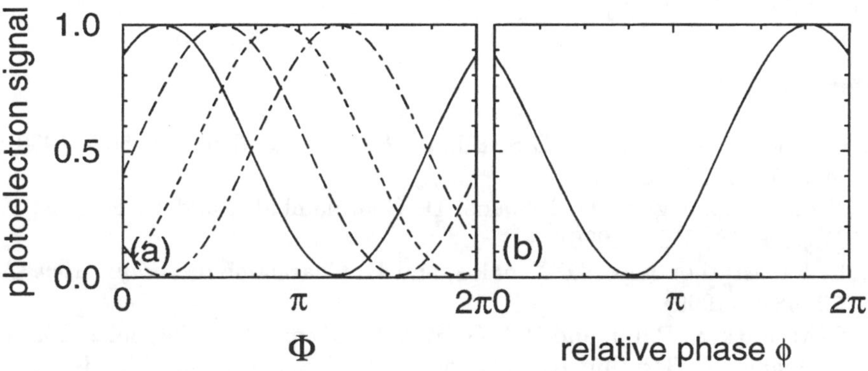

Figure 5. (a) Variation of the photoelectron signal of Na at $\Theta = \pi/2$ as a function of Φ at four values of the relative phase $\phi = 0$ (solid), $\pi/3$ (dashed), $2\pi/3$ (dotted), and π (dot-dashed) under two-color fields. (b) Variation of the photoelectron signal at $(\Theta, \Phi) = (\pi/2, 0)$ as a function of relative phase ϕ. Both lasers are *circularly* polarized. The final state energy is 0.0278 Ryd above ionization threshold for both (a) and (b).

4 Conclusions

In conclusion, we have proposed a new method for directly determining the phase difference of continua utilizing phase-controlled lasers. Although we have given specific numerical results for a light alkali-metal atom (Na), it is obvious that this method would work for any other atom as well, as long as the single-active-electron approximation is valid. The most important feature of the proposed method is that the determination of the phase difference is insensitive to the values (ratios) of radial matrix elements, which are not known with sufficient accuracy in most cases and at the same time quite influential if one is to employ the standard PAD technique for this purpose. As a consequence our method would lead to higher accuracy for determining the phase difference between continua with opposite as well as same parities. We have demonstrated that a particular pair of continuum states could be isolated through the proper choice of an initial state and laser polarization. Although the application of this technique for heavier alkali-metal atoms (Rb and Cs) would require a use of polarized atoms in the ground state and a spin-resolved photoelectron detection, the phase difference could be *unambiguously* determined with this technique, which might be compared with the report[13] based on the standard PAD method. They have made extensive speculations upon extracting the phase difference between *s* and *d* waves of Rb. Such difficulties mainly arise from the large uncertainty of radial matrix elements employed for PAD fittings.

References

1. C. Chen, Y.-Y. Yin and D.S. Elliott, *Phys. Rev. Lett.* **64**, 507 (1990); *ibid.* **65**, 1737 (1990).
2. N.E. Karapanagioti, D. Xenakis, D. Charalambidis, and C. Fotakis, *J. Phys.* B **29**, 3599 (1996).
3. E. Papastathopoulos, D. Xenakis, and D. Charalambidis, *Phys. Rev.* A **59**, 4840 (1999).
4. S.-M. Park, S.-P. Lu, and R.J. Gordon, *J. Chem. Phys.* **94**, 8622 (1991).
5. X. Wang, R. Bersohn, K. Takahashi, M. Kawasaki, and H.-L. Kim, *J. Chem. Phys.* **105**, 2992 (1996).
6. Y.-Y. Yin, C. Chen, and D.S. Elliott, *Phys. Rev. Lett.* **69**, 2353 (1992).
7. H.-L. Kim and R. Bersohn, *J. Chem. Phys.* **107**, 4546 (1997).
8. Takashi Nakajima and P. Lambropoulos, *Phys. Rev. Lett.* **70**, 1081 (1993).
9. Takashi Nakajima and P. Lambropoulos, *Phys. Rev.* A **50**, 590 (1994).

10. Takashi Nakajima, J. Zhang, and P. Lambropoulos, *J. Phys.* B **30**, 1077 (1997).
11. L. Zhu, K. Suto, J.A. Fiss, R. Wada, T. Seideman, and R.J. Gordon, *Phys. Rev. Lett.* **79**, 4108 (1997).
12. P. Lambropoulos and Takashi Nakajima, *Phys. Rev. Lett.* **82**, 2266 (1999).
13. Y.-Y. Yin and D.S. Elliott, *Phys. Rev.* A **47**, 2881 (1993).
14. C.S. Feigerle, R.N. Compton, L.E. Cuéllar, N.A. Cherepkov, and L.V. Chernysheva, *Phys. Rev.* A **53**, 4183 (1996).
15. W. Ong and S.T. Manson, *Phys. Rev.* A **20**, 2364 (1979).
16. D.W. Schumacher, F. Weihe, H.G. Miller, and P.H. Bucksbaum, *Phys. Rev. Lett.* **73**, 1344 (1994).

COHERENT CONTROL, FANO INTERFERENCE, AND NON-HERMITIAN INTERACTIONS

MARLAN O. SCULLY

Department of Physics and Institute for Quantum Studies, Texas A & M University, TX 77843
and Max-Planck-Institut für Quantenoptik, 85748 Garching, Germany

Fano interference arising from e.g. tunnel coupling of two semiconductor quantum wells can yield vanishing absorption of incident resonant radiation with nonzero stimulated emission. This is the basis for Harris-Fano lasing without inversion. It is here shown that it is interesting to analyze the problem in a basis such that the Fano decay matrix is diagonal.

1 Introduction

Quantum coherence is the basic tool of quantum control with applications to quantum optics, chemistry, and physics. One interesting approach to and generation of quantum coherence is the Fano effect[1] involving decay states or channels. For example, in tunnel coupled quantum wells, see Fig. 1a, the tunneling of electrons from the right well manifests itself as a Fano interference between the eigenstates of the tunneling interaction as per Fig. 1b.

At the conference the question was asked: "Can we rewrite the electronic interaction terms to contain the Fano interference physics?" In other words: is it possible to recast the problem of Fano interference (with driving fields), such that we shift the focus from the off-diagonal decay matrix to redefined driving potentials which include the Fano interference?

Some thought that it was not possible. In the following we show how to do this and give an example of the utility of the approach in the context of lasing without inversion[2].

In the case of two quantum wells having states $|\alpha_1\rangle$ (RHS) and $|\alpha_2\rangle$ (LHS) and coupled via a tunneling interaction, as in Fig. 1a, the time evolution of the state vector

$$|\Psi\rangle = \alpha_1|\alpha_1\rangle + \alpha_2|\alpha_2\rangle \tag{1}$$

is given by

$$\begin{aligned}\dot{\alpha}_1 &= -i\Delta\alpha_2,\\ \dot{\alpha}_2 &= -i\Delta\alpha_1 - 2\gamma\alpha_2,\end{aligned} \tag{2}$$

where Δ is the coupling rate and γ is the rate of removing from the state $|\alpha_2\rangle$.

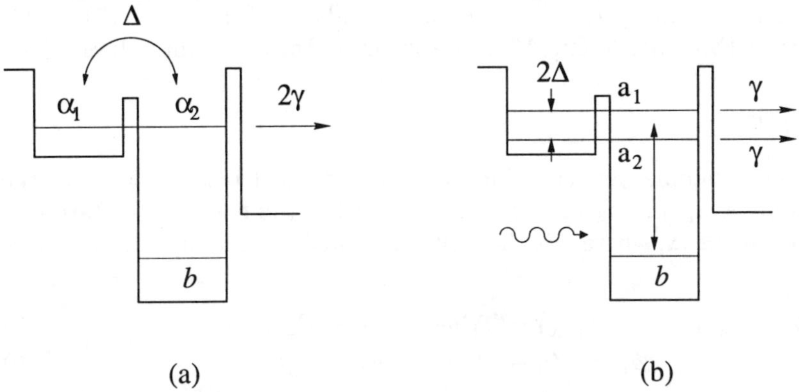

Figure 1. Typical intersubband double quantum well structure. Tunneling of electrons between levels α_1 and α_2 with coupling rate Δ.

In order to diagonalize the tunneling interaction we introduce the sum and difference states

$$|\alpha_1\rangle = \frac{1}{\sqrt{2}}(|a_2\rangle - |a_1\rangle) \tag{3}$$

$$|\alpha_2\rangle = \frac{1}{\sqrt{2}}(|a_2\rangle + |a_1\rangle) \tag{4}$$

and upon inserting (3) and (4) into (1), the state vector in the $|a_{1,2}\rangle$ representation reads

$$|\Psi\rangle = \frac{1}{\sqrt{2}}\Big[(\alpha_2 - \alpha_1)|a_1\rangle + (\alpha_2 + \alpha_1)|a_2\rangle\Big], \tag{5}$$

That is $a_1 = \frac{1}{\sqrt{2}}(\alpha_2 - \alpha_1)$ and $a_2 = \frac{1}{\sqrt{2}}(\alpha_2 + \alpha_1)$ where $|\Psi\rangle = a_1|a_1\rangle + a_2|a_2\rangle$; so that from (2) and (5) we have

$$\begin{aligned} \dot{a}_1 &= \frac{1}{\sqrt{2}}\Big[- i\Delta(\alpha_1 - \alpha_2) - 2\gamma\alpha_2\Big] \\ &= i\Delta a_1 - \gamma a_1 - \gamma a_2 \end{aligned} \tag{6}$$

and

$$\begin{aligned} \dot{a}_2 &= \frac{1}{\sqrt{2}}\Big[- i\Delta(\alpha_1 + \alpha_2) - \gamma\alpha_2\Big] \\ &= -i\Delta a_2 - \gamma a_1 - \gamma a_2 \end{aligned} \tag{7}$$

Please note the damping terms now couple a_1 and a_2 in Eq. (6) and (7), this is the essence of Fano interference[3,4] as it appears in the quantum well problem.

2 Analysis

As a specific example, let us include now a laser field between $|b\rangle$ and the excited doublet a_1 and a_2 of Fig. 1b, and writing the energies of state $a_{1,2}$ related to b as $\nu \pm \Delta$, where ν is the centrally tuned laser frequency, we arrive at

$$\dot{a}_2 = -(\gamma + i\Delta)a_2 - \gamma a_1 - i\Omega_2 b, \tag{8}$$
$$\dot{a}_1 = -(\gamma - i\Delta)a_1 - \gamma a_2 - i\Omega_1 b, \tag{9}$$
$$\dot{b} = -i\Omega_2 a_2 - i\Omega_1 a_1, \tag{10}$$

where $\Delta = \omega_2 - \nu$, $\Omega_{1,2} = \wp_{1,2}E_0/\hbar$, and $\wp_{1,2}$ are the dipole matrix elements between b and $a_{1,2}$.

Writing Eqs. (9,10) in a convenient matrix form we have

$$\frac{d}{d\tau}\begin{bmatrix} a_2 \\ a_1 \\ b \end{bmatrix} = -\Gamma \begin{bmatrix} a_2 \\ a_1 \\ b \end{bmatrix} - iV \begin{bmatrix} a_2 \\ a_1 \\ b \end{bmatrix}, \tag{11}$$

where $\tau = \gamma t$ and the Fano decay matrix is defined by

$$\Gamma = \begin{bmatrix} 1+i\tilde{\Delta} & 1 & 0 \\ 1 & 1-i\tilde{\Delta} & 0 \\ 0 & 0 & 0 \end{bmatrix}, \tag{12}$$

and probe-field interaction is given by

$$V = \begin{bmatrix} 0 & 0 & \tilde{\Omega}_2 \\ 0 & 0 & \tilde{\Omega}_1 \\ \tilde{\Omega}_2 & \tilde{\Omega}_1 & 0 \end{bmatrix}, \tag{13}$$

with $\tilde{\Delta} = \Delta/\gamma$ and $\tilde{\Omega}_{1,2} = \Omega_{1,2}/\gamma$.

We proceed by introducing the decay eigenvectors and eigenvalues such that

$$u \begin{bmatrix} 1+i\tilde{\Delta} & 1 \\ 1 & 1-i\tilde{\Delta} \end{bmatrix} u^{-1} = \begin{bmatrix} \lambda_+ & 0 \\ 0 & \lambda_- \end{bmatrix}, \tag{14}$$

where $\lambda_\pm = 1 \pm x$ with $x = \sqrt{1-\tilde{\Delta}^2}$. The eigenvectors are

$$|v_\pm) = \frac{1}{\sqrt{2}} \begin{bmatrix} 1 \\ \pm x - i\tilde{\Delta} \end{bmatrix}, \tag{15}$$

so that

$$u^{-1} = \frac{1}{\sqrt{2}} \begin{bmatrix} 1 & 1 \\ x - i\tilde{\Delta} & -x - i\tilde{\Delta} \end{bmatrix} = \begin{bmatrix} |v_+) \; |v_-) \end{bmatrix}, \tag{16}$$

where we have introduced the notation $|v_\pm)$ to distinguish it from $|v_\pm\rangle$ as discussed below.

Note that the inverse of $|v_\pm)$ must be treated with care. In the usual bra-ket vector space $\langle \; | = | \; \rangle^{-1} = | \; \rangle^\dagger$ but here $| \;)^{-1} \neq | \;)^\dagger$. The inverse eigenvectors are obtained by simply taking the inverse of u^{-1}

$$u = [u^{-1}]^{-1} = \frac{1}{\sqrt{2}x} \begin{bmatrix} x + i\tilde{\Delta} & 1 \\ x - i\tilde{\Delta} & -1 \end{bmatrix} \equiv \begin{bmatrix} (v_+| \\ (v_-| \end{bmatrix}, \tag{17}$$

which yields $(v_\pm|v_\pm) = 1$, $(v_\pm|v_\mp) = 0$.

Finally, we note that in the 3-level Hilbert space

$$U^{-1} = \frac{1}{\sqrt{2}} \begin{bmatrix} 1 & 1 & 0 \\ x - i\tilde{\Delta} & -x - i\tilde{\Delta} & 0 \\ 0 & 0 & \sqrt{2} \end{bmatrix}, \qquad U = \frac{1}{\sqrt{2}x} \begin{bmatrix} x + i\tilde{\Delta} & 1 & 0 \\ x - i\tilde{\Delta} & -1 & 0 \\ 0 & 0 & \sqrt{2}x \end{bmatrix}. \tag{18}$$

Now the transformed state vector is defined by

$$U \begin{bmatrix} a_2 \\ a_1 \\ b \end{bmatrix} = \begin{bmatrix} A_+ \\ A_- \\ B \end{bmatrix}, \tag{19}$$

that is, $A_\pm = (1/\sqrt{2}x)[a_2(x \pm i\tilde{\Delta}) \pm a_1]$, $B = b$; and the diagonal Γ operator is

$$\Gamma_t = U\Gamma U^{-1} = \begin{bmatrix} 1 + x & 0 & 0 \\ 0 & 1 - x & 0 \\ 0 & 0 & 0 \end{bmatrix}, \tag{20}$$

and the transformed interaction potential is

$$\begin{aligned} V_t &= UVU^{-1} \\ &= \frac{1}{\sqrt{2}} \begin{bmatrix} 0 & 0 & [\tilde{\Omega}_2(x + i\tilde{\Delta}) + \tilde{\Omega}_1]/x \\ 0 & 0 & [\tilde{\Omega}_2(x - i\tilde{\Delta}) - \tilde{\Omega}_1]/x \\ [\tilde{\Omega}_2 + \tilde{\Omega}_1(x - i\tilde{\Delta})] & [\tilde{\Omega}_2 - \tilde{\Omega}_1(x + i\tilde{\Delta})] & 0 \end{bmatrix}. \end{aligned} \tag{21}$$

Note that $(V_t)_{ij} \neq (V_t)^*_{ji}$, i.e. the transformed driving potential is non-Hermitian. The equations of motion in terms of $A_\pm$ and B are then found from

$$\frac{d}{d\tau}\begin{bmatrix} A_+ \\ A_- \\ B \end{bmatrix} = -\Gamma_t \begin{bmatrix} A_+ \\ A_- \\ B \end{bmatrix} - iV_t \begin{bmatrix} A_+ \\ A_- \\ B \end{bmatrix}, \tag{22}$$

to be

$$\frac{d}{d\tau}A_+ = -(1+x)A_+ - \frac{i}{\sqrt{2}x}\left[\tilde{\Omega}_2(x+i\tilde{\Delta}) + \tilde{\Omega}_1\right]B, \tag{23}$$

$$\frac{d}{d\tau}A_- = -(1-x)A_- - \frac{i}{\sqrt{2}x}\left[\tilde{\Omega}_2(x-i\tilde{\Delta}) - \tilde{\Omega}_1\right]B, \tag{24}$$

$$\frac{d}{d\tau}B = -\frac{i}{\sqrt{2}}\left[\tilde{\Omega}_2 + \tilde{\Omega}_1(x-i\tilde{\Delta})\right]A_+ - \frac{i}{\sqrt{2}}\left[\tilde{\Omega}_2 - \tilde{\Omega}_1(x+i\tilde{\Delta})\right]A_-. \tag{25}$$

3 Discussion

From Eqs. (22-24) it is clear that detailed balance has been broken. That is, for the usual two level problem, with upper level a and lower level b we have

$$\dot{a} = -(\gamma + i\Delta)\,a - i\Omega b, \tag{26}$$

$$\dot{b} = -i\Omega a. \tag{27}$$

And the symmetric driving force from $a \to b$ and $b \to a$ is determined by the single Rabi frequency Ω (taken as real) i.e. V is as given by Eq. (13) is Hermitian. Whereas in Eq. (22) the effective driving potential V_t given by Eq. (21) is not Hermitian, thus, breaking absorption-emission symmetry.

Note that in the two level case when Δ, γ are the fastest frequencies we have from (26) that $a \cong -i\Omega b/(\gamma + i\Delta)$ and inserting this into Eq. (27) yields

$$\dot{b} = -\frac{\Omega^2}{\gamma + i\Delta}b. \tag{28}$$

Hence the rate of depletion of the ground state probability, due to absorption, is given by

$$\frac{d}{dt}b^*b = -\frac{2\Omega^2\gamma}{\gamma^2+\Delta^2}b^*b. \tag{29}$$

The corresponding absorption rate in the Fano problem is found from (23) and (24) as

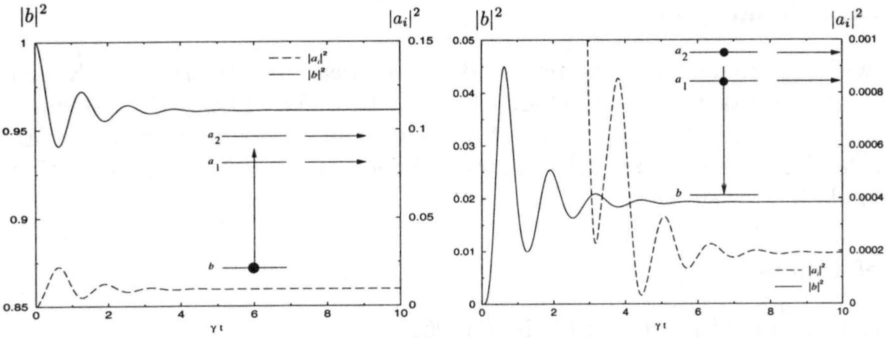

(a) (b)

Figure 2. Time evolution of the populations $|a_i|^2$ and $|b(t)|^2$ given by Eq. (3). (a) Beginning with $b(0) = 1$, a little population is induced into a_1 and a_2 yielding so called "dark state"[5]. Note that after a short time population concentrates in b, i.e., the ground state population does not change, and the absorption ceases. (b) If $a_{2,1}(0) = \pm 1/\sqrt{2}$, most of population leaves at rate γ, small fraction is trapped in dark state and the rest is recycled by removing the electrons from the RHS of Fig 1(a) and injecting then into the LH quantum well, see also Ref.6.

$$A_\pm \cong \frac{-i}{\sqrt{2}} \left[\tilde{\Omega}_2 (x \pm i\tilde{\Delta}) \pm \tilde{\Omega}_1 \right] \frac{B}{x(1 \pm x)} \tag{30}$$

and inserting this into Eq. (24) yields the Fano absorption rate

$$\begin{aligned} \frac{d}{dt} B^* B &= -\frac{1}{2} \left[\tilde{\Omega}_2 + \tilde{\Omega}_1 (x - i\tilde{\Delta}) \right] \left[\tilde{\Omega}_2 (x + i\tilde{\Delta}) + \tilde{\Omega}_1 \right] \frac{B^* B}{x(1+x)} \\ &\quad - \frac{1}{2} \left[\tilde{\Omega}_2 - \tilde{\Omega}_1 (x + i\tilde{\Delta}) \right] \left[\tilde{\Omega}_2 (x - i\tilde{\Delta}) - \tilde{\Omega}_1 \right] \frac{B^* B}{x(1-x)} + \text{c.c.} \\ &= -\frac{1}{1 - x^2} \left[(\tilde{\Omega}_2 - \tilde{\Omega}_1)^2 + i\tilde{\Delta} (\tilde{\Omega}_2^2 - \tilde{\Omega}_1^2) \right] B^* B + \text{c.c.} \\ &= -\frac{2}{\Delta^2} (\tilde{\Omega}_2 - \tilde{\Omega}_1)^2 \; B^* B, \end{aligned} \tag{31}$$

which vanishes for $\tilde{\Omega}_1 = \tilde{\Omega}_2$ (recall that we have taken $\Omega_{1,2}$ to be real). But emission does not vanish (see, Fig. 2). This is the basis of Harris-Fano lasing without inversion.

Acknowledgments

I wish to thank Y. Aharonov, B.-G. Englert, S. Harris, K. Kapale, O. Kocharovskaya, V. Kocharovsky, H. Lee, W. Pötz, G. Süssmann, H. Walther, and S. Zubairy for stimulating and useful discussions and the Office of Naval Research, the National Science Foundation, and the Welch Foundation for support.

References

1. U. Fano, *Phys. Rev.* **124**, 1866 (1961).
2. S. Harris, *Phys. Rev. Lett.* **62**, 1038 (1989).
3. H. Schmidt, K. Campman, A. Gossad, and A. Imamoglu, *Appl. Phys. Lett.* **70**, 3455 (1997).
4. F. Capasso, S. Sen, A. Cho, and A. Hutchinson, *Appl. Phys. Lett.* **50**, 930 (1987).
5. E. Arimondo, in *Progress in Optics* edited by E. Wolf, Vol. *XXXV*, p257 (Elsevier Science, Amsterdam, 1996).
6. S. Harris and J. Macklin, *Phys. Rev. A* **40** 4135 (1989).

Part III. Alignment

ALIGNMENT OF NEUTRAL MOLECULES BY A STRONG NONRESONANT LASER FIELD

HIROFUMI SAKAI
Department of Physics, Graduate School of Science, The University of Tokyo, 7-3-1, Hongo, Bunkyo-ku, Tokyo 113-0033, Japan
E-mail: hsakai@phys.s.u-tokyo.ac.jp

JAKOB JUUL LARSEN AND C. P. SAFVAN
Institute of Physics and Astronomy, University of Århus, DK-8000 Århus C, Denmark

IDA WENDT-LARSEN, KAREN MARIE HILLIGSΦE, KASPER HALD, AND HENRIK STAPELFELDT
Department of Chemistry, University of Århus, DK-8000 Århus C, Denmark
E-mail: henriks@kemi.aau.dk

We show that a strong nonresonant nanosecond laser pulse can be used to align neutral molecules. Our technique, applicable to nonpolar as well as polar molecules, relies on the anisotropic interaction between the strong laser field and the induced dipole moment of the molecules. The degree of alignment is enhanced by increasing the laser intensity or by lowering the initial rotational temperature of the molecules. We measure the alignment by photodissociating the molecules with a femtosecond laser pulse and detecting the instantaneous direction of the photofragments by ion imaging techniques. The highest degree of alignment is $\langle\langle\cos^2\theta\rangle\rangle = 0.81$.

I.Introduction

Controlling and manipulating the external degrees of freedom of atoms and molecules is one of the intriguing subjects in modern physics and chemistry. This is evident in atomic physics where lasers provide powerful means to control the velocity and position of neutral atoms,[1] and in studies of condensed matter where scanning probe microscopes enable selection and manipulation of individual atoms and molecules.[2] Chemists have aimed to control the spatial orientation of molecules because it enables them to study orientational effects in chemical reaction dynamics.[3] This kind of control can be achieved by orienting certain symmetric top molecules with elctric hexapole fields,[3,4] or by orienting molecules possessing large permanent dipole moment with strong static electric fields.[3,5,6] The advantage of these methods is that they can arrange polar molecules in a head-versus-tail order (orientation). However, there are many molecules, most notably nonpolar molecules, to which they do not apply.

Friedrich and Herschbach suggested exploiting the anisotropic polarizability interaction of an intense nonresonant laser field with the induced dipole moment of molecules.[7] The interaction creates a potential minimum for the molecules along the polarization axis of the laser field, forcing them to librate over a limited angle range instead of rotating freely with random spatial orientations. Since the interaction between the laser field and a permanent dipole moment averages out to zero, the molecules are aligned rather than oriented. This new technique can be regarded as a modern version of the optical Kerr effect[8] with the special emphasis changed from the study of electro-optic properties of liquids to alignment of isolated molecules in gas phase. Furthermore, it indicates an important new direction in molecular science because the use of strong nonresonant laser fields, in addition to alignment presented in this article, may transfer many of the successful applications of laser-induced manipulation of atoms[1] to molecules such as trapping[7,9] and focusing.[10] The convincing evidence of laser-induced alignment was first reported by Kim and Felker.[11] Here we report further experimental evidence of laser-induced alignment by using a strong, linearly polarized, infrared, nanosecond laser pulse to align a sample of neutral molecules and femtosecond laser-induced photodissociation to measure the degree of alignment. In particular, we emphasize the general applicability of the scheme by aligning I_2, ICl, CS_2, CH_3I, and C_6H_5I molecules.

The eigenstates of the molecules in the strong laser field are the so-called pendular states consisting of linear superpositions of field-free rotational states.[7] We ensure that each rotational state with quantum numbers $J|M|$ evolves into a particular pendular state by turning on the strong field adiabatically,[7] which here means slowly compared to the rotational periods of the molecules.[12] Hereby, for a given field strength, the degree of alignment can be increased by using a sample of molecules initially residing in the lowest rotational states because they are transferred into the lowest lying and most aligned pendular states. Alternatively, for molecules with a given initial rotational state distribution, the degree of alignment can be increased by increasing the intensity of the alignment pulse. Both ways of controlling the degree of alignment will be demonstrated in this article.

II. Experimental

A schematic of the experimental setup is shown in Fig. 1. To determine the degree of alignment of the molecules we measure their instantaneous orientation while the alignment laser is still on. This is necessary because the molecules evolve back into field-free rotational states when the alignment field is switched off. In the case of iodine, the measurement is performed by photodissociating the molecules[13] during the alignment pulse with a femtosecond pump pulse and observing the direction of

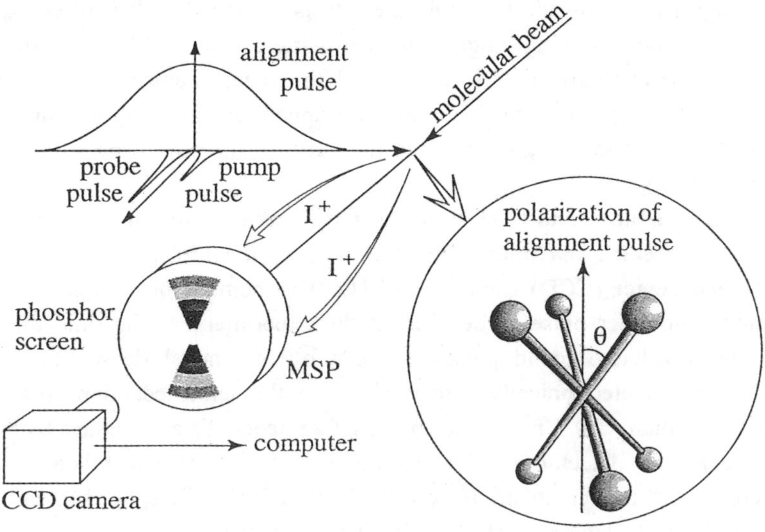

Figure 1. A schematic of the experimental setup. The polarization of the lasers are indicated by the arrows (upper left-hand side). The inset shows the aligned molecules making an angle θ with the polarization axis of the alignment laser.

the photofragments with a two-dimensional detector. Since the detector is only sensitive to ions, it is necessary to ionize the photofragments without changing their direction. This is done by firing a second, intense femtosecond probe pulse after the dissociation is complete. In the case of other molecules we use one intense femtosecond pulse to produce multiply-charged ions of the molecules.

A pulsed supersonic beam, formed by expanding the molecules in 760 Torr of argon or helium through a 0.5 mm nozzle, is crossed at 90° by the focused laser beams. The partial pressure of the molecular gas is adjusted to a few mbar. The alignment pulses come from a pulsed single-longitudinal-mode Nd:YAG laser (λ = 1064 nm) with a pulse duration of ~ 3.5 ns (full width at half maximum). The alignment pulse is focused to a spot size, ω_0^{YAG} ~ 42 μm, yielding a maximum intensity of ~ 1.4 x 10^{12} W/cm^2. The pump pulses used in the iodine experiment originate from an optical parametric amplifier and are centered at 688 nm. The pulse length is 100 fs and the peak intensity is ~ 3.0 x 10^{12} W/cm^2. The pulse energy is kept below saturation of the dissociating transition.[14] The beam waist of the probe pulse (ω_0^{probe} ~ 32 μm) is adjusted to be smaller than ω_0^{YAG} to ensure that we probe only those molecules that have been exposed to the alignment field. The pulses originate from the Ti:sapphire laser system and are centered at 800 nm.

The pulse length is 100 fs and the peak intensity is ~ 7.0 x 10^{13} W/cm^2 using linear polarization. When probing alignment with single pulse induced dissociative ionization we use circularly polarized pulses with a peak intensity of ~ 8.0 x 10^{13} W/cm^2. The laser pulses are spatially overlapped using dichroic mirrors and focused by a 30-cm focal length lens, into the vacuum chamber, where they cross a molecular beam at 90° .

The 1 kHz femtosecond laser system serves as the master clock in the present experiment. Since the maximum repetition rate of the molecular beam and the charge-coupled device (CCD) camera is 20 Hz, a homemade box triggers on every 50th femtosecond laser pulse to be used in the experiment. The time difference between the two femtosecond pulses is fixed by an optical delay line and the alignment pulse is electronically synchronized to the femtosecond pulses with a precision better than ±0.5 ns. The ionized fragments are accelerated by a static field toward an ion detector positioned on-axis with the molecular beam.[15] The detector consists of a microsphere plate (MSP) backed by a phosphor screen. Each ion detected leads to emission of a localized flash of light which is recorded by a CCD camera. Fast electronic gating of the CCD camera allows us to record mass and charge selected ion images.

III. Alignment of iodine

We measure the spatial orientation of the I_2 molecules by dissociating them using a femtosecond pump pulse as mentioned above. The pump pulse is resonant with the perpendicular $A\ ^3\Pi_{1u} \leftarrow X\ ^1\Sigma_g^+$ transition and a dissociative wave packet with a final internuclear speed of 8.9 Å/ps is created. The velocity vector of the atomic iodine fragments is a direct measure of the initial orientation of the molecule. At an optical delay of 200 ps the iodine atoms are separated by almost 2000 Å so the Coulomb energy released if both fragments are ionized is negligible compared to the dissociation energy. Hereby, we ensure that the velocity of all I^+ ions originating from ionization of the dissociated molecules is the same.

The pump pulses are polarized perpendicular to the detector surface and it is therefore mainly molecules with their internuclear axis parallel to the detector surface that are dissociated. As a result, the ionized fragments, all having the same speed, will be detected at the same radial position on the detector but at different angular positions, corresponding to the initial orientation of the parent molecule. The two-dimensional ion images of I^+ ions, shown in Fig. 2, include the basic experimental information. The circularly symmetric ion image recorded when only the femtosecond pulses are used [Fig. 2(A)] exhibits a prominent ring with a diameter, *d*, of 7.4 mm and a more diffuse distribution in the outer part of the image.

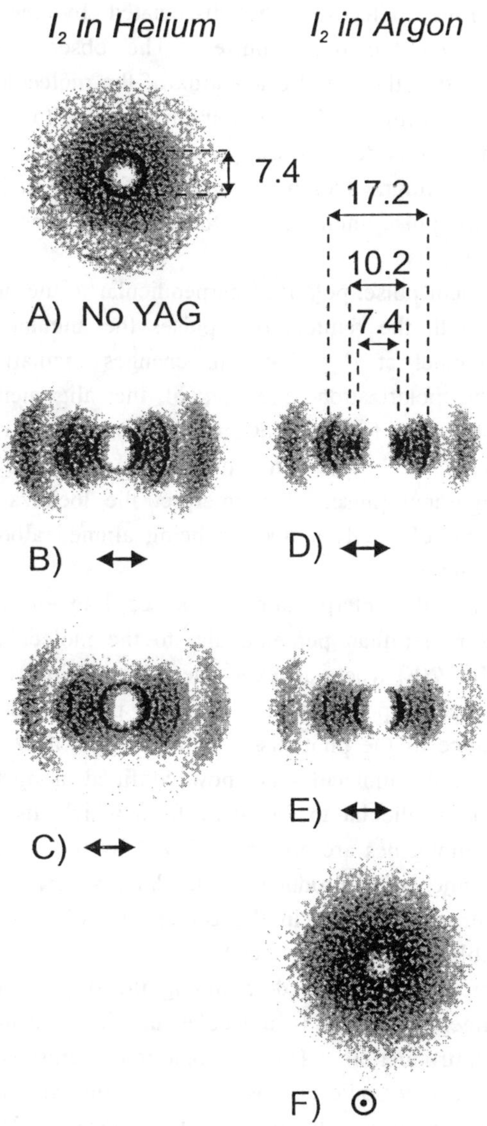

Figure 2. Ion images of I^+ fragments recorded for I_2 seeded in He (left images) or in Ar (right images), and for different intensities of the YAG pulse. (A) $I_{YAG} = 0$. (B), (D), (F) $I_{YAG} = 1.4 \times 10^{12}$ W/cm^2. (C), (E) $I_{YAG} = 5.0 \times 10^{11}$ W/cm^2. The femtosecond pulses are polarized along the molecular beam axis and the polarization direction of the YAG is shown at each image. The length scales in (A) and (D) are given in millimeters. In all images the central portion (radius $\leqq$ 2.5 mm) of the detector is not active.

The ring results from molecules photodissociated by the pump pulse and subsequently ionized by the probe pulse. The observation of a circularly symmetric ring shows that the initial orientation of the molecules is random. The deviations from this uniform angular distribution are used to identify alignment as discussed later. The I^+ ions detected away from the well-defined photodissociation channel originate from fragmentation of multiply charged molecular ions formed when the intense probe pulse interacts with those I_2 molecules that did not undergo excitation by the pump pulse.[14]

When the alignment pulse, polarized perpendicular to the molecular beam axis, is applied together with the femtosecond pulses the angular distribution of the photodissociation channel at d = 7.4 mm changes dramatically and becomes localized around the polarization direction of the alignment pulse. This is illustrated in Figs. 2(B)-2(E) for different intensities of the YAG pulse, I_{YAG}, and for I_2 seeded in Ar or in He. The pump pulse is synchronized to the maximum intensity of the alignment pulse. We interpret the localization of the angular distribution as a result of the I_2 molecules being aligned along the electric field vector of the alignment pulse.

An obvious test of this interpretation is to record an ion image with the YAG polarization parallel rather than perpendicular to the molecular beam axis. The result is shown in Fig. 2(F) where all experimental conditions are identical to those in Fig. 2(D) except that the polarization of the YAG pulse is rotated by 90°. The complete disappearance of the photodissociation channel at d = 7.4 mm confirms that the orientation of the molecules are now confined along the direction of the YAG polarization, perpendicular to the plane from which dissociation is possible. Instead, the aligned molecules are multiply ionized by the intense probe pulse and their subsequent fragmentation produces a circularly symmetric ion image with a radial distribution strongly peaked in the center of the detector. (Note that the detector has a small blind spot in the center.)

The two important parameters determining the degree of alignment are the initial rotational temperature of the I_2 molecules and the intensity of the YAG pulse, as mentioned in the introduction. The rotational temperature is controlled by using either Ar or He in the supersonic expansion because the Ar carrier gas produces I_2 molecules with lower rotational temperature than does the He carrier gas.[16] In agreement with theory,[7] we observe more localized angular distribution, i.e., stronger alignment, when Ar is used for a fixed intensity of the YAG pulse [compare Figs. 2(B) and 2(D), or 2(C) and 2(E)]. Similarly, we observe that the alignment becomes more pronounced if the intensity of the YAG pulse is increased and the initial rotational temperature is kept constant. This is seen by comparing Figs. 2(B) and 2(C) (I_2 in He), or Figs. 2(D) and 2(E) (I_2 in Ar).

We notice the appearance of two additional pairs of half rings with diameters ~ 10.2 and ~ 17.2 mm. Their radial positions reveal that they result from resonantly enhanced two-photon dissociation:

$$I_2 + \hbar\omega_{pump} + \hbar\omega_{YAG} \rightarrow I_2^* \rightarrow \begin{cases} \text{I+I}, \; d = 17.2 \text{ mm} \\ \text{I+I}^*, \; d = 10.2 \text{ mm}, \end{cases}$$

where I_2^* is an electronically excited state of the molecule, and I and I^* are the lower and upper fine structure states of the iodine atom, respectively. The absorption of an additional YAG photon can take place at the positions ① or ② in Fig. 3.[17] The transition dipole moment depends on the internuclear distance R and naturally it should be larger for R near the equilibrium distance R_e. Thus, we can expect that the absorption of an additional YAG photon is more likely at ① than at ②, which is in agreement with our observation that the 17.2-mm diameter half ring is more pronounced than the 10.2-mm diameter half ring. The two-photon dissociation channels exhibit the same angular distribution as the one-photon channel and provide an additional confirmation of the alignment of I_2 molecules.

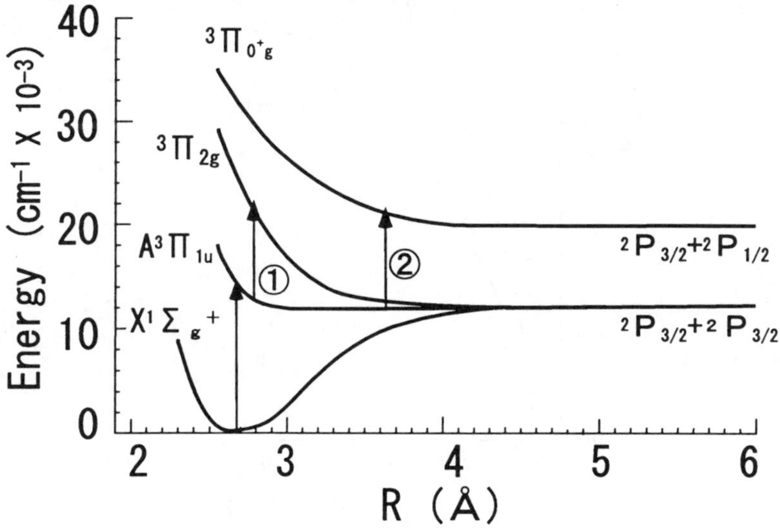

Figure 3. Schematic of a few potential curves of I_2 illustrating the appearance of two additional pairs of half rings. Although specific term labels of $^3\Pi_{2g}$ and $^3\Pi_{0^+g}$ are put on the two excited potential curves, there could be other potential curves involved as well.[17]

We can quantify the degree of alignment from the angular distribution shown in Fig. 2. For each set of (I_{YAG}, Ar/He) we have determined the average value of $\cos^2\theta$, $\langle\langle\cos^2\theta\rangle\rangle$, where θ is the angle between the YAG polarization axis and the internuclear axis. A detailed procedure to calculate the parameter $\langle\langle\cos^2\theta\rangle\rangle$ is described in Ref. 18. The results are shown in Fig. 4. We also calculated $\langle\langle\cos^2\theta\rangle\rangle$, based on the theory of Ref. 7, as a function of I_{YAG} and for different rotational temperatures, T_{rot}.[19] As seen in Fig. 4, good agreement between the experimental results and the calculations (full curves) are obtained for $T_{rot} = 3.5$ K (I_2 in Ar) and $T_{rot} = 7$ K (I_2 in He). Such rotational temperatures are consistent with the results from measurements of T_{rot} on a molecular beam similar to that reported here.[16] A more precise comparison between the measured and the calculated values of $\langle\langle\cos^2\theta\rangle\rangle$ requires an independent determination of T_{rot} in our experiment.

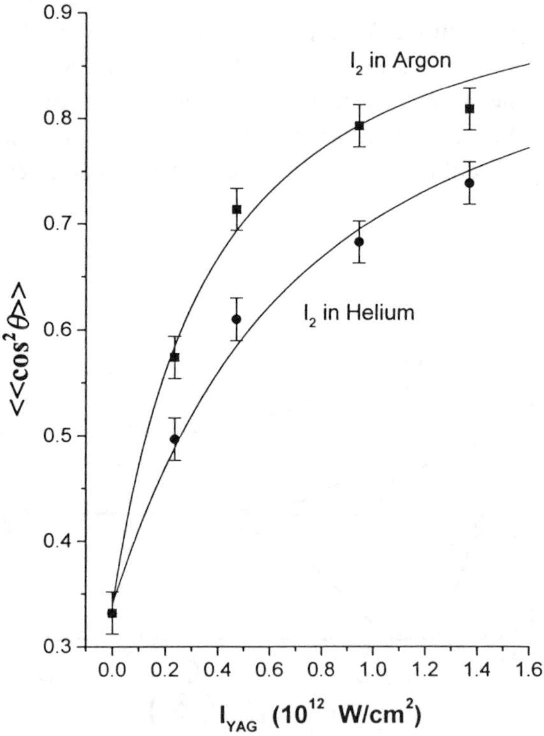

Figure 4. The observed value of $\langle\langle\cos^2\theta\rangle\rangle$ for different I_{YAG} and for I_2 in Ar (squares) and in He (circles). The full curves are the calculated values of $\langle\langle\cos^2\theta\rangle\rangle$. Note that for a sample of randomly oriented molecules the theoretical value of $\langle\langle\cos^2\theta\rangle\rangle$ is 1/3.

The initiation and the termination of the alignment when the YAG pulse is turned on and subsequently switched off can be followed by changing the delay of the pump pulse with respect to the YAG pulse. Figure 5 shows the results, represented as the $\langle\langle\cos^2\theta\rangle\rangle$ parameter, at different times. For comparison, we used the intensity profile of the YAG pulse, measured with a fast photodiode, to calculate $\langle\langle\cos^2\theta\rangle\rangle$ as a function of time (full curve in Fig. 5). The agreement between the measured and the calculated values of $\langle\langle\cos^2\theta\rangle\rangle$ confirms our scenario of the molecules evolving from field-free rotational states into pendular states and back into filed-free rotational states. The measurement of the intensity profile of the YAG pulse was performed independently of that of $\langle\langle\cos^2\theta\rangle\rangle$ and there is a possibility that the photodiode did not follow the trailing edge of the YAG pulse accurately. These may explain the small deviation of calculated $\langle\langle\cos^2\theta\rangle\rangle$ from the measured one for Time > 6 ns.

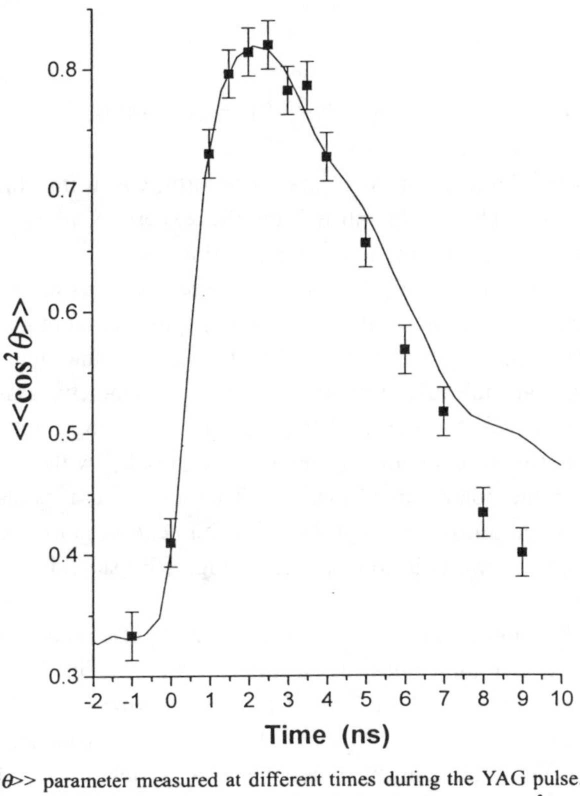

Figure 5. The $\langle\langle\cos^2\theta\rangle\rangle$ parameter measured at different times during the YAG pulse. The maximum intensity is 1.4 x 10^{12} W/cm^2 and I_2 is seeded in Ar. The full curve shows $\langle\langle\cos^2\theta\rangle\rangle$ calculated on the basis of the intensity profile of the YAG pulse.

IV. ALIGNMENT OF IODOBENZENE

To demonstrate that laser induced alignment is generally applicable to molecules other than I_2 we carried out experiments on iodobenzene (C_6H_5I). Concerning iodobenzene, we do not have enough information about an easily interpretable dissociative transition like the perpendicular $A\ {}^3\Pi_{1u} \leftarrow X\ {}^1\Sigma_g^+$ used for probing iodine.[20] Therefore, we probe alignment of iodobenzene by dissociative ionization with a single high intensity femtosecond pulse centered at 800 nm. Figures 6(A) and 6(B) show I^+ ion images recorded without the YAG laser using linearly and circularly polarized femtosecond pulses, respectively. The radial structures in the images originate from multiple ionization and subsequent Coulomb fragmentation

$$C_6H_5I + N\hbar\omega \rightarrow I^+ + C_6H_5^+ \ (d = 21.8 \text{ mm})$$

and

$$C_6H_5I + N'\hbar\omega \rightarrow I^+ + C_6H_5^{2+} \ (d = 31.2 \text{ mm})$$

The angular distributions of ions are very different using the two differently polarized probes. This is quantified by the experimental $\langle\langle \cos^2\theta_{2d} \rangle\rangle$ values[18] which are found to be 0.66 for linear polarization and 0.56 for circular polarization when measured in the $I^+ + C_6H_5^+$ channel. For comparison, a three-dimensional isotropic distribution of molecular orientations projected onto a two-dimensional detector yields $\langle\langle \cos^2\theta_{2d} \rangle\rangle = 0.50$. The difference of the angular distributions is due to the fact that molecules initially having their symmetry axis aligned with the laser polarization have a much higher probability of being multiply ionized by an intense laser pulse as compared to molecules initially with their symmetry axis perpendicular to the polarization axis.[21,22] Therefore, it is reasonable to choose the circularly polarized pulse as a probe of alignment because deviation from the (almost) circularly symmetric ion images of Fig. 6(B) should identify alignment of the molecules.

Figure 6(C) shows an ion image with the YAG pulse included, $I_{YAG} = 1.2 \times 10^{12}$ W/cm^2, where C_6H_5I molecules are seeded in Ar. As in the I_2 experiment, we observe that the angular distributions of I^+ ions are strongly peaked along the YAG polarization, being more localized when Ar is used as carrier gas. To the best of our knowledge the polarizability of iodobenzene is unknown. Assuming it to be similar to that of chlorobenzene,[23] which is anisotropic with the largest component parallel to the symmetry axis, we interpret our observations as iodobenzene being

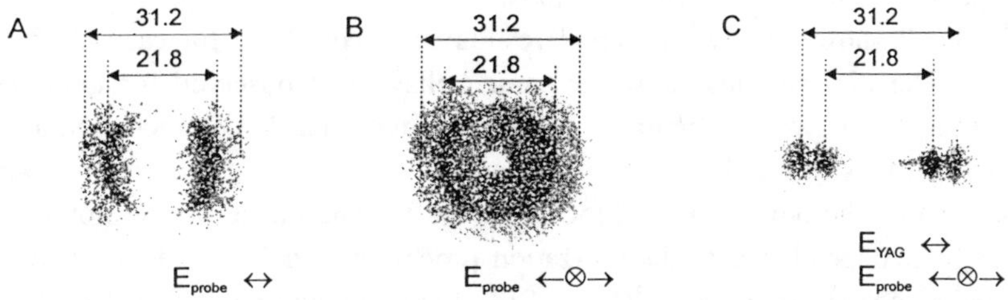

Figure 6. Ion images of I^+ fragments from single femtosecond pulse ionization and dissociation of iodobenzene seeded in Ar. (A) $I_{YAG} = 0$. The femtosecond pulse is linearly polarized. (B) $I_{YAG} = 0$. The femtosecond pulse is circularly polarized. (C) $I_{YAG} = 1.2 \times 10^{12}$ W/cm^2. The femtosecond pulse is circularly polarized.

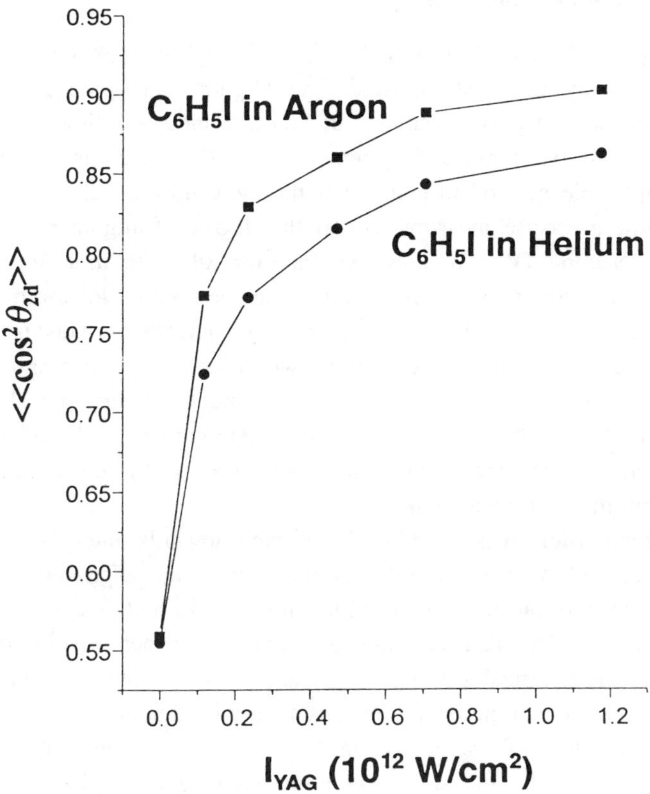

Figure 7. The observed value of $<<\cos^2\theta_{2d}>>$ for different I_{YAG} and for C_6H_5I in Ar (squares) and in He (circles). The full lines are shown to guide the eyes.

aligned along the molecular symmetry axis.

Figure 7 shows the experimental results of $\langle\langle\cos^2\theta_{2d}\rangle\rangle$ for each set of (I_{YAG}, Ar/He). The observations show the same behavior as observed for I_2 molecules. For a fixed carrier gas the alignment becomes better when I_{YAG} is increased and for a fixed YAG intensity the degree of alignment is higher for C_6H_5I seeded in Ar. In contrast to the photodissociation process in the I_2 experiment, we do not know the exact angular dependence of the ionization process of iodobenzene, so we have not converted $\langle\langle\cos^2\theta_{2d}\rangle\rangle$ to $\langle\langle\cos^2\theta\rangle\rangle$. Still, the maximum value of $\langle\langle\cos^2\theta_{2d}\rangle\rangle$ for iodobenzene, ~ 0.90, obtained for $I_{YAG} = 1.2 \times 10^{12}$ W/cm^2 and using Ar as a carrier gas, equals the maximum value obtained for iodine and indicates that an almost same degree of alignment is achieved.

V.Alignment of other molecules

We also performed experiment on ICl, CS_2, and CH_3I to show that these molecules can be aligned by the YAG pulse. The alignment is probed with circularly polarized femtosecond pulses, centered at 800 nm, inducing dissociative ionization, and by recording the ionized fragments (I^+ or S^+). This is a convenient and generally applicable method to confirm if the molecules are aligned. However, it does not allow a precise measurement of the degree of alignment, $\langle\langle\cos^2\theta\rangle\rangle$, as discussed in Section IV. We observe that for all three molecules the angular distributions of the ionized fragments become more localized around the polarization axis of the YAG pulse as I_{YAG} is increased. This is a qualitative manifestation of alignment similar to what we observe for iodine and iodobenzene. However, the observed alignment of ICl, CS_2, and CH_3I does not show the same clear dependence on the carrier gas as does the alignment of I_2 and C_6H_5I. This might be due to that the rotational temperature of these molecules was not necessarily optimized, especially in Ar.

Further extension of laser induced alignment towards other molecules than the ones demonstrated here is straightforward. We can align any molecule that possesses an anisotropic polarizability and that can be put into the gas phase and rotationally cold. The internal molecular parameters determining the degree of alignment are the polarizability anisotropy and the rotational constants[7] with larger anisotropy and smaller rotational constants favoring alignment. For a given initial rotational temperature of a sample of molecules, the maximum degree of alignment obtainable is limited by the alignment laser intensity the molecule can withstand before dissociating or ionizing. In our iodobenzene experiment YAG intensities higher than 1.2×10^{12} W/cm^2 were not used because the YAG pulse starts to ionize the molecule at this intensity. For iodine no dissociation or ionization by the YAG pulse was observed at the maximum intensity used, $I_{YAG} = 1.4 \times 10^{12}$ W/cm^2. Thus, it should be possible to increase the degree of alignment even further by increasing the laser intensity. Another way of improving the alignment is to lower the rotational temperature because alignment is very sensitive to the initial rotational temperature of the molecular beam.

VI.Discussion and concluding remarks

It should be noticed that the laser used for alignment is nonresonant[24] with respect to both the electronic and vibrational structure of the molecule because excitation of any of these internal degrees of freedom can lead to dissociation or ionization before alignment. In our experiment the YAG pulse does not cause any observable dissociation or ionization. Generally, electronic resonances are avoided by employing infrared radiation and keeping the intensity below the value where tunnel ionization becomes significant.[25] To avoid vibrational excitation it is necessary to inspect carefully the infrared absorption spectrum of the particular molecule. If the wavelength of the YAG laser accidentally coincides with a vibrational (overtone) transition and thereby causes significant dissociation, it is possible to align the molecules using a different wavelength produced by nonlinear optics methods or by employing a different laser like the CO_2 laser.

Laser induced alignment can be used to angularly confine molecules to which hexapole focusing (only symmetric tops[4]) and brute force orientation (only polar molecules[5,6]) do not apply (and also molecules to which these techniques apply). In many light induced processes and scattering experiments the distinction between orientation and alignment of the molecules in the reaction is not necessary for symmetry reasons. Therefore, it is interesting to compare the degree of alignment achievable with brute force orientation[5,6] and laser induced alignment. For pyridazine (μ = 4 D) at a rotational temperature of ~ 2 K and an orientation field of 60 kV/cm, Li *et al.*[26] estimate that half of the molecules can be restricted within a cone of 45° (half angle). Iodine, at the same rotational temperature, exposed to an aligning field of 5×10^{11} W/cm^2 will have half of the molecules restricted within a cone of ~ 12° and ~ 98% of the molecules within a 45° cone.

The reason for the better angular confinement achievable using laser induced alignment is in the much higher electric field strengths provided by pulsed lasers. In the above example, the induced dipole moment in iodine is on the order of 1 D interacting with a field of ~ 2×10^7 V/cm. This means that the interaction energy between the molecule and the laser field is approximately two orders of magnitude larger than what is possible using static fields and permanent dipole moments.

An experiment making use of laser aligned molecules must be carried out in the presence of the alignment field because the pendular states evolve back into field free states when the field is turned off. The presence of the intense laser field can influence some measurements, e.g., electronic states will be Stark shifted or additional photons can be absorbed as discussed in Section III. In the near future it might prove experimentally possible to use shorter, nonadiabatic pulses to produce field-free alignment by recurrences of rotational wave packets as discussed by Ortigoso *et al.*.[27]

Quite a few previous studies have addressed the possibility of laser-induced alignment when molecules are irradiated by intense picosecond or femtosecond laser pulses.[28] These studies differ in two important aspects from the present study. First, in the previous experiments the molecules are multiply ionized by the intense short laser pulse so a sample of neutral molecules is not maintained. Second, the indication of alignment is based on observing that the multiply charged molecular ions preferentially fragment along the laser polarization. Such an anisotropy in the angular distribution of the fragments cannot unambiguously be assigned to molecular alignment because it is now established that molecules initially oriented along the polarization axis are much more likely to undergo multiple ionization.[20,21] This ambiguity is removed from our experiment by using a nonionizing nanosecond pulse to induce alignment and a femtosecond pulse to measure the molecular orientation by photodissociation.

In conclusion, we have shown that neutral molecules can be aligned by an intense nonresonant nanosecond laser pulse.[18,29] It has importance for a number of areas in molecular science. (i) Controlling the orientation of the molecular geometry enables enhancement or suppression of photochemical events that depend on the angle between the polarization axis of the incident light and a specific molecular axis. (ii) The duration of the YAG pulse is sufficiently long that picosecond or femtosecond time-resolved studies can be performed on a sample of aligned molecules. For example, this is important for femtosecond timed Coulomb explosion studies.[12,30] (iii) Spectroscopy of pendular states as recently demonstrated by Kim and Felker in Raman experiments.[11] (iv) In addition to alignment, laser-induced dipole forces have lately been employed to deflect neutral molecules.[10] The use of nonresonant intense laser fields should allow many aspects of atomic optics to be transferred to molecules such as trapping, wave guiding, and focusing. In our recent paper,[18] some potential applications using the technique of aligning neutral molecules are discussed. They are (a) Dissociative rotational cooling, (b) Control of strong laser field ionization, (c) Selective dissociation, (d) Applications in chemical control, (e) Orientation effects in bimolecular reactions, and (f) Alignment of the molecular plane.

The present experiments were performed at Department of Chemistry, University of Århus.

References

1. For a review, see C. S. Adams and E. Riis, *Prog. Quantum Electron.* **21**, 1 (1997).
2. J. A. Stroscio and D. M. Eigler, *Science* **254**, 1319 (1991).
3. *J. Phys. Chem.* A **101**, No. 41 (1997), special issue on Stereodynamics of Chemical Reactions.
4. V. A. Cho and R. B. Bernstein, *J. Phys. Chem.* **95**, 8129 (1991); D. H. Parker and R. B. Bernstein, *Annu. Rev. Phys. Chem.* **40**, 561 (1989).
5. B. Friedrich and D. R. Herschbach, *Nature* (London) **353**, 412 (1991).
6. H. J. Loesch and A. Remscheid, *J. Phys. Chem.* **95**, 8194 (1991).
7. B. Friedrich and D. Herschbach, *Phys. Rev. Lett.* **74**, 4623 (1995); *J. Phys. Chem.* **99**, 15686 (1995).
8. J. H. Williams, *Adv. Chem. Phys.* **85**, 361 (1993).
9. J. D. Weinstein *et al.*, *Nature* (London) **395**, 148 (1998); T. Takekoshi, B. M. Patterson, and R. J. Knize, *Phys. Rev. Lett.* **81**, 5105 (1998).
10. H. Stapelfeldt, H. Sakai, E. Constant, and P. B. Corkum, *Phys. Rev. Lett.* **79**, 2787 (1997); H. Sakai, A. Tarasevitch, J. Danilov, H. Stapelfeldt, R. W. Yip, C. Ellert, E. Constant, and P. B. Corkum, *Phys. Rev.* A **57**, 2794 (1998).
11. W. Kim and P. M. Felker, *J. Chem. Phys.* **104**, 1147 (1996); **108**, 6763 (1998).
12. Our alignment pulse has a rise time of ~ 1.5 ns, which is much longer than the maximum rotational period of the I_2 molecules (~ 0.3 ns).
13. Photodissociation as a probe of the molecular orientation has been reported by other groups, see e.g., G. Bazalgette *et al.*, *J. Phys. Chem.* A **102**, 1098 (1998).
14. J. J. Larsen *et al.*, *J. Chem. Phys.* **109**, 8857 (1998).
15. A. T. J. B. Eppink and D. H. Parker, *Rev. Sci. Instrum.* **68**, 3477 (1997).
16. G. M. McClelland, K. L. Saenger, J. J. Valentini, and D. R. Herschbach, *J. Phys. Chem.* **83**, 947 (1979).
17. R. S. Mulliken, *J. Chem. Phys.* **55**, 288 (1971).
18. J. J. Larsen, H. Sakai, C. P. Safvan, I. Wendt-Larsen, and H. Stapelfeldt, *J. Chem. Phys.* **111**, 7774 (1999).
19. The calculated values of $\langle\langle\cos^2\theta\rangle\rangle$ are averaged over the different intensity regions of the focal volume of the YAG pulse.
20. H. J. Hwang and M. A. El-Sayed, *J. Phys. Chem.* **96**, 856 (1992).
21. T. Seideman, M. Yu. Ivanov, and P. B. Corkum, *Chem. Phys. Lett.* **252**, 181 (1996).

22. E. Constant, H. Stapelfeldt, and P. B. Corkum, *Phys. Rev. Lett.* **76**, 4140 (1996), J. H. Posthumus, A. J. Giles, M. R. Thompson, and K. Codling, *J. Phys.* B **29**, 5811 (1996).
23. K. J. Miller, *J. Am. Chem. Soc.* **112**, 8543 (1990).
24. A different method for achieving alignment, based on a resonant laser field, has been suggested: T. Seideman, *J. Chem. Phys.* **103**, 7887 (1995).
25. P. Dietrich and P. B. Corkum, *J. Chem. Phys.* **97**, 3187 (1992).
26. H. Li, K. J. Franks, R. J. Hanson, and W. Kong, *J. Phys. Chem.* A **102**, 8084 (1998).
27. J. Ortigoso, M. Rodriguez, M. Gupta, and B. Friedrich, *J. Chem. Phys.* **110**, 3870 (1999).
28. V. R. Bhardwaj, K. Vijayalakshmi, and D. Mathur. *Phys. Rev.* A **56**, 2455 (1997); J. H. Sanderson *et al.*, *J. Phys.* B **31**, L599 (1998), and references therein.
29. H. Sakai, C. P. Safvan, J. J. Larsen, K. M. Hilligsφe, K. Hald, and H. Stapelfeldt, *J. Chem. Phys.* **110**, 10235 (1999).
30. Ch. Ellert *et al.*, *Philos. Trans. R. Soc. London*, Ser. A **356**, 329 (1998).

EFFECT OF IR LASER PULSE ON THE UV PHOTODISSOCIATION OF CH_3I AND OCS

A. SUGITA, M. MASHINO, M. KAWASAKI, AND Y. MATSUMI[a]

Department of Molecular Engineering, Kyoto University, Kyoto 606-8501, Japan

[a]Solar-Terrestrial Environment Laboratory, Nayoga University, Toyokawa 442-8507, Japan

We have demonstrated alignment of neutral molecules and complete bleaching of electronically excited states by strong IR laser pulse. The electric field of a pulsed laser is used for alignment that is demonstrated by measurement of the control of the anisotropy of photofragments generated by polarized light. The angular distribution of $I(^2P_{3/2})$ atoms from the photodissociation of CH_3I at 304 nm was measured with the 1.06 μm laser pulse excitation:

$$CH_3I + h\nu(UV+IR) \rightarrow CH_3 + I(^2P_{3/2}).$$

The 1.06 μm laser pulse excitation of the photoprepared OCS at 230 nm bleaches completely the formation of $CO(X^1\Sigma^+, J) + S(^1D)$ and induces a new excitation pathway by upconversion:

$$OCS(v=0) + h\nu(UV) + 2h\nu(IR) \rightarrow CO(X^1\Sigma^+, J \sim 74) + S(^1S).$$

I. Introduction

Recently, there has been increasing interest in controlling chemical reactions of gas phase molecules by external optical electric fields and multiphoton processes.[1,2]

We have demonstrated control of molecular photodissociation by Nd:YAG laser pulses that are nonresonant with any vibrational excitation of molecules in the ground state. Alignment and bleaching processes with IR pulses are schematically shown in Figure 1.

Friedrich and Herschbach have proposed the use of an optical field for alignment of molecules.[3] Optical frequencies are much higher than rotational frequencies; for example, 1064 nm light has a period of 3.5 fs. The relatively slowly rotating molecule is unable to follow the rapidly fluctuating electric field, and the first order effect averages to zero. However, the second order perturbation due to the polarizability anisotropy of the molecule may be appreciable. A necessary condition for appreciable alignment is that $\mathbf{\mu}\cdot\mathbf{E}$ should be comparable to kT, where μ is the electric dipole moment of the molecule, and $\mathbf{E}$ is the optical electric field. The alignment of neutral molecules with a strong IR laser field (1.4×10^{12} W/cm^2) has been demonstrated in the photodissociation of I_2, C_6H_5I, ICl, CS_2 and CH_3I.[1] Using a sample of aligned iodine molecules, Larsen *et al.* [2] showed beautiful control of the photofragment branching ratios of the $I(^2P_{3/2})$ and $I(^2P_{1/2})$ pathways of I_2. In this paper we have demonstrated the alignment of CH_3I in the ground state with a ns Nd:YAG laser pulse:

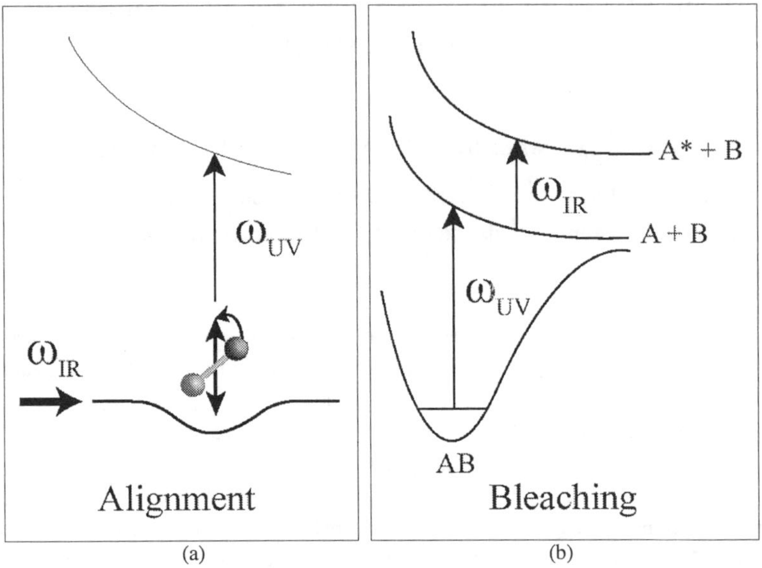

Figure 1. Schematics of IR pulse laser effect on photodissociation dynamics. (a) alignment of an electronically ground state and (b) upconversion of an electronically excited dissociative state.

$$CH_3I + h\nu(IR) \rightarrow CH_3I(aligned). \quad (1)$$

A higher excited state of $OCS(2^1\Sigma^+)$ is prepared by direct VUV excitation of the $OCS(2^1\Sigma^+ \leftarrow X^1\Sigma^+)$ transition to produce $S(^1S)$ + CO.[4] If a stepwise transition occurs by multiphoton absorption to this energy region as shown in Figure 1(b), control of the dissociation pathway can be possible by tuning the transition rate to the upper dissociative state. In this work, we have prepared the higher excited state by 1.06 μm excitation through the UV photoprepared repulsive state $OCS(1^1\Delta(2^1A'))$. Complete switching of dissociation pathway (2) to pathway (3) has been achieved by controlling the intensity of the IR laser:

$$OCS + h\nu(UV) \rightarrow CO(X^1\Sigma^+, J) + S(^1D), \quad (2)$$

$$OCS(v{=}0) + h\nu(UV{+}2IR) \rightarrow CO(X^1\Sigma^+) + S(^1S). \quad (3)$$

II. Experimental

A pulsed YAG pumped dye laser (5 ns, 0.2 mJ/pulse, 10 Hz) frequency doubled to 304.67 nm was used to dissociate CH_3I and to ionize the $I(^2P_{3/2})$ fragment by (2+1) resonance-enhanced multiphoton ionization (REMPI). The 1064 nm fundamental of the Nd:YAG laser (1-120 mJ/pulse) was used to align the parent molecule prior to dissociation. The UV and IR lasers were counterpropagated into the interaction region. Each laser was focused with a lens (f = 20 cm) on a pulsed molecular beam of CH_3I that was diluted in Ar (1%, backing pressure = 750 Torr). The rotational temperature of the CH_3I is estimated from the 7 K rotational temperature of NO, which was expanded under the same conditions. The following processes occur:

$$CH_3I + h\nu(IR) \rightarrow CH_3I(aligned), \tag{1}$$

$$CH_3I + h\nu(304.67\ nm) \rightarrow CH_3 + I(^2P_{3/2}),$$

$$I(^2P_{3/2}) + 2h\nu(304.67\ nm) \rightarrow I(^4D_{7/2}),$$

$$I(^4D_{7/2}) + h\nu(304.67\ nm) \rightarrow I^+ .$$

The iodine atoms produced by the dye laser were detected by velocity map imaging.[5,6] The apparent anisotropy parameter, β, was obtained by a least squares fit of the 3D slice to the angular distribution function,

$$I(\theta) = (1/4\pi)\{1 + \beta P_2(\cos\theta)\}, \tag{4}$$

where $I(\theta)$ is the normalized angular distribution of the I atom fragment, θ is the angle between the velocity of the I atom fragment and the electric vector of the dissociating laser beam, and $P_2(x) = (3x^2-1)/2$.[7,8]

For OCS experiment, the pulsed YAG pumped dye laser frequency doubled to 229.83-230.00 nm was used to prepare dissociative OCS in a pulsed molecular beam. The 1064 nm fundamental of the Nd:YAG laser (typically 90 mJ/pulse) was used to pump the UV photoprepared state. UV and IR laser pulses were focused with the lens on the pulsed molecular beam of OCS diluted in Ar. The photofragment $CO(X^1\Sigma^+, J)$ was ionized by the same UV laser light with additional IR pulses and was detected by velocity map imaging spectroscopy.

III. Results and Discussion

1. Alignment of CH_3I with IR laser pulse

Results

Figure 2 shows that the anisotropy parameter β increases with IR laser intensity when the IR and UV lasers have parallel polarizations and are simultaneously fired. As the IR laser intensity increases, the angular distribution becomes more anisotropic. This external field effect becomes appreciable for IR laser intensities > 2 mJ/pulse and reaches an asymptotic value at 10 mJ/pulse. Above that energy the IR laser causes a decrease in I^+ signal, although the apparent β value remains

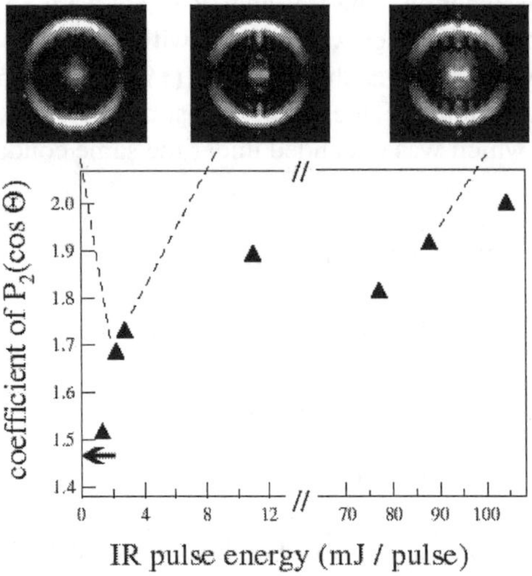

Figure 2. Change of the anisotropy parameter β with 1064 nm laser intensity for $I(^2P_{3/2})$ photofragments from the photodissociation of CH_3I at 304 nm. The arrow at zero IR energy indicates a β value averaged over several runs. The UV and IR lasers have parallel polarization. The images of the probed photofragment $I(^2P_{3/2})$ are shown.

constant. The nearly constant β value at higher pulse energies is thought to originate from CH_3I molecules situated further from the center of the region of strongest focus. In other words, in the volume where one would expect very strong alignment and even localization, the IR laser produces another dissociation pathway.

Figure 3 shows the variation of β with the delay τ between the UV and IR (10 mJ/pulse) laser pulses. An appreciable change in β is observed when the two laser pulses are temporally overlapped within ±2 ns. For $|\tau| > 2$ ns, the value of β was ~1.4. The average of β over several runs in the absence of the IR laser was 1.47±0.05, as shown by the arrow in Figure 2. Figure 4 shows the energy distribution of the CH_3 + I photofragments. No appreciable change is observed in the center-of-mass translational energy distribution with and without the nonresonant IR laser pulse. These results support the molecular alignment by nonresonant IR laser pulse as shown in Figure 1(a), and eliminate the upconversion mechanism shown in Figure 1(b).

Theory

Suppose that a) the angle between the electric fields of the dissociating and aligning lasers is χ, and b) the dimensionless parameter $\gamma = -(\alpha-\alpha_{zz})(E^2/2kT)$, where $\alpha = (\alpha_{xx} + \alpha_{yy} + \alpha_{zz})/3$ is the mean polarizability; $\Delta\alpha = \alpha_{zz} - \alpha$ is anisotropies.

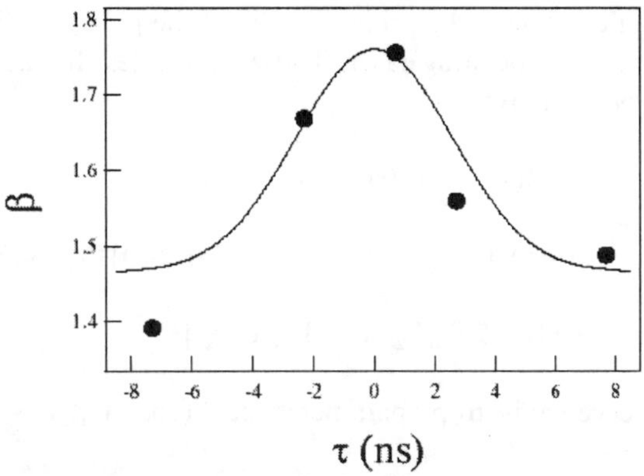

Figure 3. Temporal profile of β values for various delays between the UV laser pulse (0.1 mJ/pulse) and the IR laser pulse (10 mJ/pulse). For $|\tau| > 2$ ns, the average of over several runs in the absence of the IR laser, $\beta = 1.47$, is shown. The curve shows a temporal overlap of two Gaussian curves (3 ns and 5 ns width).

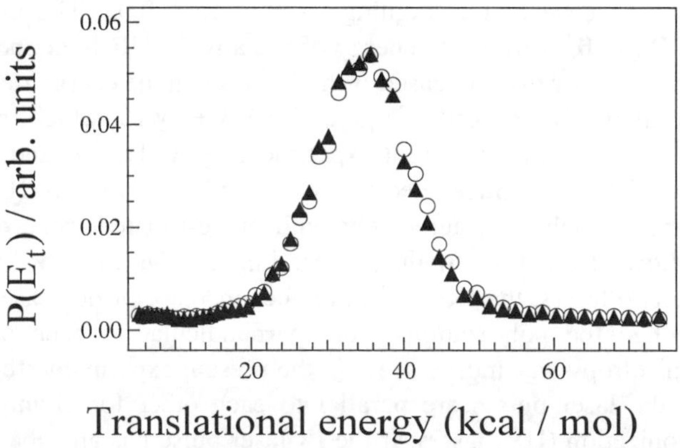

Figure 4. Center-of-mass translational energy distribution of photofragments from the photodissociation of CH_3I at 304.67 nm. The triengles were measured with the 1064 nm IR laser pulse (2-11 mJ/pulse, 5 ns duration); the circle without the IR laser pulse.

The distribution of molecular axes referred to the electric vector of the dissociating laser is given by:[9]

$$P(\theta_s) = (1/4\pi)\{1 + \gamma P_2(\cos\chi)P_2(\cos\theta_s)\}, \tag{5}$$

where θ_s is the angle between the principle axis of the polarizability tensor and the electric vector of the dissociating laser. Those molecules that absorb light have a distribution of axes given by:

$$I(\theta_s) = P(\theta_s)(1 + \beta_o P_2(\cos\theta_s))$$

$$= (1/4\pi)[1 + \{\beta_o + \gamma(1 + 2\beta_o/7 - \beta_o^2/5)\}P_2(\cos\chi)P_2(\cos\theta_s)$$

$$+ (18/35)\beta_o\gamma P_2(\cos\chi)P_4(\cos\theta_s)]. \quad (6)$$

For $\chi = 0$, the observed anisotropy parameter for $P_2(\cos\theta_s)$, β_{eff}, is given by:

$$\beta_{eff} = \beta_o + \gamma(1 + 2\beta_o/7 - \beta_o^2/5). \quad (7)$$

Analysis

The angular distribution of the I atom photofragments is fitted to Eq. (6) assuming axial recoil. The value of β_o was previously reported to be 1.5±0.1 for a dissociation wavelength of 304 nm.[6,10] In the present experiments, the value of β_o obtained in the absence of the aligning laser is 1.47±0.05. The increase of the coefficient of $P_2(\cos\theta_s)$ with pulse energy of the aligning IR laser shown in Figure 2 is fitted by values of γ that increased from 0 to 0.28 in the energy range 0 to 4 mJ. The degree of alignment, $<\cos^2\theta>$, is given by $1/3 + 2\gamma/15$, which increased from 0.333 to 0.333 + 0.037 in the present experiment. From Eq. (7) and the definition of γ, the value of the rms optical electric field is 6×10^6 V/cm or 5×10^{10} W/cm^2 at 4 mJ pulse energy, which is in agreement with our estimated value of $10^{10} \sim 10^{11}$ W/cm^2. It follows from Eq. (5) that, depending on the angle χ between the E vectors of the two lasers, the IR laser can either increase or decrease the observed anisotropy. When the polarizations were perpendicular to each other, a small decrease in anisotropy was indeed seen. In the present experiment, the polarization directions of the laser pulses are parallel to each other for alignment (IR) and dissociation/ionization (UV) lasers. If the IR laser pulse has an enhanced effect on ionization of the iodine atom, the anisotropy parameter should keep increasing with the IR laser pulse intensity. However, this is not the case in the present experiment. Thus, the IR laser pulse under the present experimental conditions does not affect the ionization process.

2. Bleaching of UV photoprepared OCS with IR laser pulse

Results

Figure 5 shows the CO images with strong IR irradiation of the YAG laser output at 1.06 μm as well as UV irradiation at λ_{UV} = 229.864 nm. Without the IR laser pulse, the REMPI image of CO(J=63) is seen from the one-photon UV

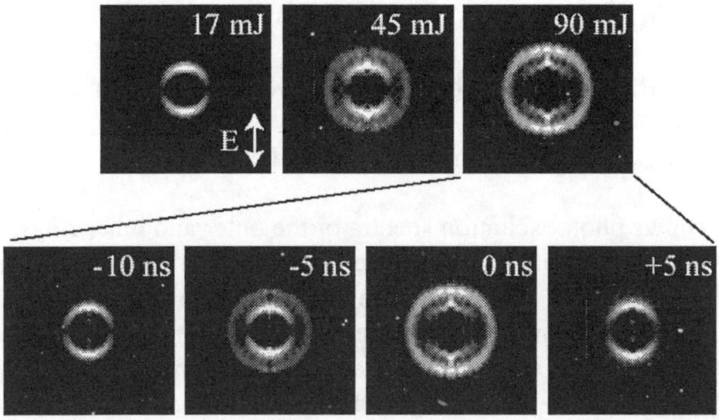

Figure 5. Effect of strong IR pulse laser irradiation on CO(J=63) images for pathway (2). Upper panels show effect of IR laser intensity at 1.06 μm. Outer ring appears for > 17 mJ. Lower panels show temporal profile of images for various delays between the IR and UV laser pulses; "- (minus)" means that the IR laser pulse comes to the interaction region prior to the UV laser pulse.

photodissociation process.[11-13] With additional IR pulses, the strong IR irradiation bleaches the inner CO image of pathway (2) and generates the outer ring of a new pathway (3). In this case, UV and IR laser pulses are temporarily and spatially overlapped. The obtained values for pathway (3), β_{out} = 0.4 - 0.9, are much smaller than β_{in} = 1.4±0.1 for pathway (2). Even when UV wavelength was shifted from the resonant wavelengths from J=54 to J=66, a small change was observed in the size of the outer rings and the β_{out} values. In other words, the UV wavelengths were not completely in resonant with the REMPI transitions of CO when the strong IR pulse were added.

The only reasonable explanation is that a) $S(^1S)$ and highly rotationally excited CO are produced after absorption of one UV and two IR photons, and b) CO in the high rotational levels, J = 70-80 (centered at ~74), is nonresonantly ionized with 2hν(UV) + nhν(IR). These J levels are estimated by the translational energies of the outer rings at various UV wavelengths. Thus, the formation of $S(^1S)$ and CO in the higher J levels is from the dissociation of bend OCS. Figure 5 shows effect of IR intensity and temporal overlap between UV and IR pulses. The intensity of the outer ring increases nonlinearly with IR intensity, in agreement with the photoexcitation mechanism of two-photon absorption of IR pulses. The outer ring disappears when temporal overlap is lost. These results eliminate involvement of vibrational excitation of the ground state OCS(v). Only a short-lived intermediate can be involved for the formation of the final repulsive state. Hence, the photoabsorption process is a sequential multiphoton process of one UV photon and two IR photons. Product CO is ionized by a multiphoton process with UV and IR:

$$OCS(v=0) + h\nu(UV) \rightarrow OCS(2^1A'\ (1^1\Delta)),$$

$$OCS(2^1A'\ (1^1\Delta)) + 2h\nu(IR) \rightarrow CO(X^1\Sigma^+, J\text{~}74) + S(^1S),$$

$$CO(X^1\Sigma^+, J\text{~}74) + 2h\nu(UV) + nh\nu(IR) \rightarrow CO^+. \quad (3')$$

Figure 6 shows photoexcitation spectra of the outer and inner rings, which were measured with and without the IR pulse as a function of dissociation (and also probe) laser wavelength. The spectrum of the inner ring without IR has the rotationally resolved Q-branch structure with slightly diffused features due to Doppler widths, which is the same as reported previously.[12] However, the excitation spectrum of the outer ring with IR is totally different from the REMPI spectrum without IR. This result also suggests that the UV wavelengths were not completely resonant with the REMPI transitions of CO. The excitation spectrum is broad because the spectral contour must reflect the nonresonant MPI process of CO as well as two different excitation pathways of OCS(UV×IR) that include the

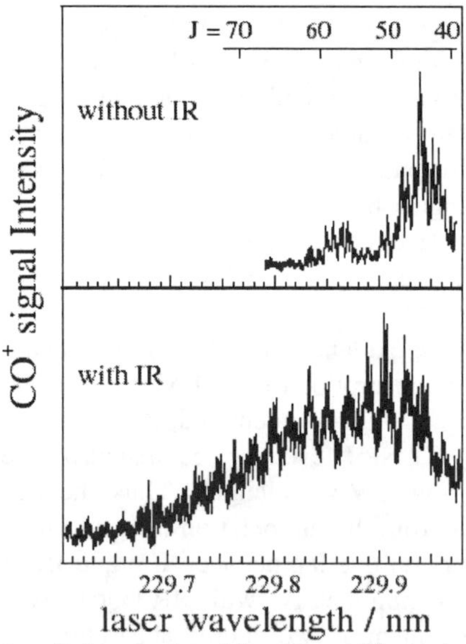

Figure 6. Photoexcitation spectra of the outer and inner rings of CO. Upper panel shows CO products from the OCS photodissociation through pathway (2) in the one-color UV laser experiment without 1.06 μm laser pulse. The structure of Q-branch heads is seen for the two-photon Q-branch transitions; $CO(X^1\Sigma^+, J) + 2h\nu(229.84\text{-}229.98\ nm) \rightarrow CO(C^1\Sigma^+, J)$. In the lower panel, OCS is excited by UV and IR pulses, and product CO through pathway (3') is nonresonantly ionized by UV + IR.

repulsive (along S-C bond) but stable (along bend) potential curve of OCS($2^1A'$ ($1^1\Delta$)).

IR excitation pathways

At 154 - 157 nm the partition of internal and translational energy was investigated for the repulsive OCS($2^1\Sigma^+$) state.[14,15] Due to the linear structure of OCS and stiffness of the C-O bond, a) 84% of the available energy is allocated to relative translational motion, 12% to vibration and 4% to rotation, and b) the angular anisotropy is as high as β = 1.8. In the present experiment, when OCS is excited simultaneously with UV and IR, S(^1S) and CO are produced via pathway (3'). The first UV excitation process of OCS is OCS($2^1A'$ ($1^1\Delta$)) $\leftarrow$ OCS($X^1\Sigma^+$). These upper states are strongly repulsive along the C-S stretching coordinate, and have short dissociation lifetimes as estimated by the β values of CO. Thus, the temporal overlap of the UV and IR is prompt as seen in Figure 5. After additional two IR photons are absorbed, OCS should cross to OCS($4^1A'$ ($2^1\Sigma^+$)) that correlated with S(^1S) + CO. However, according to Figure 7, the vertical excitation energy of this state is ~ 11 eV far above the total energy of UV+2IR (7.7 eV) under the linear geometry. However, the potential energy of OCS($4^1A'$ ($2^1\Sigma^+$)) is lowered to ~7 eV in bend coordinate, and also intersects with OCS($1^1\Sigma^+$) in linear geometry at an extended bond length at ~7.5 eV. Since the photoprepared OCS($2^1A'$ ($1^1\Delta$)) has a shallow minimum in bend mode coordinate, this dissociative state is upconverted to OCS($4^1A'$ ($2^1\Sigma^+$)) after two IR photon absorption and results in formation of S(^1S) and highly rotationally excited CO.

Since the reported β value for the VUV photodissociation of OCS($2^1\Sigma^+$) is 1.8, almost the maximum limit for a linear molecule, the final state OCS($4^1A'$ ($2^1\Sigma^+$)) has a short dissociative lifetime to generate S(^1S) + CO.[14,15] However, experimentally observed β values are 0.4 - 0.9 for the present S(^1S) pathway (3'). Let us consider, for simplicity, a sequential electronic excitation and dissociation process, OCS($4^1A'$ ($2^1\Sigma^+$)) $\leftarrow$ OCS($2^1A'$ ($1^1\Delta$)) $\leftarrow$ OCS(X $^1\Sigma^+$). The β value of the final product depends strongly on the dissociative lifetime of the final state and weakly on the β value of the intermediate state.[16,17] Because the final state OCS($4^1A'$ ($2^1\Sigma^+$)) has the large β value, it is difficult to explain the present low β values with this sequential electronic transition mechanism.

Alternative explanation is a vibrational excitation mechanism of the photoprepared bend OCS($2^1A'$) as shown by arrows in Figure 7.

$$\mathrm{OCS}(v=0) + h\nu(\mathrm{UV}) \rightarrow \mathrm{OCS}(2^1A'),$$

$$\mathrm{OCS}(2^1A') + 2h\nu(\mathrm{IR}) \rightarrow \mathrm{OCS}(2^1A', v),$$

$$\mathrm{OCS}(2^1A', v) \rightarrow \mathrm{OCS}(4^1A'\ (2^1\Sigma^+)) \rightarrow \mathrm{CO}(X^1\Sigma^+) + \mathrm{S}(^1S).$$

When the IR light intensity is so strong that the upconversion rate is much larger than the dissociation rate of the lower OCS($2^1A'$) state, bleaching is achieved. For the formation of CO($X^1\Sigma^+$) + S(^1S), the potential surface must cross over to OCS($4^1A'$ ($2^1\Sigma^+$)) that correlates with the products. During this predissociation pathway on bend coordinate, accompanied by a large angular momentum excitation of CO, the recoil direction is changed, thereby, resulting in a measurable reduction of β. Rotational motion of the parent molecule could reduce the β value but this effect is eliminated, because the supercooled molecular beam was used in the present experiment.[18]

IV. Conclusion

1. We have demonstrated that methyl iodide molecules that have an anisotropic polarizability tensor can be aligned by a ns Nd:YAG laser. Moreover, by varying the angle between the electric vectors of the aligning and dissociating electric fields within certain limits, one can control the anisotropy of the fragment velocity distribution.[9]
2. By upconversion, ns Nd:YAG laser pulses bleach completely the S(^1D) formation from the UV photodissociation of OCS and induce selective formation of S(^1S) by sequential IR multiphoton absorption via a bend vibrational excitation

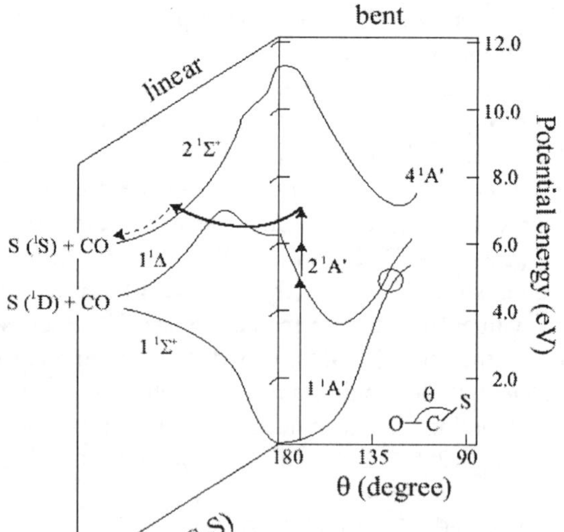

Figure 7. Schematics of the potential energy curves of OCS as a function of bond angle and internuclear distance after Ref. 14. Left-going arrow in (a) indicates the dissociation pathway via bend and stretch modes. Up-going arrows in (b) indicates the excitation process for UV + 2IR.

process of OCS.[19]

ACKNOWLEDGMENTS

M.K. thanks the Ministry of Education of Japan for support of this work by a Grant-in-Aid for a priority research field, "Molecular Physical Chemistry". A.S. thanks JSPS for fellowship. Stimulating discussion with Profs. Robert Gordon (University of Illinois) and Richard Bersohn (Columbia University) are highly appreciated.

REFERENCES

1 H. Sakai, C. P. Sufvan, J. J. Larsen, K. M. Hilligsoe, K. Hald, and H. Stapelfeldt, *J. Chem. Phys.* **110**, 10235 (1999): J. J. Larsen, H. Sakai, C. P. Sufvan, I. Wendt-Larsen, and H. Stapelfeldt, *J. Chem. Phys.* **111**, 7774 (1999).

2 J. J. Larsen, I. Wendt-Larsen, and H. Stapelfeldt, *Phys. Rev. Lett.* **83**, 1123 (1999).

3 B. Friedrich and D. Herschbach, *Phys. Rev. Lett.* **74**, 4623 (1995).

4 G. Black, R. L. Sharpless, T. G. Slanger, D. C. Lorents, *J. Chem. Phys.* **62**, 4274 (1975).

5 D. W. Chandler and P. L. Houston, *J. Chem. Phys.* **87**, 1445 (1987).

6 A. T. J. B. Eppink, D. H. Parker, *J. Chem. Phys.* **109**, 4758 (1998).

7 R. Bersohn and S. H. Lin, *Adv. Chem. Phys.* **55**, 1915 (1969).

8 R. N. Zare, *Mol. Photochem.* **4**, 1 (1972).

9 A. Sugita, M. Mashino, M. Kawasaki, Y. Mastumi, R. J. Gordon, and R. Bersohn, *J. Chem. Phys.* **112**, 2164 (2000).

10 J. R. Syage, *J. Chem. Phys.* **105**, 1007 (1996).

11 N. Sivakumar, G. E. Hall, P. L. Houston, J. W. Hepburn and I. Burak, *J. Chem. Phys.* **88**, 3692 (1988).

12 Y. Sato, Y. Masumi, M. Kawasaki, K. Tsukiyama and R. Bersohn, *J. Phys. Chem.* **99**, 16307 (1995).

13 T. Suzuki, H. Takayanagi, S. Nanbu and M. Aoyagi, *J. Chem. Phys.* **109**, 5778 (1998).

14 C. E. Strauss, G. C. McBane. P. L. Houston. I. Burak, J. W. Hepburn, *J. Chem. Phys.* **90**, 5364 (1989).

15 C. D. Pibel, K. Ohde and K. Yamanouchi, *J. Chem. Phys.* **101**, 836 (1994).

16 R. S. Sanders and K. R. Wilson, *J. Chem. Phys.* **63**, 4242 (1975).

17 M. Kawasaki, H. Sato, T. Kikuchi, S. Kobayashi, T. Arikawa, *J. Chem. Phys.* **86**, 4425 (1987).

18 H-P. Loock, J-Y. Cao, and C. X. W. Qian, *Chem. Phys. Lett.* **206**, 422

19 A. Sugita, M. Mashino, M. Kawasaki, Y. Matsumi, R. Bersohn, G. Trott-Kriegeskorte, and K-H. Gericke, *J. Chem. Phys.* **112**, 7095 (2000)

Part IV. Intense Fields

ELECTRONIC DYNAMICS AND STRUCTURE OF MOLECULES IN INTENSE LASER FIELDS

HIROHIKO KONO and ISAO KAWATA

Department of Chemistry, Graduate School of Science, Tohoku University, Sendai 980-8578, Japan

E-mail: kono@mcl.chem.tohoku.ac.jp

We theoretically investigate the electronic dynamics and structure of H_2^+ and a one-dimensional H_2 model in intense laser fields. The time-dependent Schrödinger equations for the systems are exactly solved by the dual transformation method developed by the authors. We reveal the dynamics of bound electrons and the subsequent ionization process in terms of "field-following" adiabatic states. The mechanisms of enhanced ionization in the one- and two-electron molecules are examined. For the H_2, ionic states work as a main doorway state to ionization. We also show the simple characteristic feature of molecular electronic structure in an intense field: the electronic state of each atom in the molecule can be specified in terms of adiabatic states of the atom (or its anion) in the field. Control schemes leading to electron transfer and formation of new bonds are proposed.

1 Introduction

The investigation of the interaction of atoms with ultrashort intense laser pulses has led to the discovery of new nonperturbative phenomena such as above-threshold ionization,[1, 2] tunnelling ionization,[3-6] and high-order harmonic generation.[7-11] In the high-intensity and low-frequency range, the Coulomb potential distorted by the laser electric field forms a "quasi-static" barrier through which an electron (or electrons) can tunnel.[4] Tunneling ionization occurs if the electron has enough time to penetrate the barrier(s) before the phase of the field changes. Favorable to tunneling ionization is the condition that the electric field is stronger and its period is longer. The case is classified by the Keldysh parameter $\gamma = \omega\sqrt{2I_p}/f(t)$,[3] where I_p is the ionization potential of the system, ω is the laser frequency, and $f(t)$ is the pulse envelope at time t. The quasi-static tunneling condition is given by the inequality $\gamma < 1$. High-order harmonics are generated when the electron ejected circles back to the vicinity of the nucleus (rescattering).[9] Electronic dynamics in intense fields involves such large amplitude motions. The ordinary multiphoton ionization process is regarded as the opposite case $\gamma > 1$.

For the case of molecules, nuclear motion is also involved in the dynamics of the system. Recent experiments and theories in a strong laser field case ($>10^{11}$ W/cm^2) have underscored the combined process of photodissociation and photoionization. It has been experimentally revealed that the kinetic energies of fragments are consistent with Coulomb explosions at specific internuclear distances.[12-15] The finding indicates that ionization rates at the "critical" internuclear distances exceed those near the equilibrium internuclear distance and those of dissociative fragments, and that ionization to higher-charge states occurs when the nuclei pass through the critical range.[13-20] This contradicts the prediction by simply applying the Keldysh parameter γ to diatomic molecules. With increasing internuclear distance R, γ monotonically decreases (since I_p decreases); the ionization rate monotonically increases. The fact that ionization is enhanced at critical internuclear distances suggests that strong correlation between the electronic motion and the nuclear configuration/motion exists in intense laser fields. Recent exact numerical simulations of ionization in one-electron systems such as H_2^+ have shown that the ionization rate takes a maximum at R=5-10 a.u.[14-17] It has been found that the ionization rate has a maximum (or maxima) also for two electron systems, H_2, H_4^{+2}, and H_5^{+3} in one-dimensional (1D) space.[18, 21]

In the low-frequency range, an electron or electrons adiabatically follow the laser electric field $\mathcal{E}(t)$: electron density spatially transfers from nucleus to nucleus. "Field-following" time-dependent adiabatic states can be defined as eigenfunctions $|n\rangle$ of the "instantaneous" electronic Hamiltonian $H_0(t)$ including the interaction with $\mathcal{E}(t)$. The $|n\rangle$ depend on t through the change in $\mathcal{E}(t)$. We diagonalize $H_0(t)$ in terms of bound eigenfunctions of the zero-field Hamiltonian H_{el}.[22] The extra degree of freedom arising from nuclear motion necessitates the use of "field-following" adiabatic potential surfaces. Field-induced nonadiabatic transitions, as well as nuclear-motion induced ones, are essential to describing the electronic dynamics prior to ionization.[22-24] Nonadiabatic transitions between such surfaces are enhanced by increasing the field intensity and the frequency ω: the response changes to the diabatic regime where spatial electron transfer between nuclei is suppressed. The level dynamics of adiabatic states is the key to comprehend the electronic dynamics of bound electrons and the subsequent ionization process. The ionization process can be interpreted in terms of the change in the populations of adiabatic states. If the reduction in population of an adiabatic state is not due to nonadiabatic transitions, the state is identified as a doorway state to ionization.

In this paper, we theoretically investigate the electronic dynamics and structure of a three-dimensional (3D) H_2^+ and a 1D H_2 model in intense laser fields. The former is regarded as a prototype of odd-electron molecules and the latter as even-electron molecules. In the Hamiltonian of H_2^+, the electron moves in 3D space as in the real system. When more electrons than one are involved, the 3D exact simulation is still computaionally demanding.[25] For H_2, we thus employ a 1D model in which the two electrons are allowed to move only along the molecular axis. The time-dependent Schrödinger equations for these systems are numerically solved by the dual transformation [26, 27] which can cope with awkward Coulomb potentials in a molecule. The method consists of the variable transformation of the Hamiltonian and the transformation of the wave function. The exact wave packet calculated is mapped onto adiabatic states. We demonstrate that the number of adiabatic states involved is limited because of the presence of strong radiative couplings between zero-field adiabatic (Born-Oppenheimer) states.

The electronic dynamics of H_2^+ prior to ionization are interpreted in terms of two field-following states $|1\rangle$ and $|2\rangle$ which are adiabatically connected with the lowest two electronic states $1s\sigma_g$ and $2p\sigma_u$ as the electric field changes. The net electrostatic potential for the electron has two minima. As $|\mathcal{E}(t)|$ increases, one of the potential wells descends and the other ascends. The descending and ascending wells yield the two states $|1\rangle$ and $|2\rangle$, respectively. We show that ionization proceeds via the upper adiabatic state $|2\rangle$ nonadiabatically created from the lower one $|1\rangle$. The enhanced ionization in one-electron molecules is due to electron transfer to the ascending well or spatial localization therein. The ionization barrier for $|2\rangle$ is the lowest at critical internucler distances for enhanced ionization.

The numerical calculation for the H_2 presents three types of ionization: ionization from the ionic structures H^+H^- and H^-H^+; from the covalent structure (HH); ionization due to collision between the electrons within H_2. We construct a minimum set of adiabatic states to fully analyze the ionization process of the H_2. The enhanced ionization around certain internuclear distances in two-electron molecules differs in mechanism from that in the one-electron case. In the two-electron case, excited electron transfer ionic states (e.g., H^+H^-) can cross the covalent ground state in energy. We demonstrate that field-induced transitions between a lowering ionic state and the covalent ground state are responsible for the enhanced ionization.

The final part of this paper is devoted to solve the question as to what is the characteristic feature of molecular electronic structure in an intense field. Using the 1D H_2, we numerically demonstrate that the electronic state of each atom in the molecule can be specified in terms of adiabatic states of the atom (or its anion) in the field. This can be paraphrased like "in an intense field a (homonuclear) molecule becomes atoms." The idea "a molecule becomes atoms" provides a variety of control schemes leading to electron transfer, formation of new bonds, and Coulomb explosion of a specific bond. Here we present a schematic view of bond formation in an intense field.

2 Method

2.1 Dual transformation

Although various numerical methods for electronic dynamics in laser fields have been proposed,[13,28-32] it is not an easy task to simulate, e.g., large amplitude motions of an electronic wave packet. It is even harder to include the quantum mechanical motion of nuclei. Conventional time evolution methods using grid representations,[33,34] such as split operator techniques combined with fast Fourier transform,[35,36] have been successfully applied to the nuclear wave packet dynamics in molecular physics. The success is attributed to the non-singularity that internuclear (adiabatic/diabatic) potentials are usually expressed in terms of analytic functions of distances. On the other hand, for electronic dynamics, one must cope with the Coulomb potential characterized by its long range and its singularity at the origin (at the nucleus for electron-nucleus interaction). The grid boundary in coordinate space must be chosen to be far from the origin to accommodate the wave function and grid spacings must be small to reproduce high momentum components generated near the origin. Because of those difficulties, the performance of the conventional grid methods is very poor for Coulomb systems.

Recently, we have been developing an efficient grid method to simulate electronic dynamics accurately.[27] The method is called the "dual transformation", in which both of the wave function and the Hamiltonian are consistently transformed for wave packet dynamics. No modifications of the Coulomb potentials are introduced. The choice of coordinate systems is crucial. It has been shown that for the cylindrical coordinates ρ and z the numerical accuracy is not improved by using higher order finite differences nor by using smaller grid spacings. In Ref. 27, the following three requirements are imposed on the new scaled-coordinate system (ξ and ζ) to be employed: (i) the original wave function Ψ is transformed so that it is zero at the Coulomb singular points (which ensures that the numerical difficulties concerning the singularity are avoided); (ii) the operations of the transformed Hamiltonian on the transformed wave function Φ can be well evaluated by the finite difference method even near the Coulomb singular points; (iii) the equal spacings in the new (scaled) coordinates correspond to grid spacings in the unscaled coordinates that are small near the nuclei (to cope with relatively high momentum components near the nuclei) and are large at larger distances therefrom. The transformed Schrödinger equation is discretized in space with the help of finite difference formulas and is integrated in time by the alternating-direction implicit method (ADI).[37] Among various propagation methods, the ADI method is found to be the most accurate for steep attractive potentials.

The transformation of ρ and z are expressed as $\rho = f(\xi)$ and $z = g(\zeta)$, where the functions are chosen to satisfy the three requirements (i)-(iii) described above. Transformations that satisfy the above three requirements drastically improve the accuracy and thus can reduce the required number of grid points. The transformation efficiency is also determined by the classical phase space analysis[38] based on the fact that wave packets decay exponentially in classically forbidden regions of phase space. The validity of a transformations is enforced by the fact that the missing volume in phase space decreases with decreasing grid spacings.

2.2 3D H_2^+

Here we introduce a realistic 3D model of H_2^+. The electric field of the applied laser is linearly polarized. In the model, the following assumptions are made: the molecular axis is oriented to the polarization direction[39] and the nuclear motion is restricted to vibration. Thus, the internal electronic coordinate z along the molecular axis is parallel to the polarization direction. The electron moves in 3D space. Because of the cylindrical symmetry, the z-component of the electronic angular momentum, $m\hbar$, is conserved; two cylindrical coordinates z and ρ (which is perpendicular to z) are the electronic degrees of freedom to be explicitly considered. These two electronic coordinates and the internuclear distance R can be treated quantum mechanically without using the Born-Oppenheimer (B-O) approximation.

The center-of-mass motion of the three-body system can be separated from internal coordinates such as ρ and z. Here, ρ and z are measured with respect to the center of mass of the two nuclei.[40] The Hamiltonian for the internal motions consists of the nuclear kinetic energy, electronic part H_{el}, and the dipole interaction $V_{\mathcal{E}}(z,t)$ between the molecule and the electric field $\mathcal{E}(t)$ of a laser pulse (throughout this paper atomic units are used unless otherwise noted)

$$H = -\frac{1}{m_{\rm p}}\frac{\partial^2}{\partial R^2} + H_{el}(\rho, z, R) + V_{\mathcal{E}}(z,t), \tag{1}$$

where $m_{\rm p}$ is the proton mass. The dipole interaction is given by

$$V_{\mathcal{E}}(z,t) = z\left(1 + \frac{m_{\rm e}}{2m_{\rm p} + m_{\rm e}}\right)\mathcal{E}(t), \tag{2}$$

where $m_{\rm e}$ is the electron mass. The electronic part H_{el} is

$$H_{el}(\rho, z, R) = -\frac{1}{2\mu}\left(\frac{\partial^2}{\partial \rho^2} + \frac{1}{\rho}\frac{\partial}{\partial \rho} + \frac{\partial^2}{\partial z^2}\right) + \frac{m^2}{2\rho^2} + V(\rho, z, R), \tag{3}$$

where the reduced mass μ is

$$\mu = 2m_{\rm p}m_{\rm e}/(2m_{\rm p} + m_{\rm e}), \tag{4}$$

and the potential $V(\rho, z, R)$ is the sum of the Coulomb interactions

$$V(\rho, z, R) = -\frac{1}{\sqrt{\rho^2 + (z - R/2)^2}} - \frac{1}{\sqrt{\rho^2 + (z + R/2)^2}} + \frac{1}{R}. \tag{5}$$

By using the dual transformation technique, we transform the Schrödinger equation

$$i\frac{\partial}{\partial t}\Psi(\rho, z, R) = H\Psi(\rho, z, R), \tag{6}$$

to the following one

$$i\frac{\partial}{\partial t}\Phi(\rho, z, R) = H'\Phi(\rho, z, R)\ , \tag{7}$$

where Φ is the transformed wave function that satisfies the condition (i)

$$\Phi(\xi,\zeta,R)=\sqrt{f(\xi)f'(\xi)g'(\zeta)}\;\Psi(f(\xi),g(\zeta),R), \tag{8}$$

and H' is the transformed Hamiltonian whose explicit form is given in Ref. 27. To make the time evolution scheme stable and accurate, the following normalization constraint is imposed on Φ

$$\int_0^\infty dR\int_0^\infty d\xi\int_{-\infty}^\infty d\zeta|\Phi(\xi,\zeta,R)|^2=1. \tag{9}$$

For example, for the transformation of ρ, we use the following function

$$f(\xi)=\xi\sqrt{\xi/(\xi+\alpha)}. \tag{10}$$

The parameter α designates the ρ-range where the Coulomb potentials are steep. The value $\alpha=28.3$ a.u. has been chosen. Around this value the results are insensitive to α.

2.3 "Field-following" adiabatic states

To analyze the electronic and nuclear dynamics in an intense laser field, especially to identify doorway states to ionization, we use field-following adiabatic states defined as eigenfunctions $\{|n\rangle\}$ of the "instantaneous" electronic Hamiltonian including the radiative interaction

$$H_0(t)=H_{el}(\rho,z,R)+V_{\mathcal{E}}(z,t). \tag{11}$$

For H_2^+, there are two charge resonance states $1s\sigma_g$ and $2p\sigma_u$ which are strongly coupled with each other by the dipole interaction. As will be mentioned later, it is justified to represent the dynamics of the bound component in terms of the two states.

We start with diagonalizing the Hamiltonian $H_0(t)$ in terms of the two B-O electronic states $|1s\sigma_g\rangle$ and $|2p\sigma_u\rangle$ (abbreviated as $|g\rangle$ and $|u\rangle$). The time t and R are treated as adiabatic parameters. The resulting eigenfunctions are given by[41]

$$|1\rangle=\cos\theta|g\rangle-\sin\theta|u\rangle\,, \tag{12a}$$

$$|2\rangle=\cos\theta|u\rangle+\sin\theta|g\rangle\,, \tag{12b}$$

where

$$\theta=\frac{1}{2}\arctan[2\langle g|z|u\rangle\mathcal{E}(t)/\Delta E_{ug}(R)], \tag{13}$$

with the B-O energy separation $\Delta E_{ug}(R)=E_u(R)-E_g(R)$. The eigenvalues are

$$E_{1,2}(R,t)=\frac{1}{2}\left[E_g(R)+E_u(R)\mp\sqrt{\Delta E_{ug}^2+4|\langle g|z|u\rangle\mathcal{E}(t)|^2}\,\right]. \tag{14}$$

For $H_2{}^+$, the transition moment $\langle g|z|u\rangle$ approaches $R/2$ as R increases.[42] A cross section of the instantaneous potential $V(\rho, z, R)+z\mathcal{E}(t)$ for $\mathcal{E}(t)=0.07$ a.u. (I=1.7 $\times 10^{14}$ W/cm^2) is plotted in Fig. 1 together with the energies E_1 and E_2. The transverse coordinate ρ is fixed at 0 and $R=4$ a.u. Each nucleus works as a potential

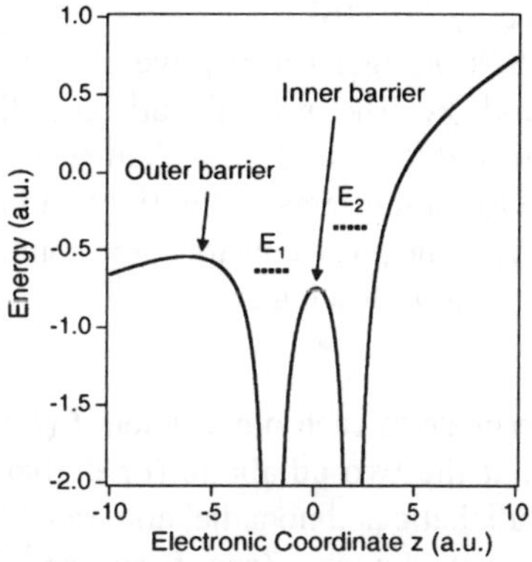

FIG. 1. The H_2^+ instantaneous potential at $\mathcal{E}(t)$=0.07 a.u. as a function of the electronic coordinate z along the molecular axis (ρ= 0 and R=4). The polarization direction is assumed to be parallel to the molecular axis. The instantaneous potential has an inner and an outer barrier. The energies of the lowest two "field-following" adiabatic states, E_1 and E_2, are denoted by horizontal dotted lines.

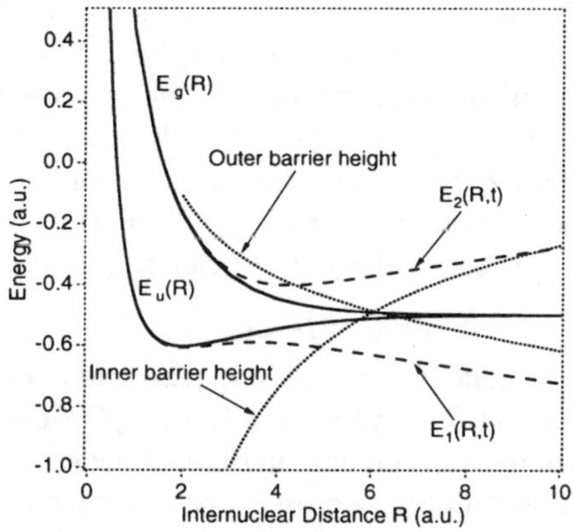

FIG. 2. The lowest two adiabatic potential energies E_1 and E_2 (broken lines) of H_2^+ and the heights of the inner and outer barriers (dotted lines) at $\mathcal{E}(t)$=0.07 a.u. as functions of R. The solid lines denote the Born-Oppenheimer surfaces of the lowest two electronic states $1s\sigma_g$ and $2p\sigma_u$.

well for the electron. The energies of the right (at $z = R/2$) and left wells are shifted, respectively, by $\pm R\mathcal{E}(t)/2$ because of the dipole interaction Eq. (2); accordingly, as $|\mathcal{E}(t)|$ increases, E_2 ascends and E_1 descends. For $\mathcal{E}(t) > 0$; the right well is the ascending one (associated with $|2\rangle$) and the left well is the descending one ($|1\rangle$). Along the polarization axis z, the instantaneous potential has an inner and an outer barrier. In Fig. 2, E_1, E_2, and the barrier heights for $\mathcal{E}(t)$ =0.045 are plotted as functions of R.

The total wave function can be expanded in terms of the two adiabatic states:

$$|\Psi\rangle = \chi_1(R)|1\rangle + \chi_2(R)|2\rangle, \tag{15}$$

where χ_1 and χ_2 are the nuclear wave functions associated with $|1\rangle$ and $|2\rangle$. Using the two localized states $|\mathrm{R}\rangle = (|\mathrm{g}\rangle + |\mathrm{u}\rangle)/\sqrt{2}$ and $|\mathrm{L}\rangle = (|\mathrm{g}\rangle - |\mathrm{u}\rangle)/\sqrt{2}$, the populations in the right and left wells associated with two nuclei, P_R and P_L, are expressed in terms of χ_1 and χ_2 as

$$P_{\mathrm{R,L}}(R) = \cos^2\left(\theta + \frac{\pi}{4}\right)|\chi_{1,2}(R)|^2 + \sin^2\left(\theta + \frac{\pi}{4}\right)|\chi_{2,1}(R)|^2 \pm \mathrm{Re}\left[\cos 2\theta\, \chi_1^*(R)\chi_2(R)\right]. \tag{16}$$

The degree of spatial localization can be estimated by Eq. (16).

Inserting Eq. (15) into the time-dependent Schrödinger Eq. (6), we obtain the coupled equations for the time evolution of χ_1 and χ_2. The nonadiabatic coupling between χ_1 and χ_2 is mainly induced by the term $\partial\theta/\partial t$ due to the change in the field $\mathcal{E}(t)$ and the other minor couplings are due to joint effects of the field and the nuclear motion. For $\mathcal{E}(t)=f(t)\sin(\omega t)$, where $f(t)$ is the pulse envelope, the solution of the coupled equations can be classified by using the following quantity[43,44]

$$\delta = \left|\Delta E_{\rm ug}^2(R)/\langle g|z|u\rangle f(t)\omega\right|, \tag{17}$$

The adiabatic energies, $E_1(R,t)$ and $E_2(R,t)$, come close to each other when $\mathcal{E}(t)=0$, e.g., at $t=n\pi/\omega$ $(n=1, 2,...)$. On condition that the two adiabatic (or diabatic) components do not interfere with each other, the adiabatic and nonadiabatic transition probabilities at a crossing point are well described by the Landau-Zener formulae,[43-45] $P_{\rm ad}=1-\exp(-\pi\delta/4)$ and $P_{\rm nonad}=\exp(-\pi\delta/4)$, respectively. For $\delta >> 1$, $P_{\rm ad}\cong 1$; the adiabatic picture of electronic dynamics works well. For $\delta << 1$, the main route is the nonadiabatic channel.

3 Electronic dynamics of H_2^+

3.1 *Interwell electron transfer*

Electron density can be transferred from well to well by an electric field. We present an example of the electron transfer between the two wells. To focus on the interwell transition, we solve the time-dependent Schrödinger equation obtained by fixing R at 4 a.u. in Eq. (7). The initial state at $t=0$ is assumed to be the first excited state $2p\sigma_u$, where the electron density is equally distributed between the two wells. The parameters for the applied pulse is as follows: $\omega=$ 0.0515 (λ=884 nm) and $f(t)=$ 0.023 (1.9×10^{13} W/cm^2). A snapshot of the packet at $t=31$ a.u. (a quarter cycle later) is shown in Fig. 3 (1 a.u. =0.0242 fs). The unequal electron distribution between the two wells results from the motion that electron density transfers from the left well to the right well. At t=31, $\mathcal{E}(t)>0$; the right well is the ascending one. The electron motion in Fig. 3 means that a part of the electron density is transferred from the descending well to

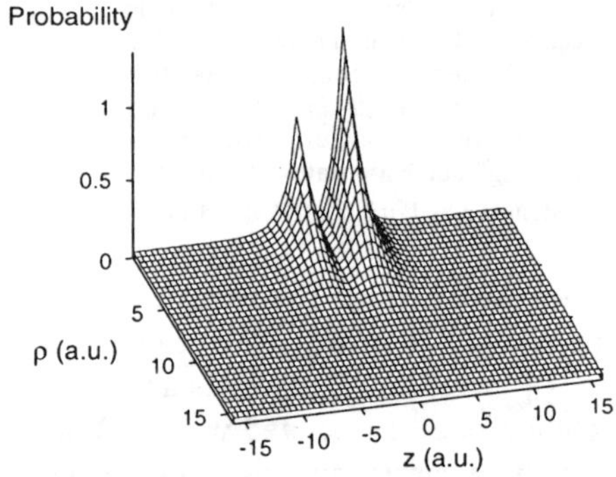

FIG. 3. The probability $|\Psi(\rho, z)|^2$ of the H_2^+ electronic wave packet in an intense field at $t=31$ a.u. (a quarter cycle later from t=0). The pulse condition is as follows: $\omega=0.0515$ a.u. and the constant pulse envelope $f(t)=0.023$ a.u. $\mathcal{E}(t)=0.023$ a.u. at t=31 a.u. R is fixed at 4 a.u. The initial state is assumed to be $2p\sigma_u$ for which the electron density is equally distributed between the two nuclei.

the ascending well (although the force due to the electric field is opposite). The ascending and descending wells change places with each other every half cycle.

Using the parameter δ, the electronic motion in an electric field can be classified into the adiabatic or diabatic regime according to whether $\delta>1$ or $\delta<1$, as mentioned in Sec. *2.3*. At small internuclear distances, the energy difference ΔE_{ug} is large and the transition moment $\langle g|z|u\rangle$ is relatively small: one can expect that $\delta>1$. The case shown in Fig. 3 for which $\delta \approx 1.6$ is nearly adiabatic. As a result, the time evolution of the excited state $2p\sigma_u$ is given by the upper adiabatic state $|2\rangle$. Staying in $|2\rangle$ means that the ascending well is more populated as in Fig. 3. If the molecule is initially in the ground state $1s\sigma_g$ and the case is adiabatic, the molecule stays in $|1\rangle$: the descending well is always more populated. From Eq. (16), when $\mathcal{E}(t) > 0$, $P_L > P_R$ for $|\Psi\rangle = |1\rangle$ and $P_R > P_L$ for $|\Psi\rangle = |2\rangle$. At larger internuclear distances, nonadiabatic transitions between $|1\rangle$ and $|2\rangle$, which occur near $\mathcal{E}(t)=0$, are dominant: the electron transfer between the two wells is suppressed.

In the near-diabatic regime, the transition rate for interwell transfer is given by the well known form $\Delta E_{ug} J_0[2\langle g|z|u\rangle f(t)/\omega]$,[43,46] where J_0 is the zeroth order Bessel function. As the field intensity increases, interwell tunneling is more suppressed. There also exist specific conditions for the Bessel function to be zero. For the zeros of the Bessel function, interwell transition is inhibited. The coherent destruction of interwell tunneling[46] is due to interference between the two adiabatic (or diabatic) components *at periodic crossing points* $t_n = n\pi/\omega$ ($n=1, 2, ...$).[43]

3.2 Enhanced ionization

The double-well nature of the effective electronic potential which leads to a different ionization mechanism from the atomic case, i.e., the existence of critical internuclear distances for enhanced ionization in molecules.[12-21] As shown in Fig. 2, the key quantities for ionization are E_1, E_2, and the inner and outer barrier heights. As R increases, the outer barrier is more suppressed by the dipole interaction. The outer barrier height V_O decreases as $-2|\mathcal{E}(t)|^{1/2} - R|\mathcal{E}(t)|/2 + 1/(R+|\mathcal{E}(t)|^{1/2}R^2)$ at large R. On the other hand, the height of the inner barrier, V_I, increases as $-3/R$. The adiabatic energies E_1 and E_2 are nearly equal to E_g and E_u respectively for the small R region, and change as $-I_p \mp R|\mathcal{E}(t)|/2$ as R increases (Here I_p is the ionization potential of H). Except in the large R region, the relative energy $E_2 - V_B$ is higher than $E_1 - V_O$, where V_B is the maximum of V_O and V_I. The upper adiabatic state $|2\rangle$ is easier to ionize than $|1\rangle$. The relative energy E_2-V_B takes the maximum value at $R=R_m$, where R_m is the position where V_O and V_I are equal to each other. We obtain $R_m \approx 1.31/\sqrt{\mathcal{E}(t)}$. Setting $V_O = V_I = E_2$, we obtain the characteristic distance for enhanced ionization given by $R_x \approx 3.86/I_p$. For H_2^+, $R_x \approx 7.7$ a.u. Similar expressions have been derived in Refs. 15 and 16.

Ionization is thus expected to be enhanced when $|2\rangle$ is populated near R_x. In

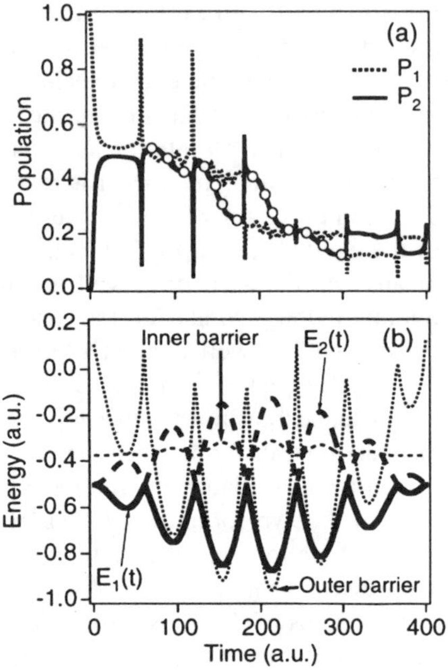

FIG. 4. (a) Change in population of the lowest two adiabatic states |1> and |2> . (b) The heights of the inner and outer barriers together with the adiabatic energies E_1 and E_2. The initial state is assumed to be $1s\sigma_g$ and R is fixed at 8 a.u. The pulse condition is as follows: ω=0.0515 a.u., the duration T=400 a.u.; the envelope $f(t)$ changes as $f_0\sin(\pi\, t/T)$ for $0<t<T$ and, otherwise, $f(t)$=0, where f_0=0.096 a.u.

the intermediate regime ($\delta\sim1$, e.g., 4<R<10), even if the initial state is $|1\rangle$, $|2\rangle$ is populated because of nonadiabatic transitions. To illustrate the enhanced ionization, we present Fig. 4 where the internuclear distance is fixed at R = 8. The pulse condition is as follows: ω = 0.0515 a.u., the duration T=400 a.u.; the envelope $f(t)$ changes as $f_0 \sin(\pi t/T)$ for $0 \le t \le T$ and, otherwise, $f(t)=0$. The maximum f_0 amounts to the peak intensity 3.2×10^{14} W/cm^2.

Analyses are made by mapping the electronic packet onto $|1\rangle$ and $|2\rangle$. The changes in population are shown in Fig. 4(a) (the initial state is $1s\sigma_g$) and the maximum heights of the inner and outer barriers are shown in Fig. 4(b) together with E_1 and E_2. As $\mathcal{E}(t)$ increases from 0, E_1 goes down and E_2 goes up. After the first level crossing at $t\approx61$ a.u., the duration for which E_2 is much higher than the barriers becomes dominant. While the electric field is near a local maximum or minimum, the population P_2 of $|2\rangle$ decreases but P_1 is nearly constant. See the reductions denoted by open circles in Fig. 4(a). This type of reduction in P_2 is the direct evidence of the ionization from $|2\rangle$ (roughly speaking, from the ascending well). Ionization is enhanced whenever P_2 is large and the barriers are low enough for the electron to tunnel from the ascending well. This is due to a combined effect of nonadiabatic transitions to $|2\rangle$ and efficient barrier suppression. For the present intense pulse, in the third and fourth half cycles, there exist periods for which E_1 is higher than the outer barrier (through which the electron tunnels from the descending well). However, the periods are much shorter than the optical cycle and the difference $E_1 - V_O$ is much smaller than $E_2 - V_I$. Thus, the ionization from $|2\rangle$ is dominant.

We have tested cases in which more than two eigenfunctions of H_{el} are used to diagonalize $H_0(t)$. For example, we diagonalize $H_0(t)$ in terms of the lowest six exact eigenstates of H_{el}, { $1s\sigma_g$, $2p\sigma_u$, $2s\sigma_g$, $3p\sigma_u$, $3d\sigma_g$, $4f\sigma_u$}. The resultant adiabatic states are numbered from $1'$ to $6'$ in the ascending order of adiabatic energy. The sum P_1+P_2 in the two-state model is nearly equal to the sum from $P_{1'}$ to $P_{6'}$ in the six-state model. This means that among the bound states of H_{el} only two states $1s\sigma_g$ and $2p\sigma_u$, or $|1\rangle$ and $|2\rangle$ in Eqs. (12), are populated before ionization. These two states are essential to describe the bound state dynamics in intense fields and the upper adiabatic state $|2\rangle$ plays as the doorway state to ionization. The electronic structure of H_2^+ is characterized by the existence of charge resonance pairs. The transition moments between the two states in a charge resonance pair is as large as $R/2$. The second adiabatic state $|2'\rangle$ arising mainly from the lowest charge resonance pair ($1s\sigma_g$, $2p\sigma_u$) exceeds in energy the lower adiabatic state arising from the charge resonance pair ($2s\sigma_g$, $3p\sigma_u$) and/or that from the pair ($3d\sigma_g$, $4f\sigma_u$). However, the nonadiabatic transitions among different charge resonance pairs hardly occur because the transition moment is much smaller between two states belonging to different charge resonance pairs than between the two state of a charge resonance pair. Hence, only two adiabatic states arising from the same charge resonance pair are essential to describe the bound state dynamics before ionization.

By including the additional degree of freedom, i.e., the molecular vibration, we can fully analyze the effect of dissociative motion on ionization. The Schrödinger equation for the 3D H_2^+, Eq. (7), is solved without using the B-O approximation. We obtain χ_1 and χ_2 by mapping the whole wave function onto $|1\rangle$ and $|2\rangle$. When $\mathcal{E}(t) \neq 0$, $|\chi_2|^2$ decreases owing to ionization. As in the R-fixed model, ionization is enhanced whenever $|\chi_2|$ is large and the barriers are low enough for the electron to tunnel. By analyzing how $|\chi_2|$ decreases in R-t space, one can tell where and when Coulomb explosions occur. The positions R_c and conjugate momenta k_c at Coulomb explosions are identified. The dissociative motion accelerates on E_1 (bond softening) and decelerates on E_2 (bond hardening). In addition to the potential energy $1/R_c$ for H_2^{2+}, the kinetic energy gained on the time-dependent surfaces by the time of Coulomb explosions contributes to the total energy of the fragment proton (more than 25 %). Results will be reported elsewhere.[47]

4 Two-electron dynamics of H_2

Ionization in two-electron systems is also enhanced around certain internuclear distances R_c. Maxima in the ionization rate have been found for two-electron systems, H_2, H_4^{+2}, H_5^{+3},[18, 21] with similar R_c's as one-electron systems [7, 10]. Enhanced ionization in two-electron molecules however differs in mechanism from that in odd-electron molecules such as H_2^+ (which is due to single-electron transfer to the ascending well or electron localization therein). In even-electron molecules, the field-following adiabatic energy of excited ionic states, e.g., H^+H^-, can cross the covalent ground state (HH). To clarify the roles of ionic and covalent states in ionization, we investigate the enhanced ionization in a 1D H_2. Doorway states to ionization are identified and the laser-induced electron transfer from a covalent to an ionic state is examined. We reveals the efficiency of electron transfer as a function of R and the origin of the R-dependence in the ionization probability. Even-electron systems are created in the course of dissociative ionization. The internuclear distance of such intermediate even-electron systems is usually larger than the equilibrium internuclear distance of their neutral parent molecules.[48] For example, after one-electron ionization from a neutral molecule of even electrons, the resultant odd-electron ion undergoes bond softening.

We begin with a simple molecular orbital (MO) picture to discuss the crossing of ionic and covalent states. For two electron systems such as H_2, one must consider at least the lowest three electronic configurations in the MO approximation, $1s\sigma_g^2$, $1s\sigma_g 2p\sigma_u$, and $2p\sigma_u^2$:[49]

$$\phi_1 = \left[a(1)a(2) + b(1)b(2) + a(1)b(2) + b(1)a(2)\right]/2, \tag{18a}$$

$$\phi_2 = \left[a(1)a(2) - b(1)b(2)\right]/\sqrt{2}, \tag{18b}$$

$$\phi_3 = \left[a(1)a(2) + b(1)b(2) - a(1)b(2) - b(1)a(2)\right]/2, \tag{18c}$$

where a and b denote the 1s-like atomic orbitals on the right and left nuclei, respectively, and 1 and 2 represent the coordinates of the two electrons. Both ground ϕ_1 and second excited states ϕ_3 are linear combinations of *ionic* (*aa*, *bb*) and *covalent* (*ab*+*ba*) configurations whereas the first excited state ϕ_2 is purely ionic. The corresponding transition dipole moments have nonzero molecular axis z-components $\langle\phi_1|z|\phi_2\rangle=\langle\phi_2|z|\phi_3\rangle= R/\sqrt{2}$.[50] The three states Eq. (18) can be thus coupled by the same radiative matrix element $\mathcal{E}(t)R/\sqrt{2}$.On condition that the radiative interaction is much larger than the energy separations between the three states at zero field, the adiabatic states obtained by the diagonalization of the 3×3 energy matrix are the covalent state $(ab+ba)/\sqrt{2}$ and the two ionic states (*aa* and *bb*). The lowering ionic state (*aa* or *bb*) which is shifted by dipole energy $-|\mathcal{E}(t)|R$ of a charge displaced from nucleus to nucleus by the field. If the dipole energy is larger than the gap between the first excited ionic state and the ground state, the lowering ionic state can be below the covalent ground state. Using the 1D H_2 model described below, we show next that the ionic state indeed becomes the ground state and that field-induced transitions between the covalent and ionic states determine the ionization process in the two electron case.

We employ a 1D model in which the two electrons are allowed to move only along the molecular axis and R is treated as a parameter. The internal coordinates of the two electrons along the molecular axis are denoted by z_1 and z_2, respectively. The Coulomb potentials of the unperturbed Hamiltonian $H_{el}(z_1, z_2)$ are regularized as in Ref. 21 by introducing the parameters ρ and ρ_e to soften the electron-nucleus Coulomb potential and the electron-electron repulsion potential, respectively:

$$H_{el}(z_1,z_2)=\sum_{j=1}^{2}\left\{-\frac{1}{2}\frac{\partial^2}{\partial z_j^2}-1\Big/\sqrt{\rho^2+(z_j-R/2)^2}-1\Big/\sqrt{\rho^2+(z_j+R/2)^2}\right\}$$

$$+\frac{1}{\sqrt{(z_1-z_2)^2+\rho_e^2}}+\frac{1}{R}\ . \qquad (19)$$

Here we set $\rho=\rho_e=1$. The potential of H_{el} has two minima at the covalent (C-) configurations around $z_1=-z_2=\pm R/2$ (*ab* and *ba*) and two saddle points at the ionic (I-) configurations around $z_1=z_2=\pm R/2$ (*aa* and *bb*). We assume that the molecule is oriented by the field, along the polarization direction.[39] The dipole interaction is then given by

$$V_{\mathcal{E}}(t)=(z_1+z_2)\mathcal{E}(t). \qquad (20)$$

The field $\mathcal{E}(t)$ is assumed to be $f(t)\sin(\omega t)$, where the envelope $f(t)$ is linearly ramped with time t so that after five cycles $f(t)$ attains its maximum f_0. The time-dependent Schrödinger equation for the system is solved by using the dual transformation method.

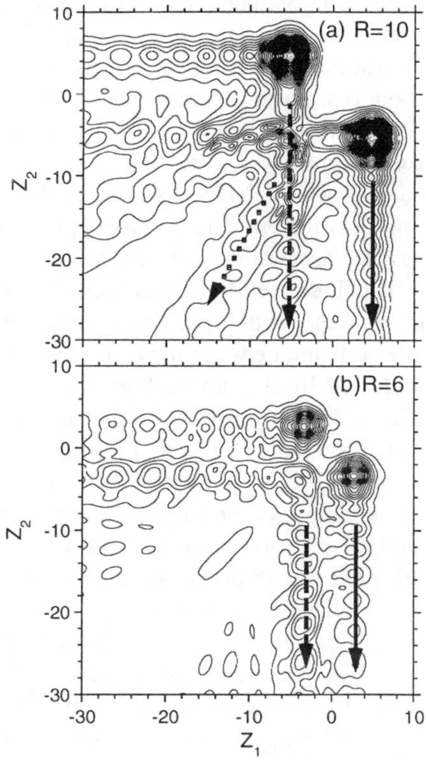

FIG. 5 Snapshots of the 1D H_2 electronic wave packet at t=642 a.u. for R=10 a.u. (a) and 6 a.u. (b). The pulse parameters used are as follows: ω= 0.0428 a.u. ; the envelope $f(t)$ is linearly ramped with t so that after five cycles $f(t)$ attains its maximum f_0=0.12 a.u. The intervals of the contour lines are constant 0.005. The meaning of each bold line is given in the text.

The present 1D model reproduces characteristic features of the 3D system. The calculated B-O potential surfaces for the lowest three electronic configurations in the 1D model are in good agreement with the 3D ones, except that the 1D potentials are all shifted down. The ionization potential of the 1D H atom for the softened Coulomb potential with $\rho=1$, 0.67 a.u., is larger than the 3D value 0.5 a.u. By changing the internuclear distance in the 1D model from its equilibrium value, one can discuss subsequent ionization processes in even-electron systems created in the course of dissociative ionization.

Snapshots of the electronic wave packet in an intense field, $|\Psi(z_1,z_2)|$, at $t=4\frac{3}{8}$ cycle (642 a.u.) are shown in Fig. 5. The R is fixed at (a) 10 a.u. and (b) 6 a.u., respectively. The field parameters are as follows: f_0=0.12 (5×10^{14} W/cm^2) and ω =0.0428 (λ=1064nm). The initial state is the exact ground state $|\phi_1\rangle$ of H_{el} which corresponds to Eq. (18a) in the MO approximation. The singlet state has the exchange symmetry $\Psi(z_1, z_2) = \Psi(z_2, z_1)$.[49] Comparing Fig. 5(a) and (b), one finds that the ionization rate is much higher at R=6 than at R=10. The population of the bound component obtained by projecting the calculated Ψ onto the bound states of H_{el} is, at $t = 4\frac{3}{8}$ cycle, 0.08 for R=6 and 0.55 for R=10. It has been demonstrated that the ionization rate has a peak (or peaks) around R=5-6 and the ionization rate at R=6 is about five times as large as that at R=10.[21]

As shown in Fig. 5(a), for R=10, the ionization process starts from the C-configurations. An electron near the descending well where $z\,\mathcal{E}(t) < 0$ (the left nucleus when $\mathcal{E}(t) > 0$) is ejected as indicated by the solid line with an arrow in Fig. 5(a) [referred to as covalent (C-) type ionization]. The other electron is left near the ascending well. Another C-type ionization which starts from the ascending well is indicated by the broken line; the electron ejected from the ascending well penetrates through the descending well without colliding with the other electron. The collision between the electrons induces the simultaneous two-electron ionization indicated by the dotted line. The electron ejected from the ascending well gains kinetic energy of $\sim R f_0$ by the time it collides with the electron in the descending well (target atom). For R=10, $R f_0 \approx 1$ is larger than I_p (H)= 0.67 for the 1D case.

For R=6, as shown in Fig. 5(b), an ionic (I-) component is created near the descending well because of the laser-induced electron transfer from the ascending well. The I-configuration has a very low ionization potential $I_p\,(\mathrm{H}^-) = 0.06$ and is regarded as a doorway state to ionization. To quantify the I-component, we define the localized I-configurations $|\mathrm{H}^+\mathrm{H}^-\rangle$ and $|\mathrm{H}^-\mathrm{H}^+\rangle$ as the ground state H^- atoms at $z = \pm R/2$, respectively. The $|\mathrm{H}^+\mathrm{H}^-\rangle$ and $|\mathrm{H}^-\mathrm{H}^+\rangle$ correspond to the ionic states *aa* and *bb* respectively, but include the electron correlation. As shown in Fig. 5(b), at $R \approx 6$, the I-component created becomes as large as $|\langle \mathrm{B}|\mathrm{H}^-\mathrm{H}^+\rangle|^2 \approx 0.2$ in the fourth cycle, where $|\mathrm{B}\rangle$ is the normalized bound component (For R=10, the peak of $|\langle \mathrm{B}|\mathrm{H}^-\mathrm{H}^+\rangle|^2 \approx 0.02$). As indicated by the broken line in Fig. 5(b), an electron is ejected from the I-configuration near the descending well [referred to as ionic (I-) type]. The ionization rate for I-type is four times as large as that for the C-type ionization denoted by the solid line. Simultaneous two-electron ionization is negligible because $R f_0 \approx 0.6 < I_p$ (H).

To examine the mechanism of the enhanced ionization, as in Sec. 3.2,[22] we map the wave packet onto field-following adiabatic states defined as eigenfunctions $|n\rangle$ of the "instantaneous" Hamiltonian $H_0(t) = H_{el} + V_{\mathcal{E}}(t)$: $H_0(t)|n\rangle = E_n(t)|n\rangle$. To diagonalize the "instantaneous" Hamiltonian, we use the lowest seven exact eigenstates of H_{el}, $\{|\phi_\ell\rangle\}$, as basis functions. The ℓth lowest state is denoted as $|\phi_\ell\rangle$. Equations (18) are approximate forms for the lowest three $|\phi_\ell\rangle$ (ℓ=1-3).

We first examine eigenfunctions of H_{el} at different R. Four singlet states at R=4.25 are drawn in Fig. 6. The probability density around the I-configurations $z_1 = z_2 = \pm R/2$ is rather high in Fig. 6(a)-(c). To quantify the I-component involved in

$|\phi_\ell\rangle$, we define $|I_g\rangle$ and $|I_u\rangle$ as

$$|I_g\rangle \equiv c_+\{|\mathrm{H}^+\mathrm{H}^-\rangle + |\mathrm{H}^-\mathrm{H}^+\rangle\}\ ,\ |I_u\rangle \equiv c_-\{|\mathrm{H}^+\mathrm{H}^-\rangle - |\mathrm{H}^-\mathrm{H}^+\rangle\}, \tag{21}$$

where $c_\pm$ are normalization constants. As shown in Table I, if R is as small as 4 a.u., the ionic characters $|I_{g,u}\rangle$ are distributed mainly among the lowest four states. On the other hand, as shown in Fig. 6(d), the ϕ_ℓ with $\ell \geq 6$, which originate from the third and higher excited states of the 1D H, are spatially "diffuse" as Rydberg states. The overlaps of the higher states ($\ell \geq 6$) with $|I_g\rangle$ and $|I_u\rangle$ are therefore very small (see Table I). As R increases, $|I_{g,u}\rangle$ are distributed among higher states. Since $\langle I_u|z|I_g\rangle$ is as large as R,[42] the ionic components involved dominate transition moments and the structures $\mathrm{H}^+\mathrm{H}^-$ and $\mathrm{H}^-\mathrm{H}^+$ which are linear combinations of $|I_g\rangle$ and $|I_u\rangle$ are easily formed in an intense field whenever $|I_g\rangle$ or $|I_u\rangle$ is involved in the dynamics.

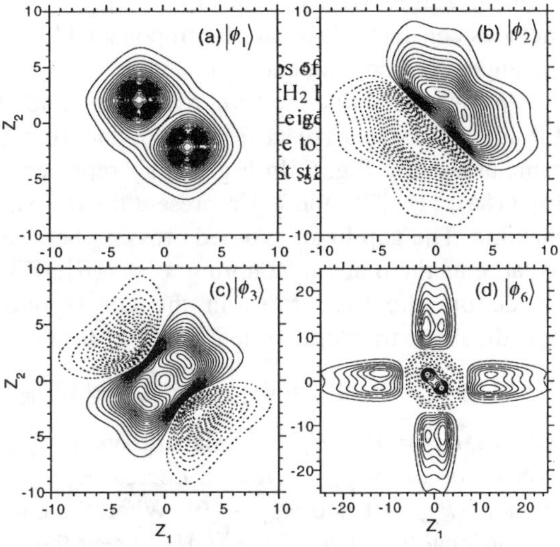

FIG. 6. Contour maps of singlet eigenfunctions of the zero-field 1D H_2 Hamiltonian at R=4.25 a.u. The sign of the eigenfunction in the solid line region is opposite to that in the dotted line region. The subscript of $|\phi >$ denotes the ℓth lowest state.

Table I. Gerade and ungerade ionic components in eigenfunctions $|\phi_\ell\rangle$ of the 1D H_2 Hamiltonian

	ℓ	1	2	3	4	5	6	7
R=4.25	$\lvert\langle I_g\lvert\phi_\ell\rangle\rvert^2$	0.45	0	0.23	0.27	0	0.01	0
	$\lvert\langle I_u\lvert\phi_\ell\rangle\rvert^2$	0	0.71	0	0	0.09	0	0.08
R=6	$\lvert\langle I_g\lvert\phi_\ell\rangle\rvert^2$	0.16	0	0.38	0.40	0	0	0.02
	$\lvert\langle I_u\lvert\phi_\ell\rangle\rvert^2$	0	0.49	0	0	0.42	0.06	0
R=10	$\lvert\langle I_g\lvert\phi_\ell\rangle\rvert^2$	0.01	0	0.03	0.79	0	0	0.15
	$\lvert\langle I_u\lvert\phi_\ell\rangle\rvert^2$	0	0.03	0	0	0.79	0.15	0

The radiative interaction $V_{\mathcal{E}}(t)$ can couple two states of different symmetry, *gerade* and *ungerade* under inversion. The transition moment $\langle\phi_1|z|\phi_2\rangle$ increases as $R/\sqrt{2}$ up to $R\approx3$ and converges to $\sqrt{2}\times$(asymtotic atomic value), which is consistent with accurate calculations for the 3D H_2.[50] The linear increase with R in the transition moment agrees with the fact that in the small R <6 region the ground state ϕ_1 includes considerable gerade I-component $|I_g\rangle$ and the first excited state ϕ_2 includes the ungerade I-component $|I_u\rangle$.

We next examine the level dynamics of adiabatic states. The eigenvalues $E_n(t)$ for adiabatic states $|n\rangle$ at R=4.25, 6, and 10 are plotted in Fig. 7. The field is the same as used in Fig. 5. In Fig. 7, $|1'\rangle$ represents the ground state near zero field as Eq. (18a); $|1\rangle$, $|2\rangle$, and $|3\rangle$ represent the lowest three adiabatic states near the field maxima. The gap in an avoided crossing between adiabatic states n and m is twice as large as the diabatic coupling $v_{\bar{n}\,\bar{m}}=\langle\bar{n}|z|\bar{m}\rangle\mathcal{E}(t)$ at the crossing, where $\bar{n}$ and $\bar{m}$ denote the two states in the corresponding diabatic representation. The nonadibabatic transition probability P_{nm} for the n-m crossing is[44, 45]

$$P_{nm} = \exp\left[-2\pi v_{\bar{n}\,\bar{m}}^2 \Big/ \left|d\Delta E_{\bar{n}\,\bar{m}}/dt\right|\right], \tag{22}$$

where $\Delta E_{ij} = E_i - E_j$; the adiabatic one is $\overline{P}_{\bar{n}\,\bar{m}} \equiv 1\text{-}P_{nm}$ [16]. As the $\bar{n}$-$\bar{m}$ spatial overlap or the $|I_{g,u}\rangle$ involved decreases, $v_{\bar{n}\,\bar{m}}$ decreases: then, the transition is diabatic ($P_{nm}\to1$ and $\overline{P}_{\bar{n}\bar{m}}\to0$). When the energy separation is large, dP_{nm}/dt is proportional to $\langle n|\partial V_{\mathcal{E}}/\partial t|m\rangle/\Delta E_{nm}$: near the field maxima, dP_{nm}/dt =0.[51]

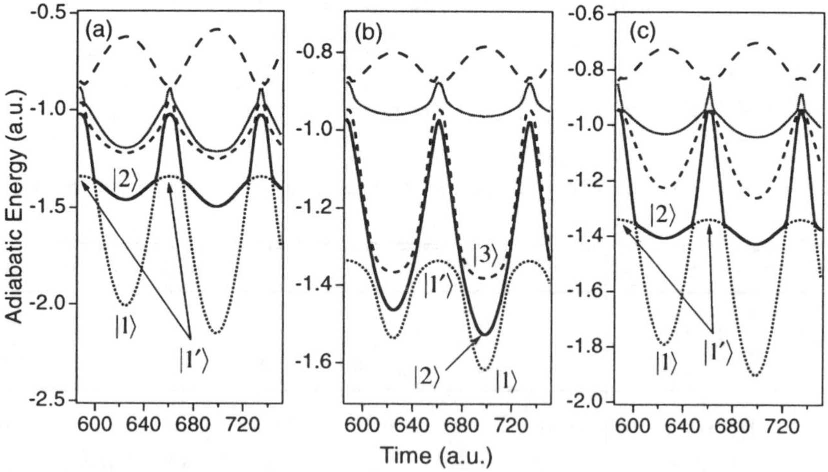

FIG. 7. The energies $\{E_n\}$ of the lowest five adiabatic states of 1D H_2 as functions of time: R=4.25 a.u. (a), 6 a.u. (b), and 10 a.u. (c). The laser field is the same as used in Fig.5. Five line species are used to denote $\{E_n\}$ for n=1-5; e.g., the solid line denotes the second adiabatic state energy E_2. |1'> represents the ground state near zero field; |1>, |2>, and |3> represent the lowest three adiabatic states near the field maxima.

In the large R region as in Fig. 7(c), the initial state $|1'\rangle$ is mainly covalent and $|1\rangle$ is ionic. At R~10, the state $|1\rangle$, whose energy changes on a large scale as $\sim R\,\mathcal{E}(t)$, becomes the ground state when $|\mathcal{E}(t)| \geq 0.05$. Population analysis, however, shows that $|1\rangle$ is little populated near the field maximum at $t\approx700$ ($|\langle 1|B\rangle|^2 \approx 0.01$): the covalent state $|1'\rangle$ is only weakly coupled with the ionic state $|1\rangle$ as indicated by the negligible avoided crossing between them (the transition dipole between the covalent state ϕ_1 and the ionic state ϕ_5 is as small as 0.13). The distance of R=10 is too large for an electron to *adiabatically* jump to the other nucleus. As a result, the main path follows the diabatic one $1' \rightarrow 2 \rightarrow 1'$ of covalent nature ($|\langle 2|H^-H^+\rangle|^2 \leq$ 0.005). The energy of the sequence changes as $\approx -a_H\mathcal{E}(t)^2$ controlled by the hydrogen atom polarizability a_H ($\approx$6 for the 1D H atom). Near the field maximum, the population of the covalent state $|2\rangle$ decreases owing to the C-type and simultaneous two-electron ionization processes shown in Fig. 5(a).

In the intermediate region ($R\approx6$) where enhanced ionization occurs, all the lowest five states have ionic characters. Whenever adiabatic energies are close, as shown in Fig. 7(b), avoided crossings through the ionic characters exist. The dominant character of $|1'\rangle$ is covalent, $|1\rangle$ is a diffuse state (< 20% ionic), $|2\rangle$ is ionic (75% near the field maximum at $t\approx700$), and $||3\rangle$ is covalent. From population analysis, the transition probabilities just after the crossing at $t\approx680$ are as follows: $P(1'\rightarrow 1)$: $P(1'\rightarrow 2)$: $P(1'\rightarrow 3)$=0:1:2. See the I- and C-components in Fig. 5(b). The adiabatic channel to the diffuse state $|1\rangle$, $1'\rightarrow 1$, is closed. Reducing the three-level crossing to two two-level problems $1'\rightarrow 1$ and $1'\rightarrow 2$ (treated as adiabatic transitions), $P(1'\rightarrow 1)/P(1'\rightarrow 2)$ can be estimated by

$$\overline{P}_{1'1}/\overline{P}_{1'2} \approx v_{1'1}^2\left(d\Delta E_{1'2}/dt\right)/v_{1'2}^2\left(d\Delta E_{1'1}/dt\right). \tag{23}$$

The coupling $v_{1'1}$ is much smaller than $v_{1'2}$ because of the smaller I-component in $|1\rangle : v_{1'1}/v_{1'2} \approx 1/6$. Energy matching is also unfavorable to the channel $1'\rightarrow 1$; i.e., $d\Delta E_{1'1}/dt > d\Delta E_{1'2}/dt$.

Energy matching and spatial overlap are most favorable to the covalent diabatic channel $1'\rightarrow 3$. The ionic state $|2\rangle$ however intrudes into the channel. At R=6, the ionic and covalent states cross for $|\mathcal{E}(t)| \approx 0.074$. The resultant avoided crossing reduces the $1'\rightarrow 3$ transition probability. Part of the population goes into the ionic state $|2\rangle$. For the present 10th half cycle, the ionization probability from $|2\rangle$ is nearly unity (I-type ionization). The population of $|3\rangle$ is reduced to 1/2 (due to C-type). The avoided crossing reduces the C-component whereas it creates an I-component of high ionization probability. It should be pointed out that maximum avoided crossing occurs at R=6. The overall ionization is therefore enhanced around R=6. The ionization probability from a pure covalent state is much smaller than the ionization probability from H^-H^+ and is nearly independent of R. The C-type ionization can be explained as the ionization of H. The enhanced ionization is hence determined by the population of H^-H^+ and the ionization probability from H^-H^+.

For R=4.25 the wave packet Ψ follows the sequence $1'\rightarrow 2\rightarrow 1'\cdots$ in Fig.

7(a). The states $|1'\rangle$ and $|2\rangle$ correspond to Eq. (18a) and H^-H^+: $|\langle 1'|I_g\rangle|^2$=0.45 at $\mathcal{E}(t)$=0 [as shown in Table I] and $|\langle 2|H^-H^+\rangle|^2 \approx$0.75 when $\mathcal{E}(t)\approx$0.12. The diabatic path alternates between $|1'\rangle$ and the more ionic state $|2\rangle$. The $|1\rangle$ which is expressed as a linear combination of ϕ_6 and ϕ_7 is a diffuse state (the transition moment between $|6\rangle$ and $|7\rangle$, 10.5 a.u., is very large). Although the lowest two states cross in energy when $|\mathcal{E}(t)|\approx$0.05, adiabatic transitions $1' \rightarrow 1$ hardly occur owing to the negligible gap $v_{11'}$($\approx 3\times 10^{-4}$). The I- and C-components in $|1'\rangle$ do not spatially overlap with the ϕ_6 and ϕ_7 in $|1\rangle$. Near the field maximum, the population of $|2\rangle$ decreases owing to I-type ionization: $|2\rangle$ is identified as the doorway state to ionization. The I-component in Ψ takes a maximum value $|\langle B|H^-H^+\rangle|^2 \approx$0.9 at the phase 0.63π for the sine electric function. Such a large I-component is however greatly sustained by the attraction of the two nearby nuclei. Although the H^-H^+ involved in the dynamics is larger at $R\approx$4 than at $R\approx$6 ($|\langle B|H^-H^+\rangle|^2 \leq$0.4 at R=6), the ionization rate is a little larger at $R\approx$6 than at $R\approx$4. As R decreases (<4 a.u.), the ionization rate decreases because the two electrons in H^-H^+ are more strongly attracted by the distant nucleus. For the 7th half cycle[$f(t)\approx$0.078], the ionization probability from H^-H^+ decreases as 1, 0.55, and 0.34 at R=6, 4.25, and 3, respectively.

In conclusion, the localized ionic configuration H^-H^+ (or H^+H^-) in the descending well is the main doorway state to ionization. When H^-H^+ is created, the ionization rate therefrom increases with increasing R owing to the decreasing nuclear attraction. On the other hand, the population of H^-H^+ decreases with increasing R. As a result, the ionization rate has a peak in the intermediate range of R. This peak correlates at R=6 with a maximum avoided crossing of adiabatic energies where the path bifurcates into the covalent and ionic ones. The avoided cros ing reduces the covalent component and increases the dominant doorway state to ionization, i.e., H^-H^+. What correlates with the internuclear distance for enhanced ionization is not just the presence of a crossing but the avoided crossing (i.e., how extensively the ionic characters are distributed among states). In the small R region, the main path is on the dominant ionic surface; in the large R region, it is the covalent diabatic one. The diffuse (Rydberg-like) state plays no role in the dynamics owing to negligible coupling with the dominant covalent and ionic states. The main coupling via the field and the electron correlation is always between the covalent and ionic states. Using only covalent and ionic states, Gibson *et al.* have analyzed charge-asymmetric dissociation in intense laser fields.[52] The interaction scheme derived here supports such a simple electrostatic consideration. The use of field-free eigenfunctions more than seven leads to the same conclusions drawn here.

5 Electronic structure in laser fields and laser-induced chemical bonds

What is the characteristic feature of molecular electronic structure in an intense field? We give a scrutiny to the electronic structure of adiabatic states. Four adiabatic states for the 1D H_2 at $\mathcal{E}(t)=0.12$ are shown in Fig. 8: (a) $|2\rangle$, (b) $|6\rangle$, (c) $|4\rangle$ and (d) $|5\rangle$. The main components of $|2\rangle$ and $|6\rangle$ are the ionic states H^-H^+ and H^+H^-, respectively. The state of each atom can be designated. For instance, in $|2\rangle$, the left atom is considered H^- and the right species is a proton.

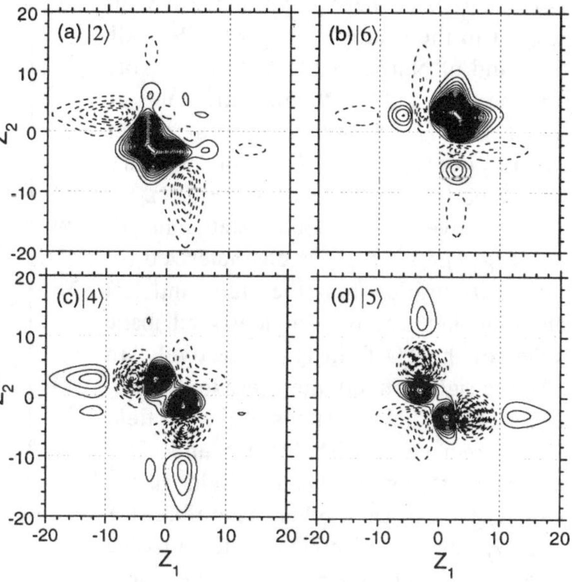

FIG. 8. Contour maps of the wave functions for four adiabatic states of 1D H_2 at $\mathcal{E}(t)$=0.12 a.u. (R=6 a.u.): (a) |2> , (b) |6> , (c) |4> and (d) |5>.

One can generally assign the state of each atom to one of diabatic states of the 1D "field" H. The lowest five adiabatic states of the 1D H at $\mathcal{E}(t)=0.12$ are drawn in Fig. 9. In this figure, the nucleus is located at $z=0$. The Hamiltonian of the 1D H in the field is diagonalized in terms of the lowest five eigenstates of the 1D field-free H (to include the diffuse component in ϕ_7). The second adiabatic state is nearly identical with the ground state of the 1D H.

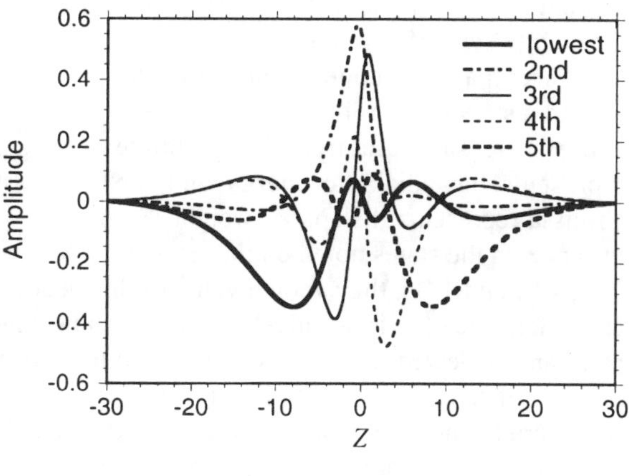

FIG. 9. Wave functions for the lowest five adiabatic states of the 1D H at $\mathcal{E}(t)$=0.12 a.u.

The interpretation of $|4\rangle$ and $|5\rangle$ in Fig. 8 is as follows. In $|4\rangle$, the right and left atoms are in the second and third adiabatic states of the 1D field H, respectively; In $|5\rangle$, the right and left atoms are in the fourth and second adiabatic states of the 1D field H, respectively. The two states

$|4\rangle$ and $|5\rangle$ are nearly symmetrical with respect to the origin $z_1 = z_2 = 0$. We call this kind of pair a counterstate pair. Note that the third and fourth adiabatic states of the 1D field H are symmetrical with respect to the nucleus of the atom and are counterparts to each other (See Fig. 9).

We have another counterstate pair $|1\rangle$ and $|7\rangle$. The state $|1\rangle$ is characterized by the configuration that the right and left atoms are in the second and lowest adiabatic states of the 1D field H, respectively. In $|7\rangle$, the right and left atoms are in the fifth and second adiabatic states of the 1D field H, respectively. The lowest and fifth adiabatic states of the 1D field H are counterparts to each other. Therefore, $|1\rangle$ and $|7\rangle$ constitute a counterstate pair. The two states in a counterstate pair are nearly symmetrical with respect to the origin $z_1 = z_2 = 0$ (definition of counterstate pair). The third adiabatic state $|3\rangle$ is mainly covalent. Both atoms are in the second adiabatic state of the 1D field H. Thus, the counterstate of $|3\rangle$ is itself.

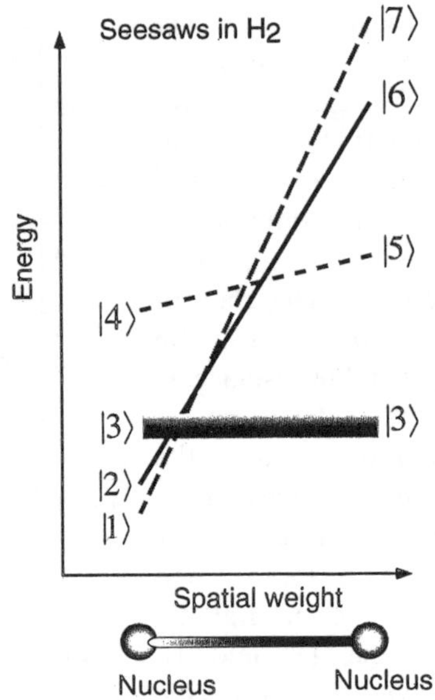

FIG. 10. A seesaw picture for seven adiabatic states $\{|n\rangle\}$ of H_2 in a laser field (R=6 a.u.). Three pairs of counterstates and the unpaired state at $\mathcal{E}(t)$=0.12 a.u. are illustrated. The ordinate represents the adiabatic energy and the abscissa represents the spatial weight. For instance, for the states noted on the left side, the average value of the electron density in coordinate space is on the left side. Each line connecting the two states of a counterstate pair can be viewed as a seesaw operated by the laser field.

Three pairs of counterstates and the unpaired state $|3\rangle$ are illustrated in Fig. 10 using a seesaw picture. The ordinate represents the adiabatic energy and the abscissa represents the spatial weight. For instance, in the states noted on the left side ($|1\rangle$, $|2\rangle$, and $|4\rangle$), the average value of the electron density in coordinate space is on the left side ($\langle z\rangle$ is negative). Each line connecting the two states of a counterstate pair can be viewed as a seesaw. The amplitude of the seesaw is determined by the product of the field amplitude $f(t)$ and the transition moment between the zero field gerade and ungerade states ϕ_n and ϕ_m constituting a counterstate pair in the field. Generally speaking, different seesaws have different amplitudes of up-and-down motion because different transitions have different transition moments. For two states of different characters, the diabatic coupling is small and the difference between the differentials of their adiabatic energies with respect to time are large: as known from Eq. (22), it is difficult to jump from seesaw to seesaw. Most of seesaws are therefore expected to be independent. For instance, the seesaw of the counterstate pair $|4\rangle$ and $|5\rangle$ is independent of the other seesaws. Two seesaws that have ionic components $|I_g\rangle$ and $|I_u\rangle$ are coupled with each other when they are close to each other. See, for example, the unpaired seesaw and the $|2\rangle$ and $|6\rangle$ pair seesaw in Fig. 10.

Nonadiabatic transitions between the two states in a counterstate pair occur when the field changes its sign. The same rule as in the H_2^+ case applies to H_2. As the seesaw amplitude, frequency, and internuclear distance increase, the coupling between the two states in a counterstate pair increases, i.e., the electron density is more localized near a nucleus.

As has been demonstrated, the state of each atom can be designated in an intense field. This can be paraphrased like "in an intense field a (homonuclear) molecule becomes atoms." The electron repulsion has an effect of confining an electron to a well and hindering the one-electron resonance between the two wells. The energy splitting between the two wells $R\mathcal{E}(t)$ further reduces one-electron resonance between them. The correlation between the electronic states in the two wells are hence weak. For the ionic counterstate pair, the two wells for the electron pair are split in energy by $2R\mathcal{E}(t)$. Two-electron resonance between the wells is negligible. Each well is regarded as an atom. The state of each atom can be designated in terms of adiabatic states of the field H (including the H^- bound state).

The idea "a molecule becomes atoms" leads to a variety of control schemes such

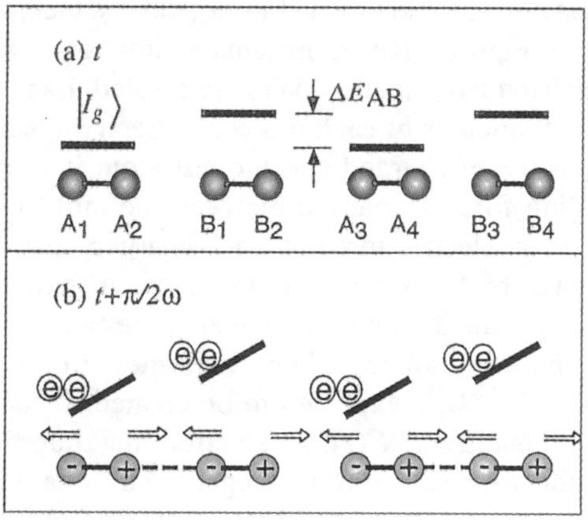

FIG. 11. A schematic view of bond formation in an intense field. The model system consists of diatomic molecules AA and BB aligned alternately (through weak bonds). Each bold line represents a seesaw in a molecule and its height represents the adiabatic energy. We assume that the zero-field adiabatic energy is higher in AA than BB by ΔE_{AB}. In an intense field, each atom is in an adiabatic state of the field atom. When $\mathcal{E}(t)>0$ as in Fig.11(b), the adiabatic states of the right atoms of AA (BB) increase in energy and those of the left atoms decrease. If ionic components are involved, the counterstates in the seesaw can be localized ionic states both for AA and BB. Then, a Coulombic binding force is exerted between the molecules AA and BB. The laser field also exerts an electric force on each charged atom. The direction of the force is denoted by the arrow with the atom. Although the two forces are opposite to each other for the atoms in a molecule, the two forces are constructive between AA and BB: a new "transient" bond is formed.

as electron transfer, formation of new bonds, and Coulomb explosion of a specific bond. We close this paper to present a schematic view of bond formation in an intense field. We consider a model system where diatomic molecules AA and BB are aligned alternately (through weak bonds) as shown in Fig. 11. Each bold line represents a seesaw in a molecule and its height represents the adiabatic energy. We assume that as shown in Fig. 11(a) the adiabatic energy at $\mathcal{E}(t)=0$ is higher in AA than in BB (the gap is denoted by ΔE_{AB}). In an intense field, each atom is in an adiabatic state of the field atom. When $\mathcal{E}(t+\pi/2\omega)>0$, the energy of the right atom in AA (BB) increases and that of the left atom decreases.

The counterstates in the seesaw can be localized ionic states of AA and BB as the states $|2\rangle$ and $|6\rangle$ in Fig. 8 (for instance, at small $R<4$, the doorway seesaw is ionic). Figure 11(b) represents a situation near a field maximum $\mathcal{E}(t+\pi/2\omega)>0$. Localized two-electron pairs are created. Each pair of two electrons are located near the left nucleus of each molecule. Each molecule is polarized so that the right atom is positively charged and the left atom is negatively charged. Then, a Coulombic binding force is exerted between the molecules AA and BB. The laser field also exerts an electric force on each charged atom. The direction of the electric force is denoted by the arrow with the atom. Although the two forces are opposite to each other for the atoms in a molecule, the two forces are constructive between AA and BB: a new "transient" bond is formed. In the course of dissociative ionization, the molecule AB is expected to be created. If the gap ΔE_{AB} is compensated with the dipole energy $2R\mathcal{E}(t)$, an electron may hop from B_1 to the positively charged A_2 around $t+\pi/2\omega$ so that (regarded as a bare proton in the field). Then, one of the two electrons on A_1 will be ejected. Consequently, three electrons are shared by the three atoms A_1, A_2, and B_1. This is a scenario of the formation of molecule AAB.

Acknowledgments

We would like to thank Prof. Y. Fujimura and Prof. A. D. Bandrauk for helpful discussions. This work was supported in part by the Department of High-Density Optical Pulse Generation and Advanced Material Control Techniques.

References

1. J. H. Eberly, J. Japanainen, and K. Rzazewski, Phys. Rep. **204**, 331 (1991).
2. M. Lewenstein, K. C. Kulander, K. J. Schafer, and P. H. Bucksbaum, Phys. Rev. A**51**, 1495(1995).
3. L. V. Keldysh, Sov. Phys. JETP **20**, 1307 (1965).
4. S. Augst, D. D. Meyerhofer, D. Strickland, and S. L. Chin, J. Opt. Soc. Am. B **8**, 858(1991).

5. M. V. Ammosov, N. B. Delone, and V. P. Krainov, Sov. Phys. JETP **64**, 1191 (1986).
6. P. B. Corkum, Phys. Rev. Lett. **71**, 1994 (1993).
7. K. Burnett, V. C. Reed, J. Cooper, and P. L. Knight, Phys. Rev. A **45**, 3347 (1992).
8. J. L. Krause, K. J. Schafer, and K. C. Kulander, Phys. Rev. A **45**, 4998 (1992).
9. T. Zuo, A. D. Bandrauk, M. Ivanov, and P. B. Corkum, Phys. Rev. A **51**, 3991 (1995).
10. A. Rundquist, G.Durfee III, Z.Chang, C.Herne, S.Backus, M. M. Murnane, and H. C. Kapteyn, Science **280**, 1412(1998).
11. M. Kakehata, H. Takada, K. Miyazaki, H. Kono, I. Kawata, and Y. Fujimura, Appl.Phys. B **70**, 219(2000).
12. P. A. Hatherly, M. Stankiewicz, K. Codling, L. J. Frasinski, and G. M. Cross, J. Phys. B **27**, 2993 (1994).
13. S. Chelkowski, T. Zuo, O. Atabek and A. D. Bandrauk, Phys. Rev. A **52,** 2977 (1995).
14. T. Zuo and A. D. Bandrauk, Phys. Rev. A **52**, R2511 (1995); A **54**, 3254 (1996).
15. S. Chelkowski and A. D. Bandrauk, J.Phys. B **28**, L723(1995).
16. T. Seideman, M. Yu. Ivanov, and P. B. Corkum, Phys. Rev. Lett. **75**, 2819 (1995).
17. M. Ivanov, T. Seideman, and P. Corkum, Phys. Rev. A **54**, 1541 (1996).
18. A. D. Bandrauk, Comments At. Mol. Phys. D**1**, 97 (1999).
19. D. M. Villeneuve, M. Yu. Ivanov, and P. B. Corkum, Phys. Rev. A **54**, 736 (1996).
20. T. D. G. Walsh, F. A. Ilkov, S. L. Chin, F. Châteauneuf, T. T. Nguen-Dang, S. Chelkowski, A. D. Bandrauk, and O. Atabek, Phys. Rev. A **58**, 3922 (1998).
21. H. Yu, T. Zuo, and A. D. Bandrauk, Phys. Rev. A**54**, 3290 (1996).
22. I. Kawata, H. Kono, and Y. Fujimura, J. Chem. Phys. **110**, 11152 (1999); Chem. Phys. Lett. **289**, 546 (1998).
23. P. Dietrich, M. Yu. Ivanov, F. A. Ilkov, and P. B. Corkum, Phys. Rev. Lett. **77**, 4150 (1996).
24. A. D. Bandrauk, *Molecules in Intense Laser Fields* (M. Dekker, NY, 1994), Chs. 1-3.
25. K. Harumiya, I. Kawata, H. Kono, and Y. Fujimura, to be published.
26. H. Kono, A. Kita, Y.Ohtsuki and Y. Fujimura, J. Comput. Phys. **130**, 148 (1997).
27. I. Kawata, H. Kono, and Y. Fujimura, J. Chem. Phys. **111**, 9498 (1999).
28. Ue-Li Pen and T.F.Jiamg, Phys. Rev. A **53**, 623 (1996).
29. M. Horbatsch, Phys. Rev. A **44**, R5346 (1991).
30. K. C. Kulander, K. J. Schafer, andJ. L. Krause, Adv. At. Mol. Phys. Suppl. **1**,

247 (1992).
31. H. Yu, A. D. Bandrauk, V. Sonnard, J. Math. Chem. **15**, 273 (1994).
32. Th. Mercouris, Y. Komninos, S. Dionissopoulou, and C. A.Nicolaides, Phys. Rev. A **50**, 4109 (1994).
33. R. Kosloff, J. Phys. Chem. **92**, 2087 (1988).
34. N. Balakrishnan, C. Kalyanaraman, and N.S athyamurthy, Phys. Rep. **280**, 79 (1997).
35. M. D. Feit, J. A. Fleck, Jr., and A. Steiger, J. Comput. Phys. **47**, 412 (1982).
36. H. Kono and S. H. Lin, J. Phys. Chem. **84**, 1071 (1986).
37. A. R. Mitchell, Computational Methods in Partial Differential Equations (Wiley, New York, 1969); N. N. Yanenko, The Method of Fractional Steps (Springer, New York, 1971).
38. E. Fattal, R. Baer, and R. Kosloff and D. Kosloff, Phys. Rev. E **53**, 1217 (1996).
39. B. Friedrich and D. Hershbach, Phys. Rev. Lett. **74**, 4623 (1995); H. Sakai, C. P. Safran, J. J. Larsen, K. M. Hilligsøe, K. Hald, and H. Stapelfeldt, J. Chem. Phys. **110**, 10235 (1999).
40. J. R. Hiskes, Phys. Rev. **122**,1207(1961).
41. F. H. Mies, A. Giusti-Suzor, K. C. Kulander and K. J. Schafer in: *Super-Intense Laser-Atom Physics,* eds. B. Piraux, A. L'Huillier and K. Rzazewski, NATO ASI Series B, Vol. 316 (Plenum Press, New York, 1993).
42. R. S. Mulliken, J. Chem. Phys. **7**, 20 (1939).
43. Y. Kayanuma, Phys. Rev. B **47**, 9940 (1993); Phys. Rev. A **50**, 843 (1994).
44. C. Zhu and H. Nakamura, J.Chem. Phys. **107**, 7839 (1997).
45. C. Zener, Proc. Roy. Soc. London, Ser.A **137**, 696 (1932).
46. F. Grossmann, T. Dittrich, P. Jung, and P. Hänngi, Phys. Rev. Lett. **67**, 516 (1991).
47. I. Kawata, H. Kono, and Y. Fujimura, to be published.
48. A. Hishikawa, A. Iwamae, and K. Yamanouchi, Phys. Rev. Lett. **83**, 1127 (1999).
49. J.C. Slater, *Quantum Theory of Molecules and Solids*, vol. I (McGraw-Hill, NY, 1963).
50. L. Wolniewicz, J.Chem.Phys. **51**, 5002(1969).
51. A. Messiah, Quantum Mechanics, Vol. 2 (North-Holland, Amsterdam, 1970) Ch. 17.
52. G. N. Gibson, M. Li, C. Guo, and J. P. Nibarger, Phys. Rev. A **58**, 4723 (1998).

LIGHT-DRESSED STATES OF H_2O IN INTENSE LASER FIELDS

SHILIN LIU, AKIYOSHI HISHIKAWA, ATSUSHI IWAMAE, AND KAORU YAMANOUCHI*

Department of Chemistry, School of Science, University of Tokyo, 7-3-1 Hongo, Bunkyo-ku, Tokyo 113-0033, Japan

The Coulomb explosion processes of H_2O, i.e., (i) $H_2O^{3+} \rightarrow H^+ + O^+ + H^+$ and (ii) $H_2O^{4+} \rightarrow H^+ + O^{2+} + H^+$, in ultrashort intense laser fields (~ 1 PW/cm^2) are investigated by the mass-resolved momentum imaging (MRMI) technique. From the analysis of these MRMI maps of the atomic fragment ions, it is found that the ∠H-O-H bond angle of the parent ions exhibits a significantly broad distribution (FWHM ~ 60°) centered at the linear configuration and the O-H bond lengths for H_2O^{3+} and H_2O^{4+} become 1.7 and 2.0 times as long as R_e = 0.958 Å, respectively. The characteristic ultrafast bond angle deformation occurring prior to the Coulomb explosion is interpreted by the formation of the light-dressed potential energy surfaces of H_2O^+.

1 Introduction

With the development of short-pulsed laser technology in recent years, the magnitude of the electric field generated by the laser light can be raised as high as the Coulombic fields in molecules. Once a molecule is exposed to such a strong laser field, interaction between molecules and the laser field leads to a variety of new phenomena such as above-threshold dissociation (ATD), bond-softening and bond-hardening,[1,2] spatial alignment of molecules[3] and formation of multiply charged ions followed by a Coulomb explosion process.[4]

In the case of polyatomic molecules, as the laser field intensity increases, the *light-dressed potential energy surfaces* (LDPESs) begin to be formed through the coupling between molecular electronic states and the laser field. On the LDPESs, the molecular structural deformation occurs within an ultrashort laser pulse. If the subsequent ultrafast electron stripping process leads to the Coulomb explosion, the fragmentation pattern is expected to carry information on the geometrical structure of molecules deformed largely through the formation of LDPESs. The structural change of polyatomic molecules in intense laser fields was previously discussed by Cornaggia and co-workers for $C_2H_2^+$,[5] $C_3H_3^+$,[6] CO_2[7] and SO_2[8] based on the coincidence patterns in a covariance map.

---

* to whom correspondence should be addressed

Since the momentum imposed on fragment ions depends sensitively on the geometrical structure of the parent ions just before the Coulomb explosion, information about the deformation of molecular skeletal structure can be derived directly by measuring the momentum vector of fragment ions produced in the Coulomb explosion process. Recently, we developed a method to visualize the momentum vector distribution of the fragment ions called *mass-resolved momentum imaging* (MRMI), [9-14] which has been successfully applied to diatomic[9-11] and triatomic[9, 12-14] molecules in intense laser fields of ~1 PW/cm^2, and has begun to be regarded as an efficient method by other groups.[15,16] As shown in our previous papers[13,14] and described briefly in section 3.2, the geometrical structures of molecules just before the Coulomb explosion in intense laser fields can be determined directly by the analysis of the MRMI maps, and enable us to discuss the formation of LDPESs within an ultrashort intense laser pulse.

In our recent studies of CO_2[13] and NO_2,[14] the analyses of the MRMI maps for the fragment ions showed that the structural deformation occurs not only along the bond-stretching coordinate but along the bending coordinate. Within the ultrashort laser-pulse duration (~100 fs), the ∠O-C-O angle distribution of CO_2 spreads significantly compared with the electronic ground state of neutral CO_2, while NO_2 evolves from its initial bent distribution towards a linear distribution. These structural deformations along the bending coordinate were attributed to the formation of LDPESs.

In the present work, we apply the MRMI method to the Coulomb explosion dynamics of H_2O in intense laser fields. In the laser-field intensity regime of ~ 1.2 PW/cm^2, we identify only two types of three-body Coulomb explosion channels, i.e., $H_2O^{3+} \rightarrow H^+ + O^+ + H^+$ and $H_2O^{4+} \rightarrow H^+ + O^{2+} + H^+$, which are called the (1, 1, 1) and (1, 2, 1) pathways, respectively. Therefore, due to this limited number of fragmentation pathways, the geometrical structures of H_2O^{3+} and H_2O^{4+} just before the explosion can be determined with high-precision from the MRMI maps of the atomic fragment ions, and it becomes possible to discuss an effect of the formation of LDPESs in intense laser fields. Recently, the structural deformation of H_2O in intense laser fields was investigated by Sanderson *et al.*[15] using the MRMI method. In the present study, we aimed to derive more precise information about the distribution of the geometrical structure of the parent ions by properly taking into account the effect of the finite detector size.

2 Experimental

The details of our experimental setup were described previously.[9-14] The femtosecond laser system consists of a Ti:sapphire mode-locked oscillator and a regenerative amplifier, which produces amplified laser pulses (100 fs, $\lambda \sim 800$ nm) with a total energy of ~1 mJ/pulse at the repetition rate of 10 Hz.

The laser beam was focused by a quartz lens into a skimmed pulsed molecular beam with the H_2O gas in the region between the extraction parallel plates of a linear time-of-flight (TOF) mass spectrometer, where the extraction electric field of 234.2 V/cm was applied. In order to achieve a high angular resolution, a plate with a 10-mmϕ aperture was placed in front of the MCP ion detector. In our TOF mass spectrum with typical mass-resolution of $m/\Delta m \sim 620$, ion species with the different mass and charge numbers were separated well with no temporal overlap. The TOF mass spectrum with a high S/N ratio was obtained by accumulating the spectra for ~ 1000 laser shots using a digital oscilloscope at a 1 GHz sampling rate. When the nozzle was not operated, the background pressure in the main chamber was 1×10^{-8} Torr. Even during the nozzle operation, the pressure in the main chamber was kept below 6×10^{-8} Torr in order to avoid the space charge effect.

The angular distribution of the mass-resolved fragment ions was measured by rotating the polarization vector of the laser light with respect to the TOF axis. A total of 36 TOF mass spectra covering the 180° polarization angle were measured to construct the MRMI maps.

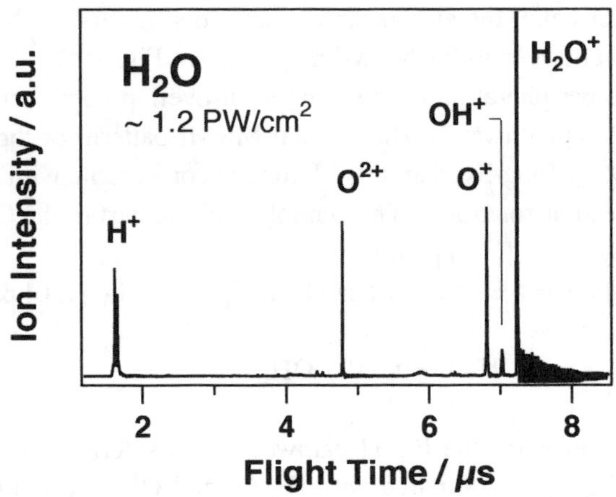

Figure 1. The TOF mass spectrum of H_2O at $\lambda = 800$ nm with the laser field intensity of ~1.2 PW/cm^2 and the laser polarization direction parallel to the TOF axis.

3 Results and Discussion

3.1 Overview of MS-TOF and MRMI maps

3.1.1 Two-body fragmentation pathway

In Fig. 1, the TOF mass spectrum of H_2O is shown, which was measured with the laser polarization direction parallel to the TOF axis. In the mass spectrum the fragment ions, H^+, O^{2+}, O^+ and OH^+, as well as the singly charged parent ions, H_2O^+, are identified.

The momentum vector distributions of the OH^+ and H^+ ions are shown in Figs. 2 (a) and (b), respectively, in a form of MRMI maps. The most probable two-body dissociation pathways to produce OH^+ are

$$[0, 1] \text{ pathway:} \qquad H_2O^+ \rightarrow H + OH^+ \tag{1}$$

and

$$[1, 1] \text{ pathway:} \qquad H_2O^{2+} \rightarrow H^+ + OH^+, \tag{2}$$

where a two-body pathway to form H^{p+} and OH^{q+} is denoted as a $[p, q]$ pathway as in our previous study of N_2O.[12] It can be seen from Fig. 2(a) that (i) the momentum distribution of OH^+ is isotropic and (ii) there are at least two contributions, i.e., the central part within the momentum circle with a radius of ~20 $\times 10^3$ amu m/s and the peripheral part with the momentum radius of ~40×10^3 amu m/s. On the other hand, the circular momentum distribution with a radius of ~40 $\times 10^3$ amu m/s can be seen in the MRMI map of H^+ in Fig. 2(b). Since the circular part in H^+ and the peripheral part in OH^+ have an overlap with each other, they are regarded as the [1, 1] pathway. The circular MRMI pattern for the [1, 1] pathway suggests that H_2O^{2+} is long-lived and its lifetime is comparable with or longer than a period of the molecular rotation. The central isotropic part in the OH^+ channel can be assigned mostly to the [0, 1] pathway.

In Fig. 2(b), the weak central part can be assigned to the [1, 0] pathway,

$$[1, 0] \text{ pathway:} \qquad H_2O^+ \rightarrow H^+ + OH, \tag{3}$$

which could compete with the [0, 1] pathway. Considering (i) the weak signal intensity of OH^+ in Fig. 1 compared with H^+, O^+ and O^{2+}, and (ii) the weak signal intensities of the [1, 0] and [1, 1] pathways in Fig. 2(b), the main dissociation

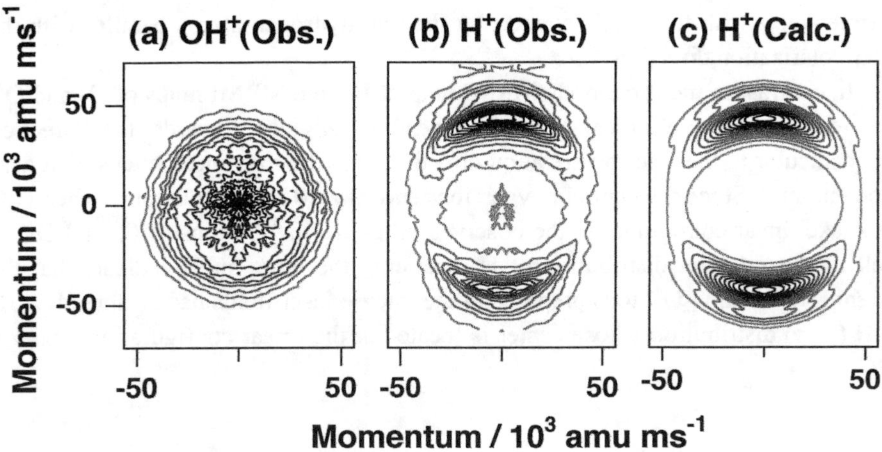

Figure 2. The MRMI maps of (a) OH^+ and (b) H^+ observed at λ ~800 nm with the laser-field intensity of ~1.2 PW/cm^2 and (c) the best-fit MRMI map of H^+.

pathways for H_2O ions in intense laser fields are the three-body fragmentation Coulomb explosion pathways.

3.1.2 Three-body fragmentation pathways

The MRMI map of H^+ in Fig. 2(b) exhibits a clear crescent-like pattern along the laser polarization direction. In the crescent pattern two components are identified, i.e., the inner and the outer crescent whose momentum centers are located at ~ 42×10^3 and ~ 56×10^3 amu m/s, respectively, which are assigned to the two three-body Coulomb explosion pathways; i.e.,

$$\text{(1, 1, 1) pathway:} \qquad H_2O^{3+} \rightarrow H^+ + O^+ + H^+ \tag{4}$$

and

$$\text{(1, 2, 1) pathway:} \qquad H_2O^{4+} \rightarrow H^+ + O^{2+} + H^+. \tag{5}$$

The anisotropic momentum distribution pattern shows clearly that H^+ ions are ejected preferentially in the direction along the laser polarization vector, indicating that H_2O ions which undergo these three-body Coulomb explosions are prepared preferentially with the a-axis connecting the two hydrogen atoms parallel with the laser polarization direction.

In contrast to the three-body MRMI map of H^+, the MRMI maps of O^+ and O^{2+} in Fig. 3 exhibit elliptical patterns extending slightly towards the direction perpendicular to the laser polarization direction with the maximum intensity at zero momentum. Since the central oxygen ions stay at zero momentum only when H_2O ions take linear configuration, the observed elliptical distributions of O^+ and O^{2+} as well as the crescent distribution of H^+ spanning the large angle indicate that the parent H_2O^{3+} and H_2O^{4+} ions prepared in the intense laser fields have a spread $\angle$H-O-H (= γ) distribution whose center is located at the linear configuration, i.e., $\gamma = 180°$.

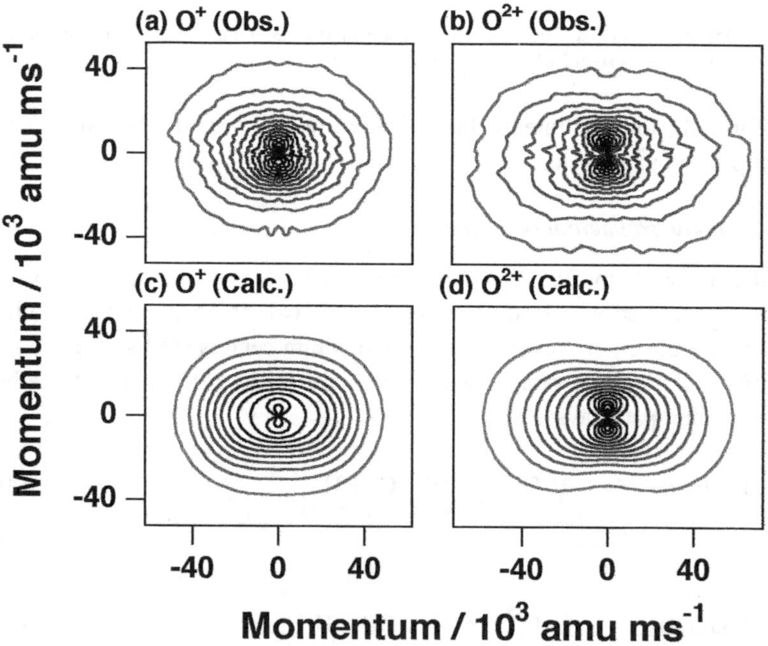

Figure 3. The observed MRMI maps of (a) O^+ and (b) O^{2+} fragment ions and the best-fit MRMI maps of (c) O^+ and (d) O^{2+}.

3.2 *MRMI analysis*

In order to derive quantitatively the information concerning the geometrical structure of H_2O^{3+} and H_2O^{4+} just before the Coulomb explosion, we performed a trial-and-error simulation of the MRMI maps of all the fragment ions generated from the (1,1,1) and (1,2,1) pathways by taking the following steps[13, 14]: (i) the released momenta of the fragment ions are calculated for a given molecular geometry expressed by the O-H bond length, R(O-H), and the bond angle γ, and they are converted into the MRMI maps of H^+, O^+ and O^{2+} for a single explosion pathway by assuming the distributions of R(O-H) and γ, (ii) the MRMI map of H^+ is synthesized by adding the relevant single-pathway MRMI maps with their weights estimated from the observed yields of O^+ and O^{2+}, (iii) the geometrical parameters R(O-H) and γ are determined by comparing the synthesized and observed MRMI maps for all the fragment ions, and (iv) the steps (i) - (iii) are repeated with a new set of the geometrical parameters until the synthesized MRMI maps agree with the observed ones. The procedure of the simulation is schematically summarized in Fig. 4.

The geometry of the H_2O ions just before the explosion is expressed by the bond length distribution $F_R(R)$[13,14] and the bond angle distribution $F_\gamma(\gamma)$[13,14,17] as,

$$F_R(R) \propto \exp\left[-\eta\left(1/\sqrt{R} - 1/\sqrt{R_0}\right)^2\right] \quad , \tag{6}$$

and

$$F_\gamma(\gamma) \propto \left\{\exp\left[-(\gamma-\gamma_0)^2/2(\sigma_\gamma/\sqrt{4\ln 2})^2\right] + \exp\left[-(\gamma+\gamma_0-2\pi)^2/2(\sigma_\gamma/\sqrt{4\ln 2})^2\right]\right\}^2 \quad , \tag{7}$$

where R_0, η, γ_0 and σ_γ are the variable parameters. The spatial angular distribution of H_2O ions is described as[3]

$$F_\theta(\theta) \propto \exp\left[-\frac{1}{2\sigma_\theta^2}\sin^2\theta\right] \quad , \tag{8}$$

where θ is the angle between the molecular a-axis and the laser polarization vector. The distribution width σ_θ is treated as a parameter representing the angular distribution of parent ions formed both by the torque given to H_2O from the laser field[3] and by the ionization enhancement[18] for a molecule whose θ is closer to zero. For given $F_R(R)$ and $F_\gamma(\gamma)$, the momentum distributions, $G(p_m, \theta_m^0)$s, of the fragment ions represented by the released momentum p_m and the ejection angle θ_m^0 of the fragment ion in the molecular fixed frame are calculated for a given explosion

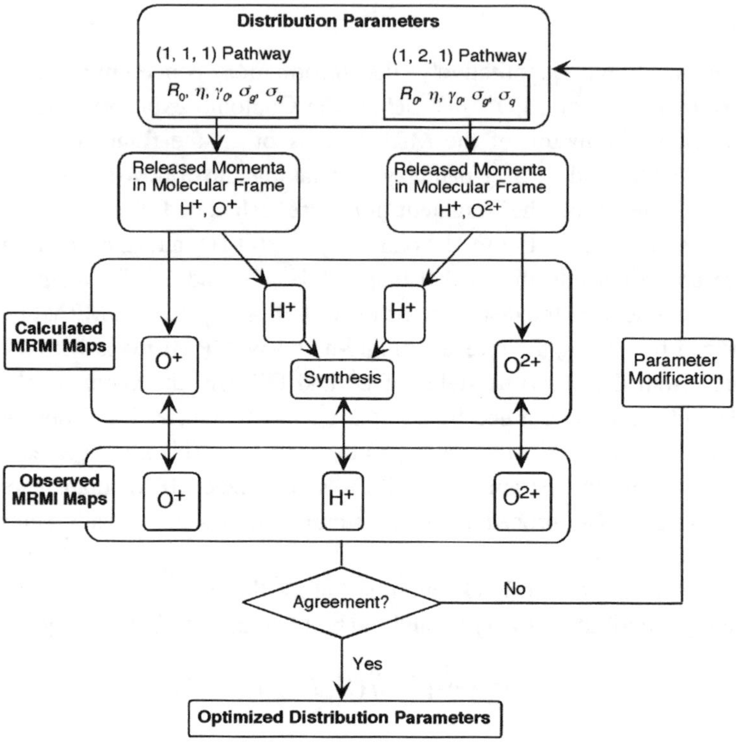

Figure 4. Schematic diagram of the simulation procedure of the MRMI maps of H^+, O^+ and O^{2+} produced after the three-body Coulomb explosion of H_2O.

pathway by numerically solving the classical equation of motion under the assumption that the repulsive forces between fragment ions are all Coulombic.

From the one-dimensional momentum distribution, the two-dimensional MRMI map is constructed in terms of the Legendre polynomials, P_{2k}'s, as

$$I(p_{//}) = \int_{|p_{//}|}^{p_{\max}} \frac{G(p_m, \theta_m^0)}{2p_m} \left[\sum_{k=0}^{\infty} a_{2k} P_{2k}(\cos\theta_m^0) P_{2k}(\cos\alpha) P_{2k}\left(\frac{p_{//}}{p_m}\right) \right] \mathrm{d}p_m \quad , \tag{9}$$

where α denotes the angle between the laser polarization vector and the TOF axis. The coefficients a_{2k}'s are determined by the Legendre expansion of the spatial angular distribution function $F_\theta(\theta)$. The $p_{\max}$ defined as

TABLE I. Optimized parameters for the geometrical and spatial-angular distributions determined from the trial-and-error simulation of the MRMI maps of H^+, O^+ and O^{2+}.[a)]

pathway	R_0 /Å	η	γ_0 /deg	σ_γ /deg	σ_θ /deg
(1, 1, 1)	1.6(1)	100(20)	180	60(10)	20(3)
(1, 2, 1)	1.9(1)	150(30)	180	60(10)	17(2)

[a)] Numbers in parentheses represent the estimated limits of error.

$$p_{\max} = \left[p_{//}^2 + \left(\frac{md}{2t_0} \right)^2 \right]^{1/2} , \tag{10}$$

represents the upper limit of the released momentum of the fragment ions that can reach the detector with an effective diameter d and the flight time t_0.[10-14] The MRMI maps for a single Coulomb explosion pathway are then synthesized for all the respective fragment ions.

The ratio of the relative yields for the (1,1,1) and (1,2,1) pathways, $I(1,1,1)$ and $I(1,2,1)$, was derived to be $I(1,1,1) : I(1,2,1) = 1.0 : 0.25$, by fitting the momentum-scaled (MS) TOF spectrum of H^+ with two Gaussian momentum distributions. The geometrical parameters, γ_0, σ_γ, R_0, η and σ_θ, were adjusted in a trial-and-error simulation of the observed MRMI maps of H^+, O^+ and O^{2+} in a manner described above so that all the MRMI maps are reproduced simultaneously. The simulation of the MRMI maps revealed that (i) the extent of the spread and that of the ellipticity of the MRMI maps of O^+ and O^{2+} are sensitive to γ_0 and σ_γ, respectively, (ii) the magnitudes of the central momentum and the width of the momentum distribution of H^+ are sensitive to R_0 and η, respectively, and (iii) the angular distribution of H^+ is sensitive to σ_θ. Therefore, these five parameters were determined with only small correlations.

The best-fit MRMI maps for H^+, O^+ and O^{2+} obtained after several trial-and-error iterations are shown in Fig. 2(c) for H^+ and in Fig. 3 for O^+ and O^{2+}, which reproduce well the corresponding observed MRMI maps. The optimized parameters are listed in Table I, and the distributions of $F_R(R)$, $F_\gamma(\gamma)$ and $F_\theta(\theta)$ are shown in Fig. 5. The corresponding distribution function obtained recently by Sanderson *et al.* [15] are also shown for comparison.

As seen in Fig. 5(a), the bond length for the parent ions tends to become larger with the charge number increases, i.e., R_0's for H_2O^{3+} and H_2O^{4+} are 1.7(1) and 2.0(1) times longer than that in the neutral electronic ground state (R_e(O-H) =

0.958Å), respectively. This type of bond elongation in intense laser fields was observed for diatomic molecules, I_2[19], N_2[10] and NO,[11] and triatomic molecules, CO_2[13] and NO_2,[14] indicating the existence of a common dynamical mechanism determining the bond elongation processes.

In contrast to the bent geometry of neutral H_2O in the electronic ground state ($\gamma = 104.5°$), both of the bond angle distributions of H_2O^{3+} and H_2O^{4+} in Fig. 5(b) exhibit a large extent ($\sigma_\gamma = 60°$) of spread whose center is located at 180°. It can be said that H_2O changes its geometry in intense laser fields from the initially bent geometry to the linear geometry.

There are two types of interaction which make the spatial angular distribution narrower; (i) the laser-induced torque which aligns a molecule along the laser polarization direction, and (ii) the enhanced ionization when the molecular *a*-axis is close to the laser polarization direction. Therefore, the determined width, σ_θ, of the spatial angular distribution is regarded as an effective parameters. However, it should be noted that the geometrical structure and its distribution represented by R_0, η, γ_0 and σ_γ are extracted free from the spatial angular distribution as far as the anisotropic distributions of the ejected fragment ions are expressed as a sum of the Legendre products in Eq. (9).

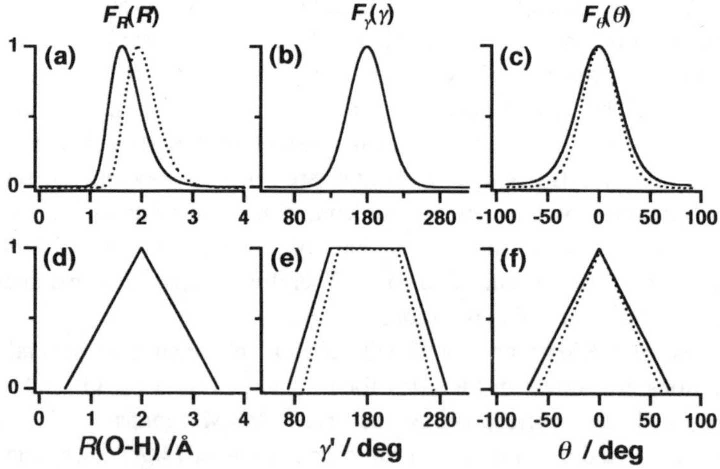

Figure 5. Optimized distributions, $F_R(R)$, $F_\gamma(\gamma)$ and $F_\theta(\theta)$, of H_2O^{3+} (solid curve) and H_2O^{4+} (dashed curve) just before the Coulomb explosion derived in the present study, (a) - (c), and in Ref. 15, (d) - (f). The same γ distributions are obtained for H_2O^{3+} and H_2O^{4+} in (b).

3.3 *Structural deformations of H_2O*

3.3.1 *Bond elongation*

From the MRMI analysis the bond length distributions for H_2O^{3+} and H_2O^{4+} were determined to have peaks at R(O-H)/R_e(O-H) = 1.7(1) and 2.0(1), respectively. It has been reported experimentally[20,21] as well as theoretically[22,23] that the ionization rate of small molecules in intense laser fields is significantly enhanced at elongated bond length and reaches a maximum at $R \sim 2R_e$. By applying the classical field-ionization model,[21] the appearance laser intensity for the generation of H_2O^{3+} and H_2O^{4+} was calculated as a function of R(O-H) and γ when the a-axis of H_2O is fixed in the direction parallel to the laser polarization vector. The derived appearance intensities are plotted as a two-dimensional map in Fig. 6.

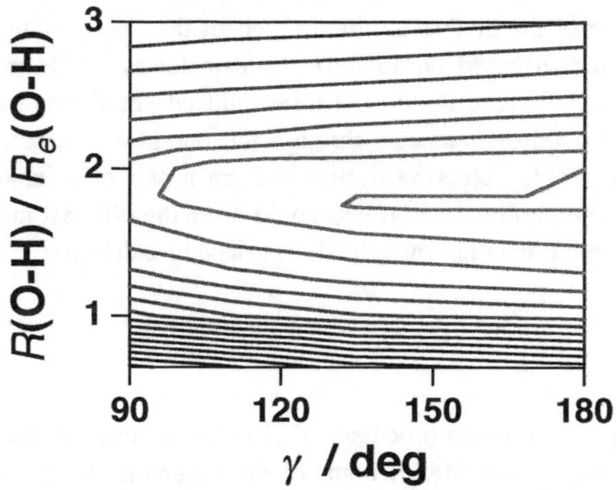

Figure 6. The calculated appearance intensities of the laser field as a function of the bond length scaled by R_e(O-H) and the ∠H-O-H bond angle, γ, for the (1, 1, 1) Coulomb explosion pathway based on the classical field ionization model.

It was found that (i) the critical geometrical parameters, R_c(O-H) and γ_c, at which the appearance intensity takes a minimum is R_c(O-H)/R_e(O-H) = 1.8 and γ_c = 180°, and (ii) the appearance intensity depends only weakly on the bending angle. Therefore, the γ_c value has a large uncertainty and is not defined well. The calculated dependence of the appearance intensity on the bond length explains well quantitatively the derived bond length distribution of H_2O^{3+} and H_2O^{4+} shown in

Fig. 5(a). This behavior of the ionization appearance intensity, i.e., its weak dependence on the bond angle and sensitive dependence on the bond length, was also identified for CO_2[13] and NO_2,[14] and is considered to be a common feature of triatomic molecules in intense laser fields.

3.3.2 Bond-angle deformation

In order to explain the derived broad bond-angle distribution shown in Fig. 5(b), the formation of LDPESs in intense laser fields was inferred, on which the ultrafast bond angle deformation occurs. As suggested by Rottke *et al.*[24] and Sanderson *et al.*,[15] the phenomenon responsible for this spread of the bond angle centered at 180° can be ascribed to the potential softening along the bending coordinate caused by the coupling between the bent ground $\tilde{X}^2B_1$ state ($\gamma_e = 120°$) and the linear excited $\tilde{A}^2A_1$ state of H_2O^+ in intense laser fields. When λ = 800 nm, the one-photon dressed state of $\tilde{A}^2A_1$ crosses with the bent $\tilde{X}^2B_1$ state near the two equivalent equilibrium positions of the double-minimum well of the $\tilde{X}^2B_1$ state as shown in Fig. 7, which was drawn based on an *ab initio* calculation by Brommer *et al.*[25] The γ distribution could spread due to the laser-induced population transfer to the excited state having a linear geometry through the one-photon crossing points at which a significant avoided crossing occurs to form a pair of adiabatic LDPESs. Then, the lower counterpart of the LDPESs could drive the ultrafast nuclear motion towards the linear structure within the ultrashort laser-pulse duration.

4. Summary

The ultrafast nuclear deformation processes of H_2O ions in intense laser fields were investigated by measuring the MRMI maps of the fragment ions produced by the two three-body Coulomb explosion processes, i.e., $H_2O^{3+} \rightarrow H^+ + O^+ + H^+$ and $H_2O^{4+} \rightarrow H^+ + O^{2+} + H^+$.

From the analysis of the MRMI maps, it was found that (i) the O-H bond lengths of H_2O^{3+} and H_2O^{4+} respectively become 1.7(1) and 2.0(1) times longer than that of the neutral H_2O in the electronic ground state, and (ii) the ∠H-O-H bond angle becomes ~180° with a Gaussian distribution width of 60°(10) for both of the two parent ions. The observed geometrical deformation was interpreted by the formation of LDPESs by the one-photon coupling between the bent ground $\tilde{X}^2B_1$ state ($\gamma_e = 120°$) and the linear excited $\tilde{A}^2A_1$ state of H_2O^+ in intense laser fields.

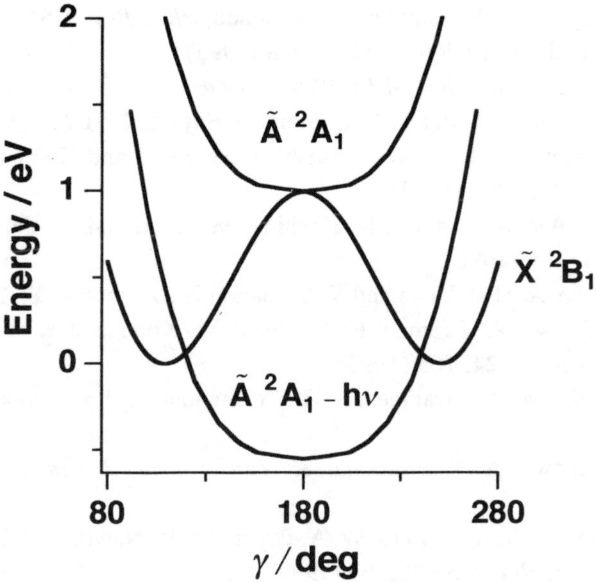

Figure 7. Potential functions along the bending coordinate of the $\tilde{A}\ ^2A_1$, $\tilde{X}\ ^2B_1$ and light-dressed $\tilde{A}\ ^2A_1$ states of H_2O^+, drawn using the theoretical potential functions (Ref. 26) of the $\tilde{A}\ ^2A_1$ and $\tilde{X}\ ^2B_1$ states derived when R(O-H) = 0.997 Å.

Acknowledgements

The present work was supported by the CREST (Core Research for Evolutionary Science and Technology) fund from Japan Science and Technology Corporation.

References

1. A. D. Bandrauk, *Molecules in Intense Laser Fields*, (M. Dekker Pub., New York, 1993).
2. A. Giusti-Suzor, F. H. Mies, L. F. DiMauro, E. Charron, and B. Yang, *J. Phys. B* **28**, 309 (1995).
3. B. Friedrich and D. Herschbach, *Phys. Rev. Lett.* **74**, 4623 (1995).
4. K. Codling and L. J. Frasinski, *J. Phys. B* **26**, 783 (1993), and references therein.

5. C. Cornaggia, M. Schimit and D. Normand, *Phys. Rev. A* **51**, 1431 (1995).
6. C. Cornaggia, *Phys. Rev. A* **52**, R4328 (1995).
7. C. Cornaggia, *Phys. Rev. A* **54**, R2555 (1996).
8. C. Cornaggia, F. Salin and C. LeBlanc, *J. Phys. B* **29**, L749 (1996).
9. A. Hishikawa, A. Iwamae, K. Hoshina, M. Kono and K. Yamanouchi, *Chem. Phys. Lett.* **282**, 283 (1998).
10. A. Hishikawa, A. Iwamae, K. Hoshina, M. Kono and K. Yamanouchi, *Chem. Phys.* **231**, 315 (1998).
11. A. Iwamae, A. Hishikawa and K. Yamanouchi, *J. Phys. B* **32**, 223 (2000).
12. A. Hishikawa, A. Iwamae, K. Hoshina, M. Kono and K. Yamanouchi, *Res. Chem. Intermed.* **24**, 765 (1998).
13. A. Hishikawa, A. Iwamae and K. Yamanouchi, *Phys. Rev. Lett.* **83**, 1127 (1999).
14. A. Hishikawa, A. Iwamae and K. Yamanouchi, *J. Chem. Phys.* **111**, 8871 (1999).
15. J. H. Sanderson, A. El-Zein, W. A. Bryan, W. R. Newll, A. J. Langley and P. F. Taday, *Phys. Rev. A* **59**, R2567 (1999).
16. J. H. Posthums, J. Plumridge, P. F. Taday, J. H. Sanderson, A J. Langley, K. Codling and W. A. Bryan, *J. Phys. B* **32**, L93 (1999).
17. T. Graber, E. P. Kanter, Z. Vager and D. Zaifman, *J. Chem. Phys.* **98**, 7725 (1993).
18. J. H. Posthumus, J. Plumridge, M. K. Thomas, K. Codling, L. J. Frasinski, A. J. Langley and P. F. Taday, *J. Phys. B* **31**, L553 (1998).
19. G. N. Gibson, M. Li, C. Guo and J. P. Nibarger, *Phys. Rev. A* **58**, 4723 (1998).
20. J. H. Posthums, A. J. Giles, M. R. Thomspon and K. Codling, *J. Phys. B* **29**, 5811 (1996).
21. E. Constant, H. Stapefelt and P. B. Corkum, *Phys. Rev. Lett.* **76**, 4140 (1996).
22. T. Seideman, M. Y. Ivanov and P. B. Corkum, *Phys. Rev. Lett.* **75**, 2819 (1995).
23. S. Chelkowski, T. Zuo, O. Atabek and A. D. Bandrauk, *Phys. Rev. A* **52**, 2977 (1995).
24. H. Rottke, C. Trump, and W. Sandner, *J. Phys. B* **31**, 1083 (1998).
25. M. Brommer, B. Weis, B. Follmeg, P. Rosmus, S. Carter, N. C. Handy, H. Werner and R. J. Knowles, *J. Chem. Phys.* **98**, 5222 (1993).

ATOMIC DYNAMICS AT LONG WAVELENGTHS

B. SHEEHY, J.D.D. MARTIN, T. CLATTERBUCK, DALWOO KIM* AND L.F. DIMAURO

Chemistry Department, Brookhaven National Laboratory, Upton, NY 11973

K. J. SCHAFER

Department of Physics and Astronomy, Louisiana State University, Baton Rouge, LA 70803

M. B. GAARDE

Lund Institute Technology, P.O. Box 118, S-22100 Lund, Sweden

K. C. KULANDER

TAMP, Lawrence Livermore National Laboratory, Livermore, CA 94551

We examine strong field atomic physics in a wavelength region (3-4 μm) where very little work has previously been done. The soft photon energy allows the exploration of one- electron atoms with low binding energies (alkali metals). We find that photoionization spectra differ from rare gas studies at shorter wavelengths due to more complex ion core potentials. Harmonic generation is studied, and we find that harmonic bandwidths are consistent with theory and the possibility of compression to pulse widths much shorter than that of the driving pulse. Harmonic yields in the visible and UV are sufficient for a complete study of their amplitude and phase characteristics.

1 Introduction

The interaction of atoms with strong optical fields has given rise to many new phenomena that have been as interesting in their fundamental physics as they are rich in potential applications. In recent years, our understanding has progressed so that many diverse phenomena may be modeled within a simple theoretical framework, while short-pulse technology has matured so that applications are more promising than ever. A great deal of the experimental work has been done in the near-infrared, visible, and UV parts of the spectrum, as this is where the development of short-pulse technology has afforded the most intense pulses. However, there are reasons to expect both interesting differences in the fundamental physics and more insight into applications by working at longer wavelengths. In this paper we explore some of these differences both theoretically and experimentally.

---

*PERMANENT ADDRESS: RIST, POHANG, KOREA

Many strong field phenomena may be understood in terms of the "rescattering model",[1,2,3] in which an electron is first promoted into the continuum by a tunneling process, and subsequently accelerated in the optical field. The electron's motion in the optical field may return it to the vicinity of the ion core, where it may radiatively recombine or scatter. This model has been very successful in predicting single- electron effects, including many features of high harmonic generation (HHG)[2,3] and photoelectron energy spectra (PES) in above-threshold ionization (ATI).[4,5] Extending experimental studies of these phenomena to longer wavelengths is appealing because the characteristic parameters in the rescattering model all scale differently with the wavelength of the light. The tunneling-ionization step may be parameterized by the Keldysh [5] adiabaticity parameter γ which scales as $1/\lambda$. This dimensionless parameter is given by $\gamma = \sqrt{I_p/2U_p}$, where I_p is the atomic ionization potential and U_p , the ponderomotive potential, is the cycle-averaged kinetic energy of a free electron moving in the optical field. The Keldysh parameter compares the characteristic time τ for tunneling through the field-suppressed Coulomb barrier to the optical period. Smaller ionization potentials or larger intensities decrease τ, while longer wavelengths increase the optical period, decreasing γ and improving the tunneling approximation. For values of γ much less than 1, tunneling ionization dominates. In comparing different atoms irradiated by different fields, it is instructive to compare experiments in which the values of γ are similar, as the physics of the ionization step is comparable. Thus, experiments on the alkalis in the mid-infrared are comparable to rare gas experiments in the near-infrared at much larger intensities. For example, at 3.5 μm, γ equals 1 at an intensity of 1.9 TW/cm^2 for potassium, while for argon in a 0.8 μm field, γ equals 1 at 130 TW/cm^2.

While γ scales as $1/\lambda$, the second dynamical step in the rescattering model, the energy acquired by the electron as it is accelerated in the optical field, scales as λ^2 At equal intensities, electron energies are an order of magnitude larger in the mid-IR than in the near-IR. Yet another scaling enters the problem when one considers the time between the tunneling ionization step and the return of the electron to the ion core. This time scales linearly with λ and is important because the electron wave packet spreads during this time, reducing its overlap with the core and the probability of scattering or recombining.

Other differences are also of interest. Consider the coupling of excited states in the argon/potassium comparison above. In argon at 0.8 μm, the first excited state lies 8 photons above the ground state, while only 11 photons are required to reach the ionization threshold. The ionization of potassium at 3.5 μm is of slightly higher order (13 photons), but the first excited state lies just 5 photons above the ground state. The role of this coupling is further explored in another paper in this volume.[6] The variation between alkali species in the ionic core potential also provides a good detailed test of the rescattering model. As shown below, small differences in the core potential can manifest themselves quite clearly in the PES by their effect on the rescattering process.

One of the most promising applications of strong-field physics is the production of short wavelength radiation through HHG. Harmonic orders as high as the 401st have been reached, producing wavelengths as short as 2 nm.[7] Recent theoretical studies[8,9] indicate that it is likely that the coherence properties of the high harmonics may be exploited in order to generate pulses on an attosecond time scale. Experimentally, it is difficult to test these theories at such short wavelengths. By using a fundamental wavelength in the mid-IR, harmonics may be generated in the visible and ultraviolet, where their coherence properties may be more easily measured. We have developed a mid-infrared source of sufficient intensity and produced visible harmonics at levels adequate to do these measurements. We have measured the spectral properties of some of these harmonics, and compared their relative abundance with theoretical predictions. We find that the spectral measurements are consistent with dramatic pulse shortening. Conclusive evidence will require the temporal and coherence measurements that are underway.

2 Mid-Infrared Source

Tunable mid-IR light is generated by three-wave mixing in a KTP crystal. A block diagram of the laser system is shown in Fig. 1. The tunable pump source includes a Spectra Physics Tsunami Ti:Sapphire mode-locked oscillator, pumped by a diode- pumped, intracavity doubled neodymium vanadate laser (Spectra Physics Millennia). The 1.5 psec pulses are stretched to 100 psec using a 2000 groove/mm grating in a stretcher of standard all-reflective design. A regenerative Ti:Sapphire amplifier, pumped by 12 mJ, 527 nm, 170 nsec pulses from a Quantronix model 527 Q-switched, intracavity-doubled, Nd:YLF laser, picks pulses from the 80 MHz oscillator train at 1 kHz and amplifies them to 2.8 mJ. After losses in the compressor, 1.6 mJ of radiation tunable from 780 - 830 nm is available for producing the mid-IR.

The signal beam is fixed in frequency at 1053 nm. A diode pumped mode-locked Nd:YLF laser (Lightwave model 131) produces 25 psec, 2.5 nJ pulses at 80 MHz. These are picked at 1 kHz and amplified to a maximum energy of 1.8 mJ in a discharge lamp pumped regenerative amplifier built in our laboratory. The parametric process requires the temporal overlap of the pulses in the mixing crystal. The synchronization requirement is somewhat

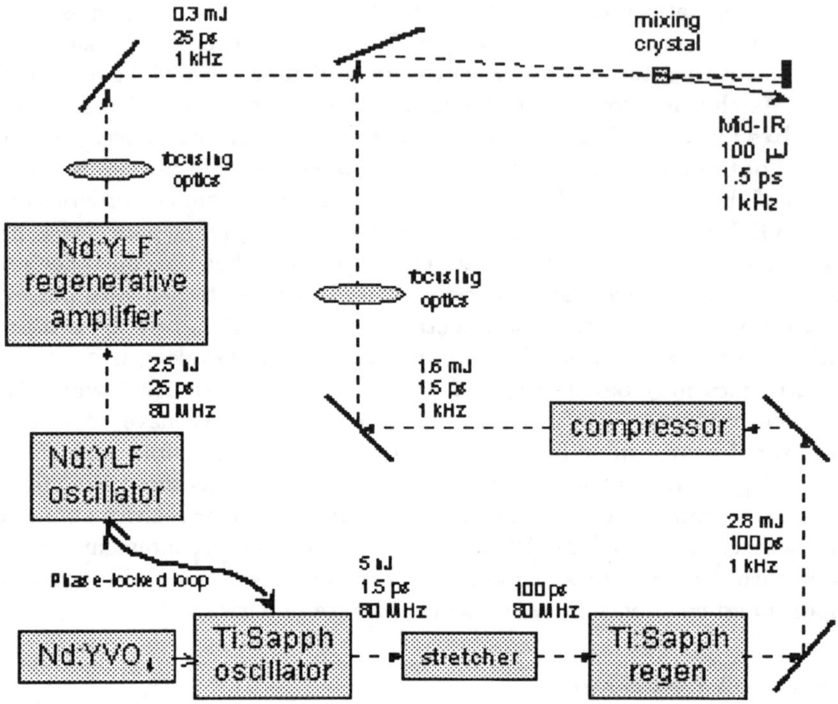

Figure 1. Block diagram of the mid-infrared source.

relaxed by using a signal seed pulse that is much longer than the pump pulse.[10] As the amplifiers introduce no additional jitter, the problem reduces to one of synchronizing the oscillators to much less than the signal pulse width. Both oscillators are locked to the same RF reference using a phase- locked loop that feeds back to the cavity lengths. The two oscillators were cross-correlated using sum-frequency generation in BBO, and the jitter found to be less than 5 psec (the resolution limit of the apparatus).

In all of the experiments described here, the three-wave mixing was done using type-II phase matching in a 5mm long KTP crystal cut in the X-Z plane at 40 degrees. KTA and $KNbO_3$ have also been used, with comparable conversion efficiency. At shorter pulse widths and longer wavelengths, the group velocity mismatch and transmission curves of the crystals dictates the best crystal choice. The pump and signal beams are focussed to 0.9 mm and 1.5 mm diameters respectively, with a 1 degree angle between them in the direction perpendicular to the phase-matching plane. The small angle facilitates the separation of the idler beam: the centers of the pump and signal beam are blocked 20 cm downstream, and residual pump and signal light is filtered out by collimating the idler with a silicon lens. The system produces 100 μJ, 1.5 psec pulses at 1 kHz, tunable between 3 and 4.1 μm, from inputs of 1.6 mJ in the pump beam and 0.3 mJ in the signal beam. The signal beam acts only as a seed, and is saturated at this level. The system's tunability can be extended to cover 2.5 - 5 μm with minor optics changes in the oscillator and regenerative amplifier. The Tsunami oscillator can also be easily configured to run at 100 fsec pulse widths. With a grating change in the stretcher/compressor, the system may then be used to produce 100 fsec pulses in the mid-IR.

3 Photoionization Experiments

The rescattering model's predictions about the gross structure of high harmonic and electron spectra follow from very simple considerations of an electron's motion in an oscillating field. An electron that never revisits the ion core can acquire a maximum drift energy of $2U_p$, if it does return, its maximum return energy is $3.2U_p$, while if it elastically scatters and is accelerated again in the field, the maximum drift energy that it can attain is $10U_p$. The well known $3U_p + I_p$ cutoff[11] in high harmonic spectra occurs at the maximum return energy. Photoelectron spectra of helium and neon in the tunneling regime show a dramatic change at $2U_p$, and the plateau shape observed in the high energy tail is given by modeling the rescattering of the returning electron wave packet.[4,5] In the case of the rare gases, the scattering potentials presented by the ion cores of the different atomic species do not differ enough to appreciably change the shape of the high-energy part ($>2U_p$) of the PES. In the alkali atoms there is more variation.

In Fig. 2, the PES of potassium and sodium excited by 3.2 μm radiation are compared. In the experiment, an atomic beam produced by a thermal source is directed through the focus of an electrostatic parabolic mirror, where it is crossed by the mid- IR laser beam. The mirror collects photoelectrons emitted over a 2π steradian solid angle, and the photoelectron energy is analyzed by time-of-flight. Using f/4 optics, the mid-IR beam may be focused to a waist diameter of $40 \pm 5\mu m$, to achieve a maximum intensity of 3 TW/cm^2. The absolute cross sections are not measured in the experiment, due to uncertainties in the beam densities and focal volumes, and the experimental curves

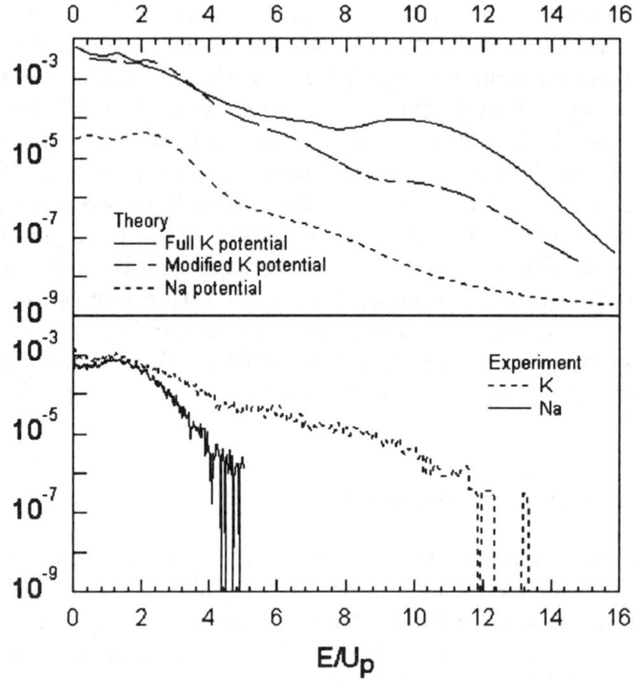

Figure 2. Theoretical and experimental photoelectron spectra for sodium and potassium excited by 3.2 μm radiation. The theoretical curves are calculated for a peak intensity of 1 TW/cm^2. The modified K potential is formed by replacing the potassium d potential with that of hydrogen. The K an Na experimental spectra are taken at 1.4(5) TW/cm^2 and 1.7(6) TW/cm^2 respectively. The experimental curves have been shifted on the y-axis to highlight the difference in shape. The electron energy is measured in units of the ponderomotive potential at the peak intensity at which the spectrum is calculated or measured.

are scaled to overlap in order to show the differences in the energy dependence.

The theoretical curves are obtained by a numerical integration of the Schrodinger equation using semi-empirically derived ℓ-dependent pseudopotentials for the ion cores,[12] and is discussed elsewhere.[6] The interesting difference between the potentials lies in an attractive well in the d potential of the potassium core, while the sodium is nearly hydrogenic. The s and p potentials of both atoms are more repulsive than the analogous hydrogen potentials and for $\ell > 2$, the potential is dominated by the centrifugal term. The attractive well has a dramatic effect on the shapes of the PES. In both the theory and experiment, the electron yield in sodium falls precipitously above $2U_p$, while in potassium it falls off much more slowly. The interpretation of this is that the attractive well greatly enhances the probability that an electron returning to the core will elastically scatter and contribute to the high-energy part of the spectrum. This is borne out in the theory by the fact that when the d potential in potassium is replaced with that of the hydrogen atom [Fig. 2(a)], the high- energy electron yield drops, and the PES resembles that of sodium.

4 High Harmonics and Attosecond Pulses

The possibility that the harmonic series produced in HHG might yield very short pulses was suggested[13] soon after HHG was first observed. Since HHG is a high order nonlinear process, it is to be expected that individual harmonics might be shorter than the driving pulse, but in addition, the pulse width might be further reduced if many harmonics could be coherently combined. The effectiveness of this of course depends on whether the harmonics are in phase, but it appears that theoretically there are circumstances in which this is the case. Antoine et al.[8] proposed a mechanism in which macroscopic phase matching could select only in-phase contributions to harmonics in the plateau region and create a terahertz train of subfemtosecond pulses. Schafer and Kulander [9] calculated that harmonics at the high energy end of the plateau have spectral phase chirps and relative phases that would permit their compression and combination to form a single subfemtosecond pulse. Evaluating and exploiting mechanisms such as these requires an experimental characterization of the coherence properties of the harmonics. Since this is difficult at the short wavelengths produced in HHG experiments driven by near-infrared sources, it seems useful to study these properties in the visible and UV by using a driving source in the mid- infrared.

Linearly polarized mid-IR light is focused by f/4 optics near the exit aperture of an alkali oven. There is no collimation of the atomic beam, and the width of the $\cos^3\theta$ density profile where it is intersected by the laser beam is 1mm. The atomic density is varied with the oven temperature over a range of 0.1-2 torr. The mid-IR beam is focused to $40 \pm 5\mu m$, which implies a confocal parameter of $\sim$3 mm. The interaction region is imaged with unit magnification onto the entrance slit of a 0.18 meter flat field air monochromator which has a gated, intensified CCD camera at its image plane. The collection system's absolute efficiency is calibrated at the third and fifth harmonics by

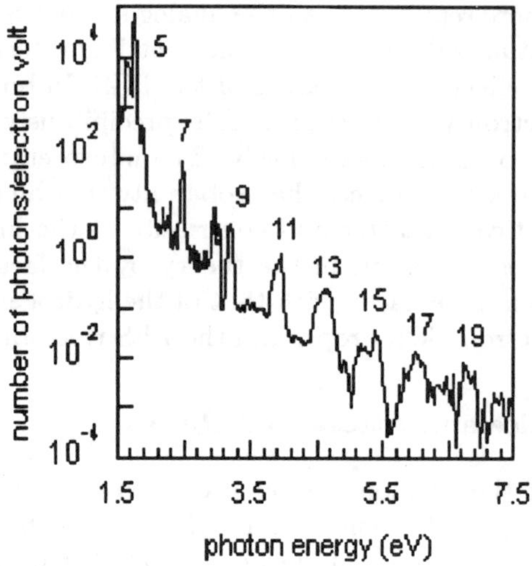

Figure 3. **The harmonic spectrum of Rubidium irradiated with a 3.6 μm, 1.9 psec pulse (peak intensity 2 TW/cm^2). Unlabeled lines are due to atomic fluorescence. The spectrum is not corrected for the spectral response of the collection system.**

comparison with a calibrated photodiode, and the relative throughput at other wavelengths determined from the emission spectrum of a mercury discharge lamp.

Figure 3 shows a sample spectrum of Rubidium irradiated by a 3.6 μm, 1.9 psec pulse. The distribution covers 9 orders of magnitude in signal, and is not adjusted for the spectral response of the collection system. The peak intensity of 2 TW/cm^2 is roughly twice the saturation intensity of 1.1(4) TW/cm^2, which was determined by the intensity dependence of the total photoionization and harmonic yield. There is a clear odd-harmonic series extending to the 19th harmonic (190 nm) which is the short- wavelength cutoff of our collection system. The $3U_p + I_p$ harmonic cutoff[11] would occur at the 23rd harmonic under these conditions.

We observed that at low atomic densities, the harmonic yield scaled with the square of the density, which implies macroscopic coherence. Near field images confirm that the emission is confined to within a beam waist of the interaction volume, and the harmonic yield shows a weak dependence on the

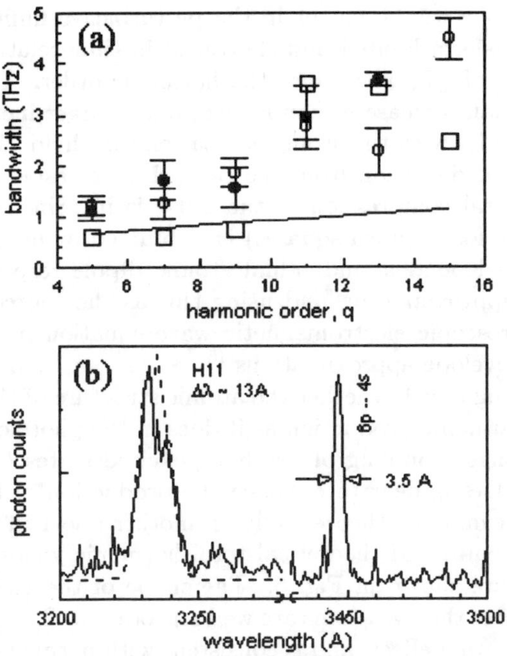

Figure 4. (a) Spectral linewidths of harmonics for Rb excited by 3.2 μm radiation (open circles) and K in a 3.6 μm field (filled circles). The open squares are the calculated linewidths of K at 3.6 μm. The solid line represents the line width associated with a purely perturbative broadening proportional to $\sqrt{q}$. (b) A typical measured (solid line) and calculated (dotted line) harmonic lineshape in K. The instrumental resolution is shown by the nearby $6p \rightarrow 4s$ fluorescence line.

position of the focal spot relative to the atomic distribution, suggesting that we are close to the optimal geometry of a loosely focused beam in which the confocal parameter is approximately equal to the length of the medium [14]. We find that, depending on the harmonic order, rubidium produces harmonics more efficiently than potassium, for the same atomic number density, by a factor of 3-6.

The lineshape of a high order harmonic (H11) produced in K is shown in Fig. 4(b). The bandwidth is nearly 4 times the instrumental bandwidth, which is shown by a nearby fluorescence line. The bandwidth as a function of harmonic order is shown in Fig. 4(a), for both Rb and K, at 3.2 and 3.6 μm respectively, near the saturation intensity. For comparison, we also show the bandwidth that one would expect in the perturbative limit, i.e. if the harmonics were transform-limited and shortened in time relative to the driving pulse by a factor of $\sqrt{q}$, where q is the harmonic order. The low harmonic orders show a small increase in bandwidth, but a larger increase occurs near H11, which coincides with the atomic ionization threshold. The high harmonics show a marked deviation from the perturbative result and suggest that significant temporal compression of the pulse is in principle possible. The theoretical values for K (open squares) are obtained by numerically calculating the intensity-dependent individual atomic dipole response in the single active electron approximation, and using this as the source term in a solution of the macroscopic electromagnetic wave equation in the paraxial and slowly-varying envelope approximations.[15]

It is interesting to ask whether the atomic structure of the alkalis plays as large a role in harmonic production as it does in the photoionization spectra. In fact, the stronger coupling of low-lying excited states (in comparison to the rare gases) does appear to enhance the production of harmonics in the plateau. This is examined theoretically in another paper in this volume.[6]

The experimental and theoretical total harmonic yields as a function of harmonic order are shown in Fig. 5. The shape of the yield curve and the relative efficiency of the two atoms are well reproduced. The Z-dependent efficiency is a single-atom effect, and is consistent with harmonic[14] and electron[5] distributions in rare gases. In the absolute photon yields, however, there is a large discrepancy (roughly a factor of 1000) between theory and experiment. Our calculations indicate that imperfections in the transverse mode of the mid-IR beam can dramatically reduce the harmonic yield. Our shot-averaged analysis of the beam spot quality does indicate a deviation from TEM_{00}, but not enough to completely explain this difference.

5 Summary

Strong field studies in the mid-IR show a number of interesting contrasts to studies at shorter wavelength. In photoionization, the electron spectra show a greater variation across atomic species. This may be explained by differences in the ion cores, which present very different potentials to returning electrons. The effects of these differences on the photoelectron spectra provide a detailed test of the rescattering model. The alkali atomic structure also plays a role in harmonic generation, where the stronger coupling of low-lying excited states enhances harmonic production at higher orders. We have produced harmonics of a mid-IR field in the visible and UV up to the 19th order, and found that spectral measurements are consistent with theory and with the possibility of

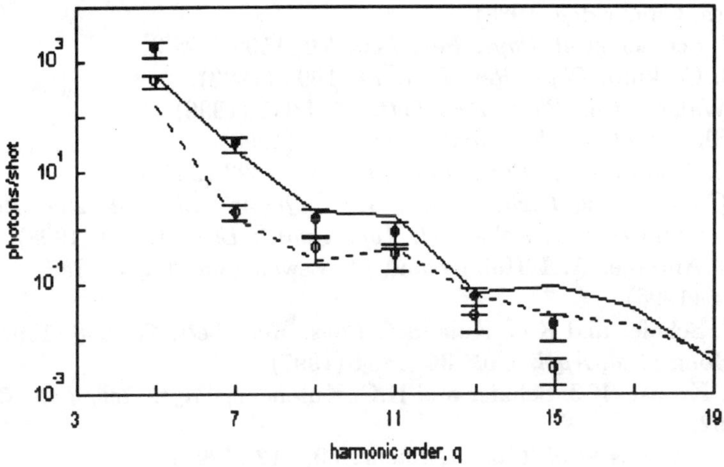

Figure 5. Photon yields as a function of harmonic order for K (open circles) and Rb (filled circles) in a 3.2 μm, 1.2 TW/cm^2 optical field. The calculated yields, scaled by a factor of 10^{-3} are shown by the solid (Rb) and dotted (K) lines.

pulse compression. The photon yield of harmonics is now sufficiently large to make temporal measurements, as well as measurements which yield complete amplitude and phase information, such as frequency-resolved optical gating.[16]

Acknowledgments

The experiments were carried out at Brookhaven National Laboratory under contract No. DE-AC02-98CH10886 with the U.S. Department of Energy and supported by its Division of Chemical Sciences, Office of Basic Energy Sciences and BNL/LDRD No. 99-56. P. A. acknowledge travel support from NATO under Contract No. SA.5-2-05(RG910678). K. J. S. and M. B. G. acknowledge support from the Louisiana State Board of Regents through Grant No. LEQSF96-99-RD-A-14, and from the Swedish National Science Research Council.

References

1. H B van Linden van den Heuvell and H G Muller, in *Multiphoton Processes-1988*, ed. S J Smith and P L Knight (Cambridge University

Press, Cambridge, 1988).
2. K.J. Schafer *et al*, *Phys. Rev. Lett.* **70**, 1599 (1993).
3. P.B. Corkum, *Phys. Rev. Lett.* **71**, 1994 (1993).
4. B. Walker *et al*, *Phys. Rev. Lett.* **77**, 5031 (1996).
5. B. Sheehy *et al*, *Phys. Rev. A* **58** 3942 (1998).
6. M.B. Gaarde *et al*, *Phys. Rev. Lett.* **84**, 2822 (2000).
7. Z. Chang *et al*, *IEEE J. Sel. Top. Quant. Elec.* **4**, 266 (1998); C. Spielmann *et al*, *IEEE J. Sel. Top. Quant. Elec.* **4**, 249 (1998).
8. Ph. Antoine, A. L'Huillier and M. Lewenstein, *Phys. Rev. Lett.* **77**, 1234 (1996).
9. K.J. Schafer and K.C. Kulander, *Phys. Rev. Lett.* **78**, 638 (1997).
10. J. Hong *et al*, *Appl. Opt.* **36**, 1894 (1997).
11. J.L. Krause, K.J. Schafer and K.C. Kulander, *Phys. Rev. Lett.* **68**, 3535 (1992).
12. W.J. Stevens *et al*, *Can. J. Chem.* **70**, 612 (1992).
13. S.E. Harris, J.J. Macklin and T.W. Hansch, *Opt. Commun.* **100**, 487 (1993); G. Farkas and C. Toth, *Phys. Lett. A* **168** 447 (1992).
14. A. L'Huillier *et al*, *J. Nonlinear Opt. Phys. Mater.* **4**, 647 (1995).
15. M.B. Gaarde *et al*, *Phys. Rev. A* **57** 4553 (1998).
16. R. Trebino *et al*, *Rev. Sci Instrum.* **68**, 3277 (1977).

NEW WAY OF CONTROLLING MOLECULAR PROCESSES BY LASERS

YOSHIAKI TERANISHI* AND KUNINOBU NAGAYA
Department of Functional Molecular Science,
The Graduate University for Advanced Studies

HIROKI NAKAMURA
Department of Functional Molecular Science,
The Graduate University for Advanced Studies,
and
Department of Theoretical Studies,
Institute for Molecular Science,
Myodaiji, Okazaki 444-8585, Japan

Two new ways of controlling molecular processes are proposed. One is to sweep laser frequency and/or intensity at avoided crossings among dressed states to control nonadiabatic transitions there. The second is to use the intriguing phenomenon of complete reflection in the nonadiabatic tunneling type transition in the time-independent framework. The newly completed semiclassical theory of nonadiabatic transitions can give a nice analytical formulation for these.

1 Introduction

Control of molecular processes by lasers has attracted much attention recently,[1,2] and several schemes have been proposed such as coherent control, pump-dump method, pulse-shape driven method, adiabatic rapid passage with linear charp, and π-pulse.[3,4,5,6,7,8,9,10,11,12,13] Considering the importance of nonadiabatic transitions due to curve crossing not only in ordinary molecular processes but also in molecular processes induced by lasers, we have proposed a new way of controlling nonadiabatic transitions by sweeping laser frequency and/or intensity at curve crossings among the dressed (or Floquet) states.[14,15] By using the newly developed time-dependent theory of nonadiabatic transitions,[16] we can analytically formulate appropriate control conditions. On the other hand, the intriguing phenomenon of complete reflection in the nonadiabatic tunneling (NT) type transition in the time-independent framework has been utilized to propose a new mechanism of molecular switching[17,18,19] and a new possibility of controlling molecular photodissociation.[20] These new ideas are briefly reviewed in this report.

*PRESENT ADDRESS; RIKEN, HIROSAWA 2-1, WAKO 351-0198, JAPAN

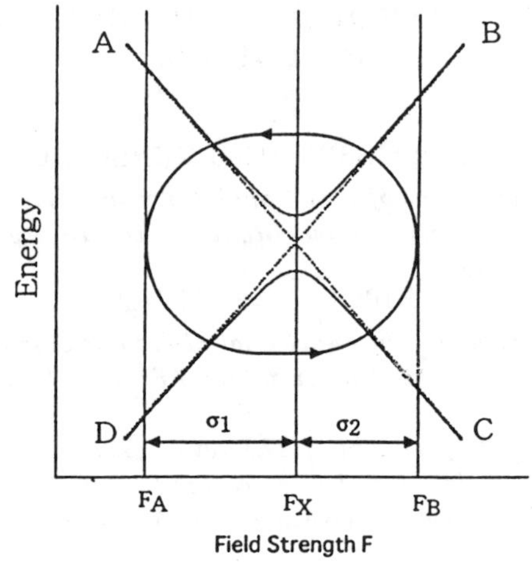

Figure 1. Schematic two diabatic (dotted lines) and two adiabatic (solid lines) potentials as a function of time-dependent field parameter F.

2 Time-Dependent Theory of Nonadiabatic Transition

Based on the complete solutions of the time-independent Landau-Zener-Stueckelberg type curve crossing problems obtained by Zhu and Nakamura,[21,22] we have formulated a new time-dependent theory[16] which can cover the cases of two closely lying crossings, tangential touching, and diabatically avoided crossing of potentials (see Fig.2).

The basic quantity is the following I-matrix which describes the transition amplitude from $F_A = F(t = t_A)$ to $F_B = F(t = t_B)$ in Fig.1, where F is the time-dependent field parameter,

$$I = \begin{pmatrix} \sqrt{1-p}e^{i(\varphi+\sigma_1/2+\sigma_2/2)} & \sqrt{p}e^{i(\phi-\sigma_1/2+\sigma_2/2)} \\ -\sqrt{p}e^{-i(\phi-\sigma_1/2+\sigma_2/2)} & \sqrt{1-p}e^{-i(\varphi+\sigma_1/2+\sigma_2/2)} \end{pmatrix}, \tag{2.1}$$

where

$$\sigma_1 = \int_{t_A}^{t_X} \Delta E(t)dt, \quad \sigma_2 = \int_{t_X}^{t_B} \Delta E(t)dt, \tag{2.2}$$

$$\varphi = \phi_s = \frac{\delta}{\pi}\ln\left(\frac{\delta}{\pi}\right) - \frac{\delta}{\pi} - \arg\Gamma\left(i\frac{\delta}{\pi}\right) - \frac{\pi}{4}, \tag{2.3}$$

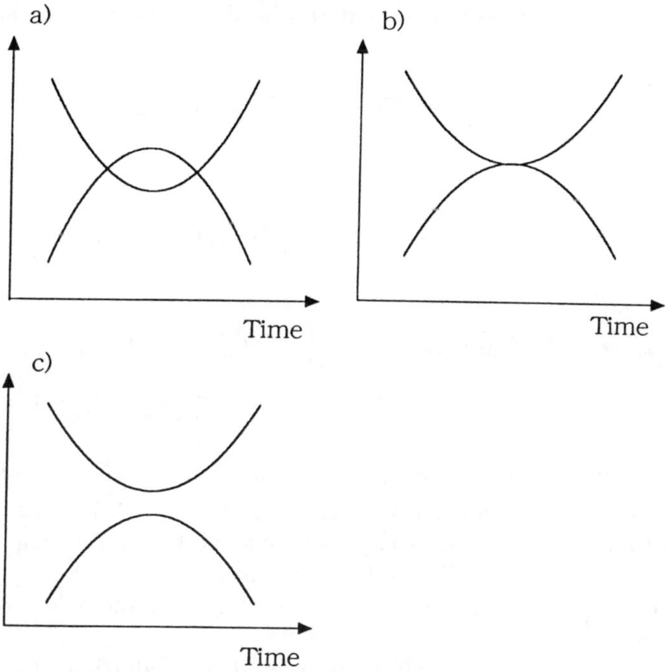

Figure 2. Schematic diabatic potentials. a) two crossings, b) tangential touching, c) diabatically avoided crossing.

$$p = \exp\left[-\frac{\pi}{4\sqrt{\alpha\beta}}\left(\frac{2}{1+\sqrt{1+\beta^{-2}(0.4\alpha+0.7)}}\right)^{1/2}\right], \qquad (2.4)$$

$$\phi = \sigma_0, \;\; \sigma_0 + i\delta = \frac{1}{2\sqrt{\alpha}}\int_0^i \left(\frac{1+\zeta^2}{\zeta+\beta}\right)^{1/2} d\zeta, \qquad (2.5)$$

and $\Delta E(t)$ represents the adiabatic energy difference. The parameters α and β are defined from the original diabatic potentials ($\varepsilon_1 = \alpha_1 t^2 + \beta_1$ and $\varepsilon_2 = \alpha_2(t-\gamma)^2 + \beta_2$) and the diabatic coupling V_0 as

$$\alpha = \frac{(\alpha_1 - \alpha_2)\hbar^2}{8V_0^3}(>0), \;\; \beta = \frac{1}{2V_0}\left(\beta_2 - \beta_1 + \frac{\alpha_1\alpha_2\gamma^2}{\alpha_1 - \alpha_2}\right). \qquad (2.6)$$

In general, these parameters α and β can be determined from the shape of $\Delta E(t)$ as

$$\alpha = \frac{\hbar^2}{(\Delta E_m)^3 t_m^2}\sqrt{(\Delta E_M - \Delta E_m)(\Delta E_M + \Delta E_m)} \tag{2.7a}$$

and

$$\beta = \frac{1}{\Delta E_m}\sqrt{(\Delta E_M - \Delta E_m)(\Delta E_M + \Delta E_m)} \tag{2.7b}$$

with

$$\Delta E_M = \Delta E(t_M),\ \Delta E_m = \frac{1}{2}\left(\Delta E(t_m^{(+)}) + \Delta E(t_m^{(-)})\right),$$
$$t_m = \frac{1}{2}\left(t_m^{(+)} - t_m^{(-)}\right), \tag{2.8}$$

where t_M corresponds to the local maximum position of $\Delta E(t)$ and $t_m^{(\pm)}(t_m^{(+)} > t_m^{(-)})$ are the two minimum positions. These expressions (2.7a) - (2.8) are more convenient than Eq.(2.6), since the parameters can be directly evaluated from adiabatic potentials even if the diabatic potentials are not exactly quadratic. Furthermore, Eqs.(2.7a) and (2.7b) can actually give more accurate results than Eq.(2.6).

The transition matrix T after one period of oscillation ($F_A \to F_X \to F_B \to F_X \to F_A$) is given by

$$T = I^t I = \begin{pmatrix} [p + (1-p)e^{2i\psi}]e^{-i\sigma} & -2i\sqrt{p(1-p)}\sin\psi \\ -2i\sqrt{p(1-p)}\sin\psi & [p + (1-p)e^{-2i\psi}]e^{i\sigma} \end{pmatrix}, \tag{2.9}$$

where

$$\psi = \varphi + \phi + \sigma_2,\ \ \sigma = 2\phi + \sigma_2 - \sigma_1. \tag{2.10}$$

This transition corresponds to the one passage of the two closely lying crossings ($\beta > 0$). In the case of diabatically avoided crossing($\beta < 0$), Eq.(2.9) still holds with the proper replacement of some quantities. We do not go into the details here.

In the case of Rosen-Zener type of transition, we have

$$p = \frac{1}{1 + \exp[\delta_{RZ}]}, \tag{2.11}$$

$$\varphi = \gamma(2\delta_{RZ}) - \gamma(\delta_{RZ}), \tag{2.12}$$

$$\phi = -\varphi + \sigma_0, \tag{2.13}$$

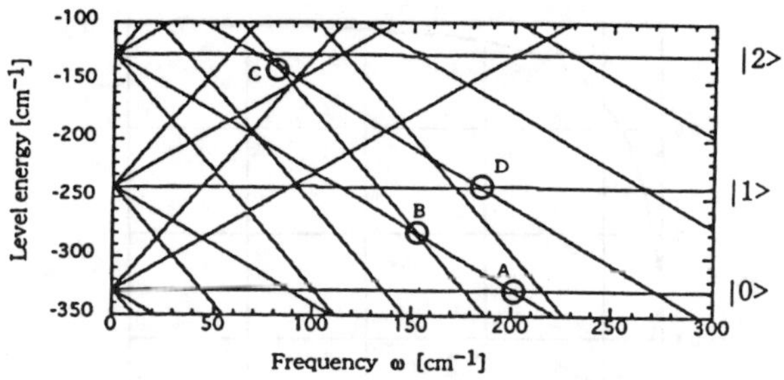

Figure 3. Floquet-state diagram, i.e. vibrational levels of Fig.5 as a function of laser frequency.

$$\delta_{RZ} = \pi\Delta/\hbar\lambda, \tag{2.14}$$

where Δ is the constant diabatic energy difference, λ is the exponent of the diabatic coupling $\mathrm{sech}(\lambda t)$ and $\gamma(X)$ is given below.

In some cases it is useful to use the exponential potential model given by

$$\begin{pmatrix} U_1 + V_1 e^{-\kappa t} & V e^{-\kappa t} \\ V e^{-\kappa t} & U_2 + V_2 e^{-\kappa t} \end{pmatrix}. \tag{2.15}$$

In this case we have

$$p = e^{-\pi\delta_2} \frac{\sin\mathrm{h}(\pi\delta_1)}{\sin\mathrm{h}[\pi(\delta_1 + \delta_2)]} \tag{2.16}$$

$$\varphi = -\gamma(\delta_1 + \delta_2) + \gamma(\delta_2) \tag{2.17}$$

$$\phi = \gamma(\delta_1) - \gamma(\delta_1 + \delta_2) - 2\left[\sqrt{(\delta_1 + \delta_2)\delta_2} + \frac{\delta_1}{2}\ln\frac{\sqrt{\delta_1 + \delta_2} - \sqrt{\delta_2}}{\sqrt{\delta_1 + \delta_2} + \sqrt{\delta_2}}\right], \tag{2.18}$$

where

$$\delta_{1(2)} = \frac{U_1 - U_2}{2\hbar\kappa}\left[1 + (-)\frac{(V_1 - V_2)/2V}{\sqrt{1 + [(V_1 - V_2)/2V]^2}}\right] \tag{2.19}$$

and

$$\gamma(X) = X\ln X - X - \arg\Gamma(iX). \tag{2.20}$$

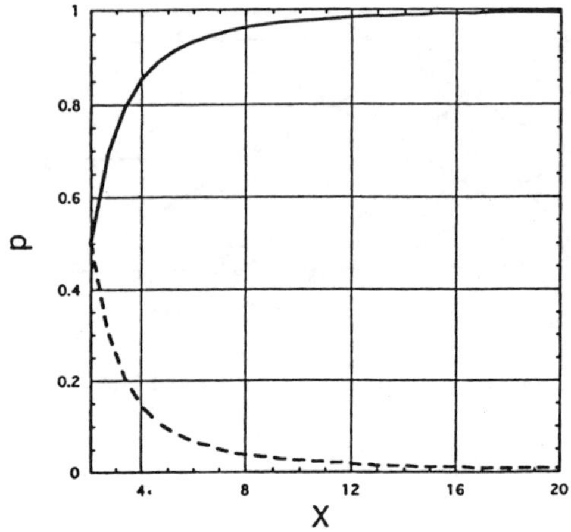

Figure 4. The range of p to fulfill the complete control by n-times of field oscillation. $X = 2n$ for n-period ($2n + 1$ for n and half-period).

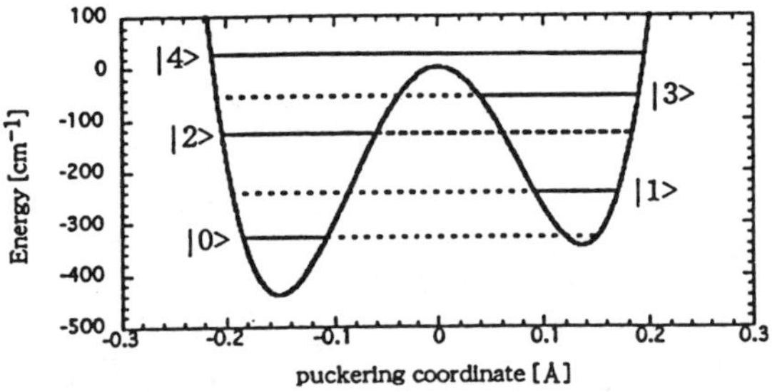

Figure 5. A model double well potential of the puckering of trimethylenimine.

3 Control of Molecular Processes by Sweeping Lasers

We start with the dressed (or Floquet) state picture as a function of laser frequency ω (see Fig.3), in which there appear many avoided crossings, the separations of which are determined by the laser intensity and the transition dipole moment. The horizontal lines represent the energy levels without any absorption or emission of photons; and the lines with positive (negative) slopes correspond to the states which absorbed (emitted) photons. In each avoided crossing we have a situation like in Fig.1. Our basic idea is to control the nonadiabatic transition there as we desire by sweeping the laser frequency periodically. If the energy separation ΔE between the adiabatic Floquet states is large enough at the avoided crossing point, the nonadiabatic transition probability p for one passage of the crossing point is small and thus we can make a transition by slowly following one of the adiabatic states. This is nothing but the adiabatic rapid passage method, and naturally requires a strong laser. By sweeping the laser periodically, we can relax this condition very much. After n periods of oscillation, the overall probability of the transition $1 \to 2, P_{12}^{(n)} = |(T^n)_{12}|^2$, is given by

$$P_{12}^{(n)} = 4\frac{\sin^2(n\xi)}{\sin^2\xi}p(1-p)\sin^2\psi, \tag{3.1}$$

where

$$\cos\xi = (1-p)\cos(2\psi - \sigma) + p\cos\sigma. \tag{3.2}$$

Since $P_{12}^{(n)} = 1$ means $P_{12}^{(2n)} = 0$, we can have the following two conditions for $P_{12}^{(n)} = 1$ to be satisfied:

$$\sin^2(n\xi) = 1 \tag{3.3}$$

and

$$4p(1-p)\sin^2\psi = \sin^2\xi. \tag{3.4}$$

For a given n, ξ is obtained from Eq.(3.3) and ψ can be determined according to Eq.(3.4) for given ξ and p. From the definitions of ψ and σ, we can find the proper values of σ_1 and σ_2 which can be adjusted by changing F_A and F_B. We can formulate similar conditions for n and half periods of oscillation, in which case we end up with the state B or C in Fig1. The number of the oscillation is better to be set as small as possible for a given p. This condition is shown in Fig.4. The range between solid and dashed lines represents the range in which the above mentioned complete control is achievable by n-times of field oscillation. The abscissa X equals to $2n$ in the case of n-period ($2n+1$ for n and half period). The one period of oscillation with $F_B > F_X$, $F_B = F_X$, and $F_B < F_X$ corresponds to the two crossings, tangential touching, and diabatically avoided crossing in Fig.2, respectively.

As an application, we consider the one-dimensional model of puckering isomerization of trimethylenimine.[8] Fig.5 is the model double well potential and Fig.3 is the corresponding Floquet diagram. Since the transition dipole moment between $|0>$ and $|1>$ is too small, it is better to use the consecutive transitions $|0> \rightarrow |2> \rightarrow |4> \rightarrow |3> \rightarrow |1>$; that is, we use the avoided crossings marked in Fig.3. The numerical results are shown in Fig.6, in which the constant laser intensity and the one period of frequency sweeping are used. If the constant laser intensity is difficult to maintain, we can use the exponential potential model, Eqs.(2.15)-(2.20), the results of which are shown in Fig.7.

Instead of frequency sweeping, we can sweep the laser intensity and use the Rosen-Zener type transition, in which the diabatic energy difference is constant and the coupling (laser intensity) is an exponential or a sech function. Or more generally, we can combine them and find the best way. Since nonadiabatic transitions play important roles in various molecular processes in external fields, our idea of controlling nonadiabatic transitions would give a versatile control scheme of molecular processes. In the case of dynamic processes like photodissociation, we have to carry out wave packet propagation calculations, but the analytical estimates of the various parameters can still be done.

4 Complete Reflection Phenomenon and Control of Molecular Processes

4.1 Complete Reflection

In the case of nonadiabatic tunneling (NT) type of transition in the time-independent framework in which the two diabatic curves cross with different signs of slopes as a function of spatial coordinate and a potential barrier is created, complete reflection occurs at certain discrete energies.[21,22] The semiclassical theory explicitly gives the transmission probability as

$$P_{trans.} = \frac{4\cos^2\phi}{4\cos^2\phi + p^2/(1-p)}, \tag{4.1}$$

where p and ϕ represent, respectively, the nonadiabatic transition probability for one passage of the crossing point and the phase along the upper adiabatic potential between the two turning points with a correction due to nonadiabatic coupling. The complete reflection ($P_{trans.} = 0$) occurs at energies which satisfy $\phi = (n + \frac{1}{2})\pi (n = 0, 1, 2)$.

4.2 Molecular Switching

In a periodic system of the NT-type potential units we can utilize the complete reflection phenomenon to propose a new type of molecular switching.[17,18,19] In a periodic potential system, we always have complete

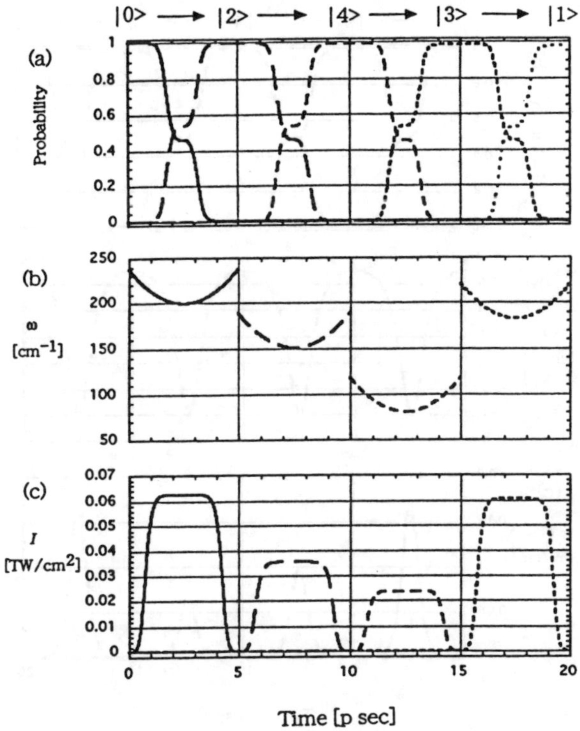

Figure 6. Control of the puckering isomerization process $|0\rangle \rightarrow |2\rangle \rightarrow |4\rangle \rightarrow |3\rangle \rightarrow |1\rangle$. The constant intensity and the quadratic variation of frequency are used.

transmission at certain discrete energies. Thus, by somehow modifying the potentials of one or more units in the system so that the new complete reflection position coincides with the original complete transmission position, we can, in principle, switch on and off the transmission. Fig.8 demonstrates this switching in one-dimensional model system.[23] Two NT-type of potential units are placed at the origin and the wave packet is sent in from the right. The one unit is composed of two linear potentials coupled by a quadratic potential, and the coupling strength in one of the two units is appropriately modified to create the complete reflection condition.[18] The total length of the potential system is $4a_0$. The mean energy of the wave packet is initially set to the complete transmission position. After the switching at $t = t_s$, the residual two-thirds of the wave packet are completely reflected back. The switching was demonstrated

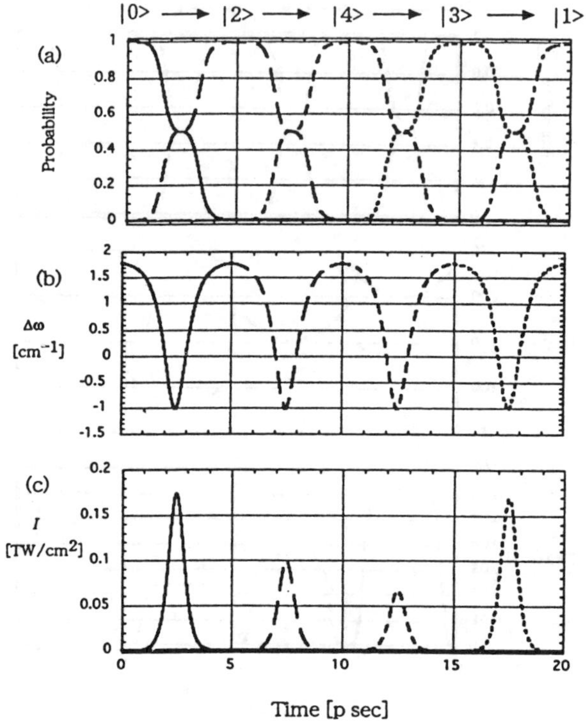

Figure 7. The same as Fig.6. The exponential model is employed.

to be possible even in a two-dimensional constiriction model.[19]

4.3 Control of Molecular Photodissociation

The complete reflection phenomenon may be used to control molecular photodissociation by using the NT-type potential surface crossing between a dressed up ground and an appropriate excited electronic state. In this case we simply apply the stationary laser field and try to create the complete reflection condition in a certain dissociation channel so that we can dissociate the molecule to the other channel as we desire. If this were possible, we could dissociate a molecule into such a channel that can never be reached by the ordinary photodissociation because of a potential barrier in that channel of the excited electronic state. Fig.9 shows numerical results for a two-dimensional

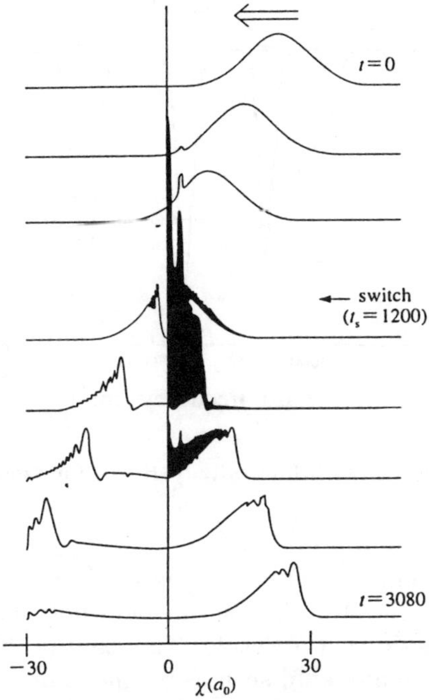

Figure 8. One-dimensional model of molecular switching. The potential system is put at the origin.

model of CH_3SH which usually does not dissociate to CH_3+SH because of the transition state of excited electronic state located in this dissociation channel.[20] The dissociation probability, which is actually the dissociation flux accumulated over one psec time scale, depicts dips at certain laser frequencies where the complete reflection is expected to occur. This kind of control is, unfortunately, not possible for the ground vibrational state, and actually Fig.9 is the case of 123rd vibrational state which corresponds to $v = 8$ local mode in the S-H bond. In this model there is no mode coupling in the ground electronic state and the wave packet stays inside. If there is a mode coupling, however, the dissociation to CH_3+SH is possible via the ground electronic state.

5 Concluding Remarks

Our scheme of sweeping lasers to control molecular processes is expected to be quite versatile, actually presenting a generalization of the adiabatic rapid

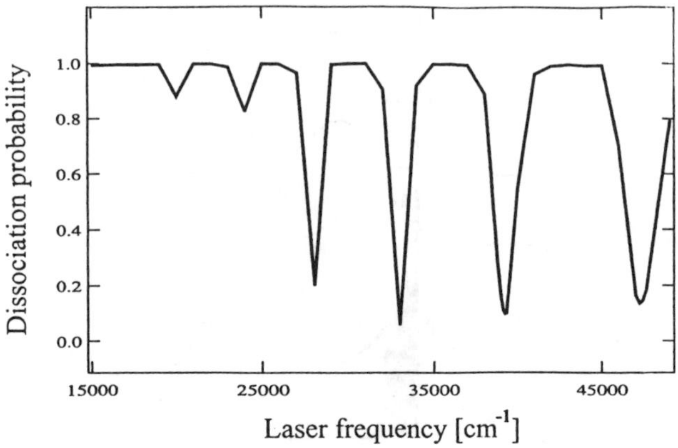

Figure 9. Dissociation of CH_3SH into CH_3S+H as a function of laser frequency.

passage and the chirping. This method is based on the phase interference between the phases induced by lasers, but does not require any information about the phase intrinsic to a molecule. The external field is not necessarily a laser field, but can be any kind such as magnetic or electric field.

The phenomenon of complete reflection is always possible in one-dimensional system, but is quite sensitive to the potential energy surface topography in multidimensional system, unfortunately. The phenomenon is, however, quite unique and presents an interesting possibility of control, if the conditions are satisfied.

References

1. R. J. Gordon and S. A. Rice, *Annu. Rev. Phys. Chem.* **48**, 601 (1997).
2. A. D. Bandrauk, Molecules in Laser Fields, (Marcel Dekker, New York, 1994).
3. P. Brumer and M. Shapiro, *Annu. Rev. Phys. Chem.* **43**, 257 (1992).
4. D. J. Tannor and S. A. Rice, *J. Chem. Phys.* **83**, 5013 (1985).
5. Kosloff, S. A. Rice, P. Gaspard, S. Tersigni, and D. J.Tannor, *Chem. Phys.* **139**, 201 (1989).
6. D. Neuhauser and H. Rabitz, *Acc. Chem. Res.* **26**, 496 (1993).
7. B. Kohler, J. L. Krause, F. Raksi, K. R. Wilson, V. V. Yakovlev, and R. M. Whitnell, *Acc. Chem. Res.* **28**, 133 (1995).
8. M. Sugawara and Y. Fujimura, *J. Chem. Phys.* **100**, 5646 (1994).

9. S. Chelkowski and A. D. Bandrauk, *J. Chem. Phys.* **99**, 4279 (1993).
10. M. M. T. Loy, *Phys. Rev. Lett.* **32**, 814 (1974).
11. J. S. Melinger, S. R. Gandhi and W. S. Warren, *J. Chem. Phys.* **101**, 6439 (1994).
12. S. Guerin, *Phys. Rev. A.* **56**, 1458 (1997).
13. K.Mishima and K.Yamashita, *J. Chem. Phys.* **109**, 1801 (1998).
14. Y. Teranishi and H. Nakamura, *Phys. Rev. Lett.* **81**, 2032 (1998).
15. Y. Teranishi and H. Nakamura, *J. Chem. Phys.* **111**, 1415 (1999).
16. Y. Teranishi and H. Nakamura, *J. Chem. Phys.* **107**, 1904 (1997).
17. H.Nakamura, *J. Chem. Phys.* **97**, 256 (1992).
18. S. Nanbu, H. Nakamura and F. O. Goodman, *J. Chem. Phys.* **107**, 5445 (1997).
19. H. Nakamura, *J. Chem. Phys.* **110**, 10253 (1999).
20. K. Nagaya, Y. Teranish, and H. Nakamura, (in preparation).
21. H. Nakamura, Dynamics of Molecules and Chemical Reactions, edited by R. E. Wyatt and J. Z. H. Zhang (Marcel Dekker, New York, 1996), p.473.
22. H. Nakamura and C. Zhu, *Comments At. Mol. Phys.* **32**, 249 (1996).
23. H. Nakamura, *Butsuri* **51**, 829 (1996) [in Japanese].

Part V. Condensed Phases

COHERENT CONTROL OF MOLECULAR EXCITATIONS IN THE CONDENSED PHASE USING OPTICAL HETERODYNE DETECTED RAMAN INDUCED KERR EFFECT SPECTROSCOPY

VALERIA D. KLEIMAN, DALE McMORROW AND JOSEPH S. MELINGER
Naval Research Laboratory
Code 6820, 4555 Overlook Ave. Washington D.C. 20375
e-mail:valeriak@ccs.nrl.navy.mil

We explore vibrational coherences in condensed phase molecules which are created by simple phase modulated femtosecond laser pulses. Experiments are performed using optical heterodyne detected Raman-induced Kerr effect spectroscopy, which independently detects the real (birefringence) and imaginary (dichroism) components of the third order molecular response. For an excitation pulse with a specific duration and phase modulation, we find the experimental signals arising from the transient birefringence and dichroism to be qualitatively different. The experimental results are interpreted using a wavepacket picture.

1. Introduction

The manipulation of molecular dynamics using phase-modulated fields has experienced significant interest in recent years.[1] One strategy used to control molecular reactions, first proposed by Rice and Tannor,[2] is based on the creation of non-stationary wavepackets using ultrafast excitation pulses. The dynamics of the excited wavepackets can be manipulated by the use of characteristic phase-modulated fields in the excitation process.

It has been demonstrated that population transfer between electric-dipole coupled electronic states can be controlled in a way that depends on the sign of the chirp of an excitation pulse.[3,4] This naturally leads to questions regarding the conditions under which the excitation of specific vibrational coherences, either on the excited or ground state potential energy surfaces, can be controlled using phase modulation techniques.

Theoretical calculations and experimental measurments have been very fruitful in the gas phase,[3-7] but relatively few experiments have been performed on molecules in solution, where most chemical reactions occur. Recently, there has been a growing interest in utilizing ultrafast phase-modulation techniques to manipulate excitations in the condensed phase. [8-10] Tailoring an excitation pulse so that a specific vibrational mode can be enhanced or suppressed is of vital interest when studying molecular systems in solution.

Phase properties of ultrashort pump pulses can greatly influence the excitation of molecules in solution.[4,9,11,12] For phase-modulated excitation pulses, an impulsive picture of the field-matter interaction can yield an incorrect interpretation of the molecular response since phase modulation (chirp) in the pump pulse changes the arrival time of each of the individual frequency components. Consider excitation with a transform-limited pulse: all colors of the broadband ultrafast pulse arrive at the sample simultaneously. In turn, all components of the ground-state wave function are simultaneously transferred to the excited-state surface. This simple picture breaks down for excitation with phase-modulated pulses. When excitation is performed with a chirped pulse, the different components of the wavepacket are created at different times. As the system evolves, interference among those components can greatly influence the shape of the wavepacket, and thus, the region of the potential energy surface it accesses at a given time. The interpretation of the system-response must include a dynamical wavepacket that is evolving in time while it is being created.

Controlling wavepacket dynamics using ultrashort phase-modulated pulses might prove to be a useful method to selectively excite vibrational coherences in either the ground or electronically excited manifold. Previous experimental work[8,13] has shown that a resonant, positive-chirp excitation pulse preferentially creates a wavepacket on the excited state surface. An equivalent negative-chirp pulse returns a fragment of the wavepacket amplitude back to the ground electronic state manifold, thus creating ground state vibrational coherences.

In this work we explore how chirped ultrafast pulses excite vibrational coherences in condensed phase molecules. We use optical heterodyne detected Raman-induced Kerr effect spectroscopy (OHD RIKES) to detect the vibrational coherences created. This experimental arrangement has the ability to detect selectively the real (birefringence) and imaginary (dichroism) components of the third-order susceptibility. We show that the transient birefringence and dichroism signals generated by chirped pulse excitation can yield qualitatively different signatures relating to the vibrational coherences produced. These qualitative differences add further insight into the wavepacket dynamics arising from ultrashort pulse excitation.

2. Experimental Section

The experimental arrangement involves measuring the transient anisotropy induced by phase-modulated femtosecond laser pulses. The sample consists of the dye molecule IR 125 (Exciton) dissolved in ethylene glycol with an optical density of approximately 1.0 near its absorption maximum (790 nm). The sample solution was circulated through a 250 μm thick nozzle.

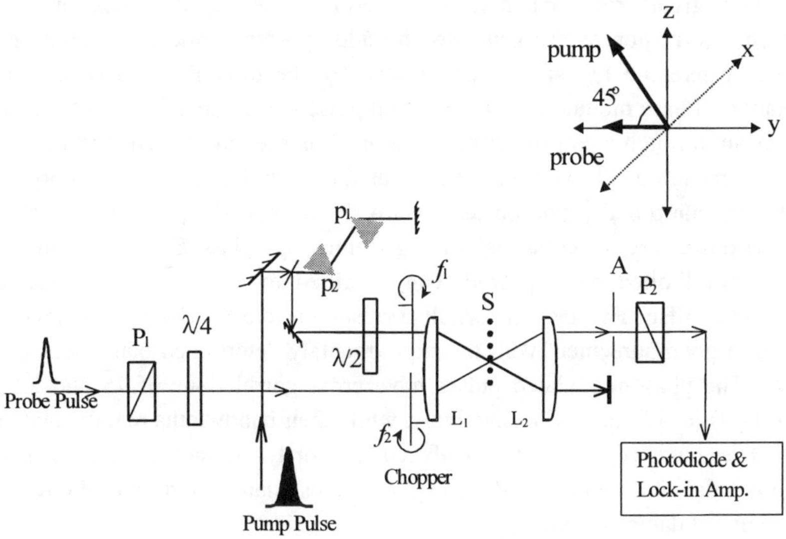

Figure 1. Experimental setup. P_1,P_2: crossed Glan-Taylor polarizers; p_1, p_2: prisms for phase modulation; λ/4, λ/2: wave-plates; L_1, L_2: lenses; S: 250μm sample jet; A:aperture; f_1,f_2: chopper frequencies.

Experiments are performed using the compressed output of a homemade self mode-locked Ti:Sapphire laser that generates sub-30 fs (FWHM) pulses centered at 790 nm at a repetition rate of 82 MHz. The OHD-RIKES experimental setup is shown in Figure 1, and has been used extensively to interrogate the low frequency intermolecular dynamics of transparent liquids.[14,15] The sample is located between crossed polarizers (P_1 and P_2), with the pump beam polarized 45° relative to the probe. Pump and probe pulses are overlapped at the sample using a 65 mm focal length achromatic lens. Pulse excitation energies at the sample position do not exceed 0.7 nJ/pulse.

The excitation pulse induces a time-dependent anisotropy in the sample absorption coefficient and refractive index. The anisotropy's temporal evolution is measured by the transmission of a weak probe pulse as a function of the optical delay between pump and probe pulses. Pump and probe beams are independently chopped, and the signal is detected at the difference frequency (Δf) using a photodiode and lock-in amplifier combination. A slight rotation of P_1 (P_2) introduces a local oscillator (LO) that is 90° out-of-phase (in-phase) with the probe field. This LO field coherently mixes with the signal field at the detector, giving rise to a heterodyne term in the signal expression.[14] Pure heterodyne responses are generated by adding scans collected with each LO sense, where the LO sign is determined by the direction of rotation of the polarizer. Phase modulation on the pump pulse is introduced by double passing the beam through a pair of prisms positioned on the pump beam pathway. Near transform-limited (TL) pulses are generated by minimizing the cross correlation between pump and probe pulses. Positively chirped [PC] (negatively chirped [NC]) pulses are generated by adding (removing) glass from the beam. Near transform limited pulse profiles can be fitted to a $sech^2$ pulse shape. The resulting 35 fsec FWHM pulsewidth corresponds to a FWHM bandwidth of 535 cm^{-1}, in good agreement with the experimentaly determined bandwidth of 515 cm^{-1}. The phase-modulated pulses have cross correlations of 75 and 82 fsec (FWHM) for PC and NC respectively, while their bandwidths remain unaltered.

The data collected are analyzed by first subtracting the long time exponential contributions and then fitting the oscillatory portion of the response to a sum of damped sinusoids:

$$S(t) = \sum_i \sin(\omega_i t + \theta_i) * e^{-t/\tau_i} ,$$

where ω_i is the central frequency, θ_i is the phase, and τ_i is a lifetime. The sinusoidal waves are fitted one at the time, starting with the frequency having the longer decay time, until the remaining data lie within the experimental noise level. When glass is introduced to (removed from) the pump beam pathway, the temporal position of the beams is changed and any absolute phase information is lost. Thus, it is only possible to compare phase values of individual sinewaves by referencing to one of the frequency components[‡].

3. Results and Discussion

The electronically resonant OHD-RIKES experiment contains several contributions to the third order molecular response function. Each contribution is manifested in the decay of the anisotropy of the birefringence and dichroism.

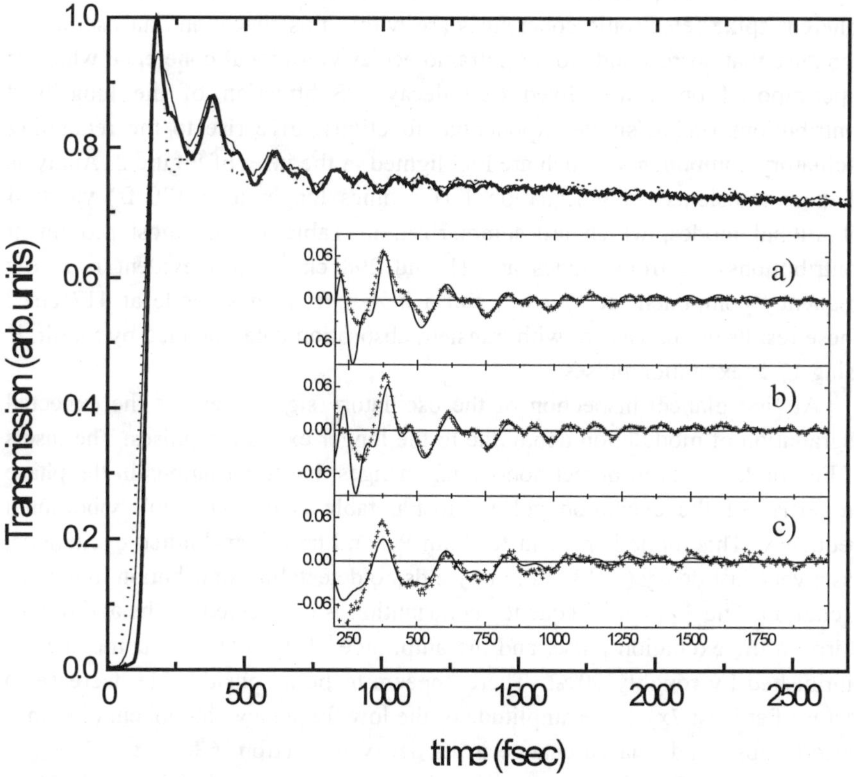

Figure 2. Transient Birefringence. a) —— Positive chirp; b) ▬ No chirp; c) .... Negative chirp of IR125 in ethylene glycol. Insert shows the oscillatory components with their respective fits (see text).

One component arises from the electronic hyperpolarizability, and it will be observed only when excitation and probe pulses are present simultaneously. Another component originates from the reorientation of electronically excited molecules. We are concerned in this work with the signal contributions that arise from intramolecular vibrational degrees of freedom which yield the vibrational coherences. The presence of an electronic resonance facilitates coherently excited vibrational modes of the ground and/or electronically excited state to be created through Raman interactions.

Figure 2 presents the normalized transient birefringence response of IR125 in ethylene glycol excited with transform limited, positively-chirped, and negatively-chirped pulses, respectively. Each data set shows a fast rise time, a coherent spike (electronic coherences are within this time), and an oscillatory signature that corresponds to the intramolecular vibrational coherence which is superimposed on a long-lived (ns) decay. Subtraction of the long-lived contributions (using single exponential functions) give rise to the remaining oscillatory components, which are highlighted in the inset of Figure 2. Analysis of these transients using Equation 1 (for times longer than 130 fs) yields 4 vibrational modes, which are summarized in Table 1. The most prominent contributions are from modes at 141 and 298 cm^{-1}. Also evident is a low frequency component at 17 cm^{-1}, and a higher frequency mode at 417 cm^{-1}. These results are consistent with transient absorption data obtained by Smolrz[16] using 17 fs excitation pulses.

At first glance, inspection of the oscillatory signals reveals the expected degradation of modulation depth due to the longer excitation pulses. The insets of Figure 2, however, depict non-trivial changes due to variations in the phase properties of the excitation pulse. Most notable is the 417 cm^{-1} vibrational frequency. This mode is present for both PC and transform-limited excitations, has a very fast decay (τ=130 fs), but yields no detectable contribution to the NC excitation. The 298 cm^{-1} frequency contribution is unaffected by the addition of chirp on the excitation pulse, and the amplitude of the 141 cm^{-1} component is diminished by roughly 30%. There appears to be a considerable decrease (a factor of at least 7x) in the amplitude of the low frequency component (17 cm^{-1}; period ~2ps), and a large change in its decay time (from 680 fs to 12 ps). A decay time of <700 fs corresponds to less than one vibrational period before this low frequency oscillation fully disappears, suggesting that we may not be able to obtain a reliable fit for this slow component.

The interference process that diminishes the 417 cm^{-1} contribution suggests the possibility that it is associated with the excited-state potential energy surface. This may be understood by considering that NC pulses tend to focus the wavepacket in the excited state. Further, the negatively chirped excitation pulse (blue-to-red frequency sweep) is properly ordered to maximize the probability of coupling back to the ground state. Thus, when frequency components corresponding to excited state vibrations are excited with negatively chirped pulses, they should appear as weaker contributions to the total signal. This argument suggests that the 417 cm^{-1} oscillation is an excited state mode, a conclusion that is consistent with its very fast decay (τ=130 fs). The 141 cm^{-1} and 298 cm^{-1} oscillations may correspond to modes present on both ground and excited surfaces.

Table 1. Fitting parameter for birefringence due to excitation with a positively-chirped (PC), transform-limited (TL), or negatively-chirped (NC) pulse.

Excitation Pulse	Frequency (cm^{-1})	Amplitude	Decay time (fsec)
PC	17	0.0039	12000
TL		0.0038	12000
NC		0.029	680
PC	141	0.14	320
TL		0.14	320
NC		0.09	370
PC	298	0.02	900
TL		0.02	900
NC		0.02	700
PC	417	0.32	130
TL		0.35	130
NC		-----	-----

Transient dichroism signals generated by near transform-limited, positively-chirped, and negatively-chirped excitation pulses are shown in Figure 3. The oscillatory component of the dichroism signal (shown in the insets of Figure 3) appears similar to the birefringence, though, notably, the relative amplitude of the oscillatory components to the total signal is significally smaller. Fitting parameters extracted from these transients are listed in Table 2 and shown in the inset of Figure 3 (note the differences in scale between insets in Figure 2 and 3). In contrast to the birefringence, the dichroism data exhibits only two frequency components: the 141 cm^{-1} and 298 cm^{-1} modes, with the 298 cm^{-1} mode showing a faster decay time. The oscillatory contributions to the NC data set are nearly identical to those obtained for excitations with TL pulses. Strikingly different are the results for the excitation with the positively chirped pulse. Here the oscillations are nearly completely attenuated. The 141 cm^{-1} mode produces no contribution to the signal, and the 298 cm^{-1} contribution is diminished by an order of magnitude.

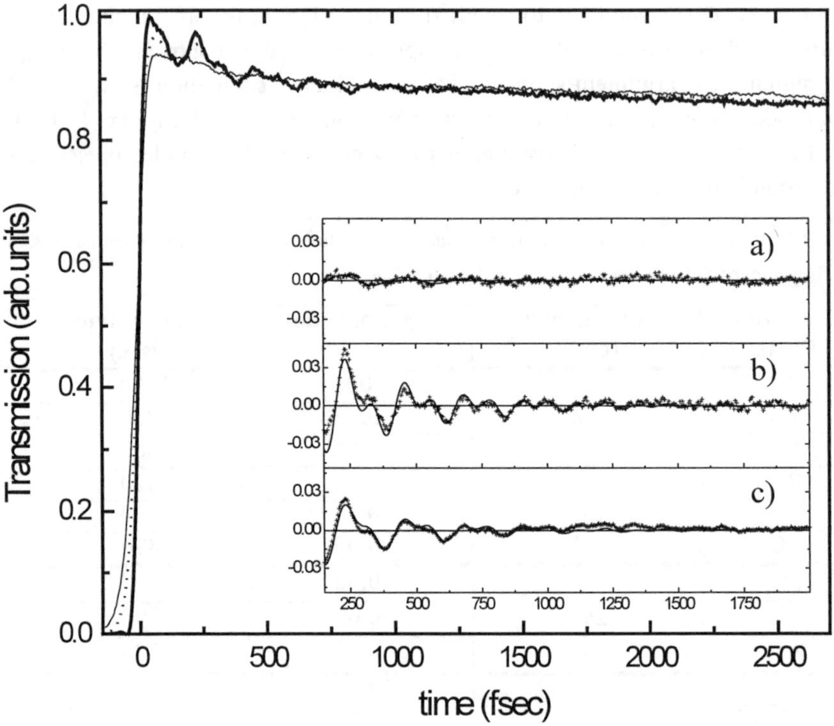

Figure 3. Transient Dichroism. a) ——Positive chirp; b) ▬ No chirp; c)·········· Negative chirp. Insert shows the oscillatory components with their respective fits (see text).

Table 2. Fitting parameters for dichroism due to excitation with a PC, TL or NC pump pulse.

Excitation Pulse	**Frequency (cm^{-1})**	**Amplitude**	**Decay time (fsec)**
PC	141	0.007	500
TL		0.05	320
NC		0.04	320
PC	298	-----	----
TL		0.025	500
NC		0.011	500

Initially, this appears to be a surprising result, considering that two oscillatory components are present in the TL data, and also in the TL and PC data from the birefringence.

Excitation with a PC pulse can be pictured as a stepwise process in which the red end of the excitation pulse transfers the initial components of the wavepacket to the excited state. As the blue end of the pulse arrives, the wavepacket has begun to move away from the initial region of the excited state surface. The formation of a wavepacket in a stepwise manner generates a broader wave function, which, in turn, is expected to reduce the amplitude of the coherent oscillations detected in the signal. Such behavior has been seen by Kohler *et al.* in their experiments on I_2 in the gas phase[12] and Bardeen *et al.* in a low-temperature matrix.[17]

For an electronically resonant excitation, it might be expected that vibrational coherences associated with both the ground and excited-state manifolds appear in each of the transient dichroism and birefringence responses. It follows that the oscillatory components of both the birefringence and dichroism should be qualitatively similar. Yet, for an ultrashort excitation pulse of finite time duration and chirp parameter, we have observed that the two responses yield notable qualitative differences. First, for the same TL excitation pulse, the percentage of the total signal that is due to vibrational coherences is at least a factor of two larger in the birefringence than in the dichroism. For the transform-limited pulse, four modes are observed in the birefringence response, and only two in the dichroism. Finally, for excitation with positively-chirped pulses, the vibrational coherences are nearly completely attenuated in the dichroism, while remaining prominent in the birefringence.

One possible explanation for these results is that the real (birefringence) and imaginary (dichroism) contributions to the molecular susceptibility preferentially sample different regions of the molecular potential energy surfaces coupled by the laser field. For example, the dichroic component should have a frequency response that is similar to that of a transient absorption/stimulated emission process. For shifted potential energy surfaces, the peak of the frequency response should occur in the region where wavepacket broadening is at a maximum, whereas the high and low frequency wings will sample the wavepacket near the turning points. Since we measure the dichroic response integrated over all probe frequencies one may expect this measurement to have preferentially sampled the central region of the coupled surfaces, thus, leading to an attenuation of the coherent oscillations. The birefringence component (being related to the real part of the suscpetibility) should have a different frequency response, possibly with larger weighting near the turning points of the coupled surfaces. If this were the case one would expect a larger contribution from the coherent oscillations to the total signal. In addition, the observed change in the decay time for the 298 cm^{-1} mode also suggests that both measurements probe different regions of the potential energy surface. It also illustrates the mode-sensitive nature of the experiment, since the decay time for the 141cm^{-1} mode remain unchanged.

Such an explanation remains to be tested by theoretical calculations, and with experiments whereby frequency dispersing the probe beam we can obtain independent information for different regions of the coupled surfaces. We are currently pursuing such experiments.

4. Conclusions

We have used the OHD RIKES experimental configuration to measure the resonant transient birefringence and dichroism of condensed phase molecules induced by chirped laser pulses. The data illustrate that the vibrational coherences are sensitive to the phase characteristics of the resonant excitation pulses. The different experimental signatures present in the transient birefringence and dichroism data suggest that the two experimental configurations are sensitive to the wavepacket dynamics over different regions of the coupled energy surfaces. It is possible that detection of both the real and imaginary components of the third-order molecular response may help discriminate between vibrational coherences created on the ground- and excited-state manifolds. Dispersing the transmitted probe and thus, selectively probing different regions of the coupled potential energy surfaces may lead to a more complete picture of the source of each vibrational coherence.

Acknowledgement

We acknowledge the generous support of the Naval Research Laboratory.

References

1. R. J. Gordon and S. A. Rice, *Ann. Rev. Phys. Chem.* **48**, 601 (1997).
2. S. A. Rice and T. D. J., *Adv. Chem. Phys.* **70**, 441 (1988).

3. B. Kohler, J. L. Krause, F. Raksi, K. R. Wilson, V. V. Yakovlev, R. M. Whitnell, and Y. Yan, *Acc. Chem. Res.* **28**, 133 (1995).
4. J. S. Melinger, G. S. R., D. Goswami, and W. S. Warren, *J. Chem. Phys* **101** (8), 6439 (1994).
5. G. Gerber and T. Baumert, *Is. J. Chem.* **34**, 103 (1994).
6. B. Kohler, V. V. Yakovlev, J. Che, J. L. Krause, M. Messina, K. R. Wilson, N. Schwentner, R. M. Whitnell, and Y. J. Yan, *Phys. Rev. Lett.* **74**, 3360 (1995).
7. J. L. Krause, R. M. Whitnell, K. R. Wilson, Y. J. Yan, and S. Mukamel, *J. Chem. Phys* **99**, 6562 (1993).
8. C. J. Bardeen, Q. Wang, and C. V. Shank, *Phys. Rev. Lett.* **75**, 3410 (1995).
9. V. D. Kleiman, S. M. Arrivo, M. J. S., and H. E. J., *Chem. Phys. Lett.* **233**, 207 (1998).
10. V. D. Kleiman, E. J. Heilweil, D. McMorrow, and M. J.S., *Ultrafast Phenomena XI* (Springer, 1998) Vol:63; 648
11. S. H. Ashworth, T. Hasche, M. Woerner, E. Riedle, and T. Elsaesser, *J. Chem. Phys* **104** (15), 5761 (1996).
12. B. Kohler, V. Yakovlev, J. Che, J. L. Krause, M. Messina, and K. R. Wilson, *Phys. Rev. Lett.* **74** (17), 3360 (1995).
13. C. J. Bardeen, Q. Wang, and C. V. Shank, *J. Phys. Chem.* **102**, 2759 (1998).
14. D. McMorrow, W. T. Lotshaw, and G. A. Kenney-Wallace, *J. Quant. Elect.* **24** (2), 443 (1988).
15. W. T. Lotshaw, D. McMorrow, N. Thantu, J. S. Melinger, and R. Kitchenham, *J. Ram. Spec* **26**, 571 (1995).
16. S. Smolrz and F. Wise .
17. C. J. Bardeen, J. Che, K. R. Wilson, V. V. Yakovlev, V. A. Apkarian, C. C. Martens, R. Zadoyan, B. Kohler, and M. Messina, *J. Chem. Phys* **106** (20), 8486 (1997).

---

‡ In essence, this methodology is equivalent to linear prediction single value decomposition (LPSVD). LPSVD analysis could not be used because it was extremely sensitive to initial conditions (cutoff frequency, initial time, etc) and it produced a variable number of additional waves with relatively similar contributions. The use of LPSVD yielded an unrealistically large number of waves, making it impossible to discern between true and spurious oscillations. Instead, fitting sine waves sequentially gave a more physical picture including only a minimum number of independent sinusoidal waves. Fourier transformation of the experimental data yielded similar results, reproducing the main features, but failing to resolve a fast-decaying high frequency component.

DYNAMICAL INTENSITY BORROWING IN PORPHYRIN J-AGGREGATES REVEALED BY SUB-5-FS SPECTROSCOPY

TAKAYOSHI KOBAYASHI, HIDEAKI KANO AND TAKASHI SAITO
Department of Physics, University of Tokyo, Hongo 7-3-1, Bunkyo, Tokyo 113-0033, JAPAN
E-mail: kobayashi@phys.s.u-tokyo.ac.jp

We observed coherent molecular vibration coupled to the Frenkel exciton in J-aggregates. The coherent oscillation in the induced absorbance change is explained by the modulation of the transition dipole moment through the 244cm^{-1}-ruffling mode.

1 Introduction

Molecular J-aggregates have attracted much attention because of their several characteristic features of spectroscopic properties.[1] Intermolecular interaction between transition dipole moments results in the coherent delocalization of excitons over an aggregate.[2] Recently it was found that the aggregate has a hierarchical structure, which is described by a macroaggregate composed of many mesoaggregates.[3] The aggregation forces in the latter and former are due to the van der Waals interaction and static dipole-diople ($\mu_{\rm st}-\mu_{\rm st}$) interaction. The transition-dipole-transition dipole ($\mu_{\rm tr}-\mu_{\rm tr}$) interaction suffers much from thermal fluctuation because of weaker long range dependence ($1/r^6$) than the static dipole-dipole interaction ($1/r^3$). The thermal fluctuation of the order of phonons in the aggregate system with the wavelength $\lambda_{\rm ph}$ disturbs the $\mu_{\rm tr}-\mu_{\rm tr}$ interaction, but it does not deteriorate the $\mu_{\rm st}-\mu_{\rm st}$ interaction. The growth of the mesoaggregate stops at the aggregate size of $r \sim (kT/\mu_{\rm tr}^2)^{1/2}$, where kT is the thermal energy. The mesoaggregates then have a large static dipole moment accumulated, and further aggregation of mesoaggregates forms a macroaggregate, since the former is not much disturbed by the thermal fluctuation up to the size level of $r \sim (kT/N\mu_{\rm st})^{1/3}$. It is very important to well characterize the above hierarchic structure to discuss the experimental results obtained by coherent interaction with a coherent laser pulse. Hereafter the name "J-aggregate" is used as a mesoaggregate.

The excitonic transition of the J-aggregates is characterized by a sharp absorption band, called the J-band, of which transition energy is lower than that of monomers by the value of twice the dipole-dipole interaction.[4,5] Among J-aggregates composed of various aromatic or macrocyclic compounds, tetraphenylporphine tetrasulfonic acid (TPPS) J-aggregates are of special interest since they are model substances for aggregates of the light-harvesting antenna chlorophyll and of primary charge-separation systems in photosynthesis. The absorption spectrum of TPPS aggregates shows a relatively weak band in the visible region and a strong peak in the near ultraviolet, denoted as Q- and B-bands, respectively, composing a quasi-three-band Frenkel exciton system. The former and latter originate from the lowest and second lowest singlet excited state (S_1- and S_2-state), respectively. A number of investigations have been performed using linear and nonlinear electronic spectroscopies[6–16] which elucidated the optical properties characteristic of the excitonic transition in porphyrin J-aggregates. Resonance Raman spectroscopy has also been applied to J-aggregates as well as to the monomer,[7,9,17] and aggregation-induced Raman scattering is observed at 242 and 316cm^{-1}.[9] However, no studies have ever been made on the vibrational properties of porphyrin J-aggregates by the method of real-time spectroscopy, because very high time-resolution is required for direct measurement of intramolecular vibrations in the time domain. Using excitation pulses which are shorter than the characteristic frequency of some vibronically active modes, it is possible to create vibrational wavepackets which move on the electronic ground and excited state potential surfaces. Several experimental studies of dye molecules and bacteriorhodopsin have been performed to study the wavepacket motion by real-time spectroscopy after 6-fs laser pulses were developed by Shank and coworkers.[18,19,20] Recently, we developed a novel optical parametric amplifier (OPA) which generates sub-5-fs visible pulses. The parametric amplifier is in the noncollinear geometry with the pump pulse in a pulse-front tilting configuration. The noncollinear optical parametric amplifier (NOPA) has also two sets of chirped mirrors to compensate for the group velocity dispersion between the signal and idler before and after the OPA.[22] High-frequency multimode wavepacket motions associated with the three carbon-carbon stretching modes and chain-deformation modes have been observed in polydiacetylenes on a sub-5-fs time-scale. Real-time observation of mode-coupling between the two stretching modes $C-C$ and $C=C$ through the $C-C=C$ bending mode has been reported.[23] This is the first observation of real time mode coupling. Recently there was a report which claims the observation of quantum beats due to mode coupling between stretching modes from the experimental result of CARS.[24] However, that result can be explained simply by the classical interference between them.

In the present study, we applied sub-5-fs real-time pump-probe spectroscopy to a TPPS J-aggregate system and observed for the first time the oscillation of transmittance change due to the modulation of the intensity of induced absorbance by the 244cm^{-1}-ruffling mode. In particular, the phase of the oscillation showed a π-phase difference between the bleaching and the induced absorption of the multi-exciton states (MES). The transition corresponding to the induced absorption can be expressed as $|n+1> \leftarrow |n>$ $(n \geq 1)$, where n is the number of excitons in a single mesoaggregate.[26] The coherent oscillations in the transient signals are explained by the modulation of the transition dipole moment due to transfer of the oscillator strength from the intense B-band to the weak Q-band by the intensity borrowing mechanism through the vibronic interaction.

2 Experimental

2.1 Sample Preparation

Tetraphenylporphine tetrasulfonic acid (TPPS: Tokyo Kasei) and polyvinylalcohol (PVA: Kanto Chemical) were used without further purification. After 200-mg PVA was dissolved into 2-ml distilled water at $\sim 100°C$, 20-mg TPPS was added into the hot solution, The concentration of TPPS was 5×10^{-3}mol/l, which is high enough for aggregate formation. A unidirectionally oriented film of porphyrin J-aggregates was prepared using a vertical spin-coating technique developed by our group some years ago.[27] Figure 1 shows a linear absorption spectrum of porphyrin J-aggregates. Sharp absorption peaks at 2.53eV (S_2-state: B_x-band or Soret-band) and 1.75eV (S_1-state: Q_x-band) correspond to B- and Q-excitons, respectively, composing a quasi-three-band Frenkel-exciton system. Fine structures in the absorption bands indicate counterparts of the two bands, namely, the B_y- and Q_y-bands, which have peaks around 2.94 and 1.86eV, respectively. From the polarization dependence of the absorption, it can be stated that the x and y components originate from the transition dipole moments which are parallel and perpendicular, respectively, to the aggregate axis defined as a direction connecting the centers of neighboring molecules composing the aggregates.

2.2 Real-Time Pump-probe Experimental Apparatus

The experimental setup of the sub-5fs time-resolved pump-probe system is described elsewhere. Here only the important parameters of the experimental system are described. The probe pulse intensity after the sample is spectrally dispersed by a 30cm monochoromator (Ritsu, MC-30) and measured with a Si photodiode. The spectral resolution of the whole system is about 3nm. The broad spectrum of the probe pulse enables the investigation over the wide spectral range from 1.6 to 2.3eV. All measurements are performed at room temperature. Although the pulse duration is shorter than 5fs, the FWHM of

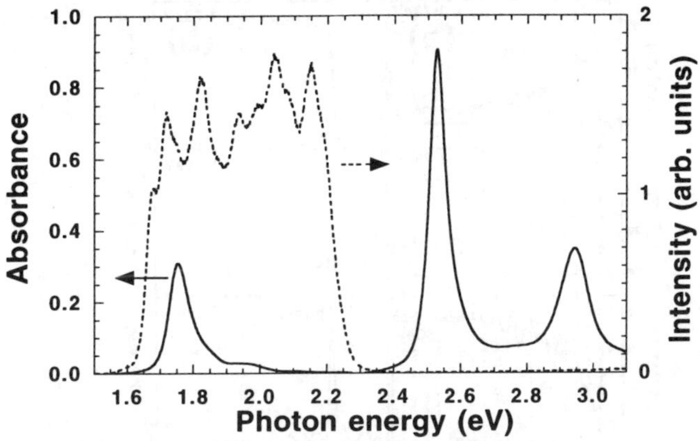

Figure 1. Linear absorption spectrum of porphyrin J-aggregates (solid line) and laser spectrum (dashed line)

the cross-correlation trace is broadened to about 8fs because of geometrical smearing due to the noncollinear angle between the pump and probe beams to avoid an artificial signal caused by the interference between them. The pump pulse energy and photon density are about 14nJ and 5.9×10^{14}photons/cm^2, respectively. Both the polarization of the pump and probe are parallel to the aggregate axis.

3 Results & discussion

3.1 Real-time Spectra

The time dependence of the transmittance change in porphyrin J-aggregates at 3 different probe photon energies is shown in Fig. 2 (a). The laser spectrum covers the whole width of the Q-band of the TPPS J-aggregates. On the other hand, the B-band is outside of the laser spectrum. It can be safely assumed that Q-exciton is created by the one-photon resonant transition process and no B-exciton can be produced at the excitation photon flux of 1.2×10^{29}photons/sec $\cdot$ cm^2. The predominant feature commonly appearing in the traces is high-frequency oscillations in addition to the underlying slow-dynamics transient components. The oscillations persist for delay times longer than 2ps. To the best of our knowledge, this is the first observation of the coherent molecular vibration in porphyrin J-aggregates. The oscillatory signal size is drastically reduced at the probe-photon energy of 1.77eV.

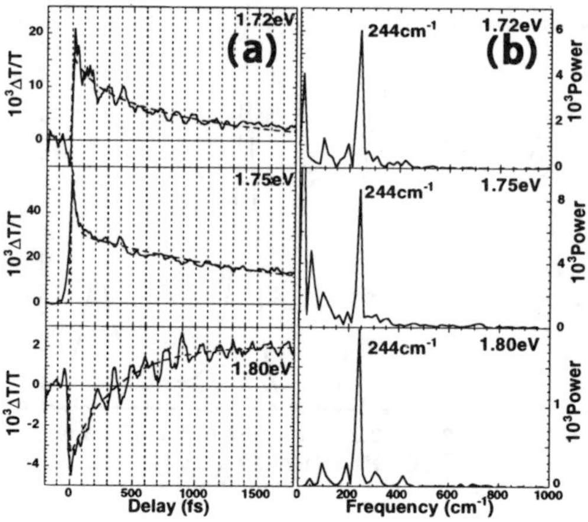

Figure 2. (a) Time-dependence of the transimttance change in porphyrin J-aggregates at 3 photon energies marked on the right. (b) Fourier-power spectra of real-time spectra integrated from 100fs to 1.8ps

It is noted that the phase of the transient oscillation does not change with probe-photon energy.

The overall slow-dynamics of transmittance changes originate from the dynamics of the excited states of J-aggregates by photoexcitation. The positive signal observed over the whole range of delay times up to 1.8ps for $E_{\text{probe}} \leq 1.77\text{eV}$ is attributed to bleaching between the Q-band and ground state. At probe energies higher than 1.77 eV, the signal becomes negative due to the induced absorption (IA) which is associated with the transition to multi-exciton states (MES). Additional complex oscillating features observed both at a negative delay and around the time origin are due to the pump-perturbed free induction decay,[28] and cross-phase modulation,[29] and are also partly due to the interference left over between pump and probe pulses.

Measured decay curves of the transmittance change are fitted by the sum of two exponential-decay functions, which is convoluted with the cross-correlation function. Because nearly chirp-free pulses are used for both pump and probe, the same cross-correlation function can be used for convolution with the response function over the whole range of the probe photon energy.

3.2 Dynamics of the Electronic States

The time dependent transmittance change is composed of the electronic and molecular vibration contributions. The former and the latter correspond to a slow-decaying component and an oscillating component, respectively. The oscillating component is observed only around the Q-band and is not observed at the probe-photon energies higher than 1.82eV.

The signal is composed of three components, namely the bleaching due to $|1, S_1> \leftrightarrow |G>$, IA due to the transition to MES ($|n+1, S_1> \leftarrow |n, S_1>(n \geq 1)$), and IA due to the transition to higher excited exciton states ($|1, S_n> \leftarrow |1, S_1>, n > 1$). Here $|n, S_1>$ is the n-exciton state of S_1-exciton, and $|1, S_n>$ is one-exciton state of S_n-exciton, namely the excited state of an exciton originated from an S_n-excited state in a TPPS molecule. The positive signal in the difference transmittance $\Delta T/T(t)$ around the Q-band is associated with bleaching, while the negative signal originates from two sources. The IA with transient oscillations is observed only at the higher photon energy side of the Q-band. Since the IA due to the transition to MES is also observed at similar photon energy due to Pauli exclusion principle, the IA with transient oscillations is attributed to the IA due to the transition to MES. On the other hand, the IA without transient oscillations has a broad spectrum far above the Q-band, and it is ascribed to the IA due to the transition to higher excited exciton states, because the absorption spectra of S_n-exciton states ($n \gg 1$) are expected to be broad.

Next we shall focus on the origin of the slow-dynamics component. The decay time constant is determined to be around 2ps. The lifetime of the Q-exciton is reported to be ~50ps.[10,13,16] Therefore, the observed fast response cannot be attributable to the Q-exciton lifetime. In our previous study, the decay time constant of the ~ 2-ps component increases with the excitation power and the signal amplitude is super-linearly dependent on pump power.[16] From the results, it was attributed to the S_2-exciton state photogenerated by two-photon absorption. In our more recent report, it was found that the spectrum of the ~ 2-ps component in IA shifts to the blue as the delay time increases in the time-resolved pump-probe experiment using a 100-fs-pump pulse.[30] Similar behavior is also observed in the IA ($|1, S_n> \leftarrow |1, S_1>, n > 1$) signal in the present experiment. The decay time constant at higher photon energy than 1.75eV increases up to ~1ps with probe-photon energy, which means blue-shifted IA. These results indicate the slow-dynamics component is not to be attributed to the two-photon generated S_2-exciton. The slow-dynamics component is, therefore, ascribed to the non-thermal S_1-excitons (Q-excitons) which thermalize with time constant of about 2ps. There is another time constant shorter than 2ps, especially at the higher photon energy side of the Q-band. This is explained by the fast relaxation process from higher MESs ($|n+1, S_1> \leftarrow |n, S_1>$ $(n >> 1)$), which is expected to be faster than the recovery of the bleaching. As a conclusion of the assignment, the bleaching ($|1, S_1> \leftrightarrow |G>$), IA ($|n+1, S_1> \leftarrow |n, S_1>$ $(n \geq 1)$), and IA ($|1, S_n> \leftarrow |1, S_1>$) are observed in the present study. Each component can be spectrally separable. In the following, the bleaching and IA ($|n+1, S_1> \leftarrow |n, S_1>$ $(n \geq 1)$) signal with transient oscillation will be considered.

3.3 Analysis of Coherent Oscillation

For the discussion of the molecular vibration, the oscillating components must be separated from the transient curves. The slow-dynamics decay component is subtracted from the transient signal. Figure 2 (b) shows the Fourier-power spectra calculated from the traces within 100fs $\sim$ 1.8ps in Fig. 2 (a). The lower limit is constrained by the exclusion of any coherent effects around the zero delay time. The resolution is determined to be 16cm^{-1} (FWHM) by the width of the delay-time window (1.7ps) in the Fourier transformation analysis. The Fourier spectra clearly show an intense peak at 244 ± 8cm^{-1} in the whole spectral region of the Q-band. The amplitude of the oscillating signal with a 137fs oscillation period corresponds to 244cm^{-1}. The intensity of the Fourier spectrum at 244cm^{-1} is drastically reduced at 1.77eV, near which the sign of the amplitude is changed. A peak at 241cm^{-1} is reported in the stationary Raman spectrum.[7,12,13] This peak is attributed to a ruffling mode.[13] The Raman signal of this mode is drastically enhanced by aggregation by a factor of more than 30 because change in the molecular configuration associated with the vibrational mode is expected to strongly influence the dipole-dipole interaction. The 317cm^{-1} doming mode, which is another aggregation-enhanced intense Raman mode in the stationary Raman spectrum, is not observed in real-time spectra. Both modes are reported in the Raman scattering experiment with resonant excitation of the B-band. On the other hand, our experiment is performed in resonant excitation of the Q-band. It is an open question whether the 317cm^{-1} doming mode is detectable by a pump-probe setup. It is also of interest to study experimentally the stationary resonance Raman spectrum under the Q-band resonance condition.

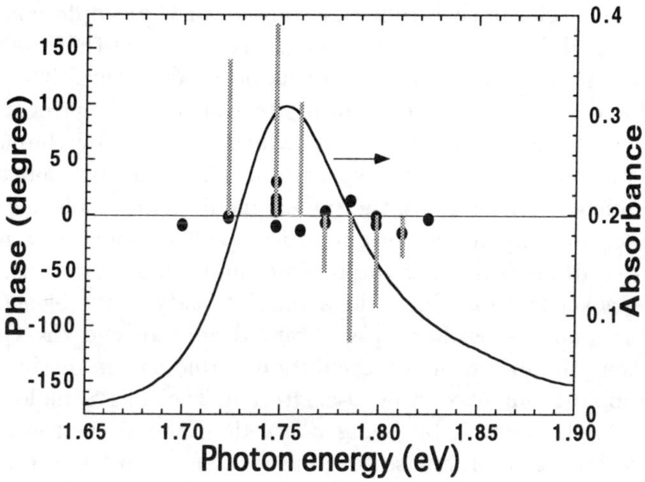

Figure 3. Probe-photon energy dependence of the phase (dots) and amplitude (bar) of the oscillation in transient signal

3.4 Probe-photon Energy Dependence of Phase of the Coherent Oscillation

The phase of the oscillation can be evaluated by the complex Fourier transformation which has already been performed in previous section. Figure 3 shows the probe-photon energy dependence of the phase and amplitude of the 244cm^{-1}-ruffling mode. It clearly shows that the oscillations are in-phase in terms of the values of $|\Delta T/T|$ in the whole spectral region of the Q-band and that the sign of the amplitude is reversed at 1.77eV, which is slightly higher than the peak photon energy of the Q-band. At 1.77eV, the transient signal changes from the bleaching (positive signal) to the IA (negative signal), and the oscillation amplitude is extremely decreased. These results indicate that the bleaching and IA increase (decrease) at the same time.

Several reports based on numerical calculation on the transmittance change in medium-size molecules associated with ultrashort pulse generation have attributed transient oscillation to the dynamics of localized vibrational wavepackets which move on the electronic ground and excited potential energy surfaces.[18–21] In these cases, the phase of the transient oscillations in the signal is expected to be shifted continuously by changing the probe-photon energy, because the transition energy depends on the nuclear position, and different photon energies of the probe pulse are absorbed at different nuclear positions. On the other hand, our experimental result shows that the phase keeps almost constant value. In addition to this peculiar feature, it is well known that the J-band associated with excitonic transition in J-aggregates interacts only weakly with molecular vibrations, which is shown by the small Stokes shift in the stationary fluorescence spectrum. This indicates that the vibronic coupling is much smaller than in the typical medium-size molecules. For these reasons, the oscillations do not originate from conventional vibronic coupling between the excitonic transition and relevant molecular vibrations represented by the wavepacket motion. Since the signal intensity of the bleaching and IA vary in-phase, and the oscillation is observed only around the Q-band, it is considered that the origin of the oscillation is due to the modulation of the transition dipole moment of the Q-exciton in the TPPS molecules, which becomes originally allowed by being coupled to the B-exciton through the ruffling-mode vibration. Increase (decrease) in the transition dipole moment is reflected by the increase (decrease) in the signal intensity both of bleaching and IA. The in-phase oscillation in terms of $|\Delta T/T|$ between the bleaching and IA is therefore due to the modulation of the Q-exciton transition dipole moment.

3.5 Evaluation of an amplitude and frequency modulation

The experimental results can be interpreted by assuming a modulated transition dipole moment of the molecule in the J-aggregates. We shall discuss the origin of the modulation in this subsection. Although the transition to the Q-band is originally forbidden because of the symmetry of the molecular structure, the transition becomes allowed because of mixing with the B-band due to configuration interaction.[31] We assume that re-coupling between the Q- and the B-band is induced by the pump pulse through the 244cm^{-1}-ruffling mode, and the mixing is enhanced or reduced depending on the phase of the vibrational motion of the TPPS molecules. The mixing of two states results in transfer of oscillator strength from the intense B-band to the weak Q-band. First, the amount of the modulation of the transition dipole moment is evaluated. The normalized modulation of the transition dipole moment of the molecule is expressed as $\delta(\mu)/\mu$, in which μ represents the transition dipole moment of TPPS molecule, and $\delta\mu$ is the change of μ induced by the ruffling vibration. Here we should notice that the modulation of μ also changes the intermolecular dipole-dipole interaction, J, which is given by

$$J = \mu^2(1 - 3\cos\theta)/\hbar a^3. \tag{1}$$

Here all transition dipoles in the TPPS molecules composing a mesoaggregate are assumed to be parallel and making an angle θ with the aggregate axis; the molecules are equidistant with lattice constant a. Since the transition energy from the ground state to the first excited state of the aggregate is originally red-shifted by $2J$ in comparison with the one of the monomer, the modulation of J gives rise to a spectral shift (frequency modulation). As a consequence, the modulation of μ results not only in amplitude modulation of the induced absorption change, but also in a frequency modulation through the modulation of J. We shall estimate the amount of the ampitude and frequency modulations due to $\delta\mu$ and δJ as follows. The observed signal is composed of the bleaching and IA due to the transition to MES. We assume that the spectral shapes of the bleaching and IA are similar to the stationary absorption spectrum, $A(\omega)$ and $A(\omega - \Delta E)$, respectively. Here ΔE results from the Pauli exclusion principle based on the Frenkel exciton theory in a one-dimensional aggregate.[25,26] The experimentally obtained nonlinear absorption spectrum, $S(\omega)$, is given by

$$S(\omega) = -\alpha A(\omega) + \beta A(\omega - \Delta E). \tag{2}$$

Here α and β are arbitary coefficients which should be determined by fitting the experimental result using Eq. 2. Assuming that the amplitude and frequency modulations are given respectively by

$$A'(\omega) = A(\omega)(1 + \delta A \cos(\omega_M t)) \tag{3}$$

and

$$\omega' = \omega + \delta\omega \cos(\omega_M t) \tag{4}$$

the modulated signal, $S'(\omega)$ is modified as

$$S'(\omega) = (-\alpha A(\omega + \delta\omega \cos(\omega_M t)) + \beta A(\omega + \delta\omega \cos(\omega_M t) - \Delta E))(1 + \delta A \cos(\omega_M t)) \tag{5}$$

Here δA, $\delta\omega$, and ω_M represent the size of the amplitude and frequency modulations and modulation frequency (244cm^{-1}), respectively. Both δA and $\delta\omega$ are assumed to be independent of ω. Equation (5) is approximately expressed as

$$\begin{aligned} S'(\omega) &\simeq (\beta - \alpha)(A(\omega) - \beta \frac{dA}{d\omega} \Delta E) \\ &+ (\beta - \alpha)(A(\omega)\delta A + \frac{dA}{d\omega}\delta\omega) \cos(\omega_M t). \end{aligned} \tag{6}$$

Since the experimentally obtained transmittance change is composed of the slow-dynamics component, $(\Delta T/T)_{\mathrm{slow}}$, and the oscillating component,

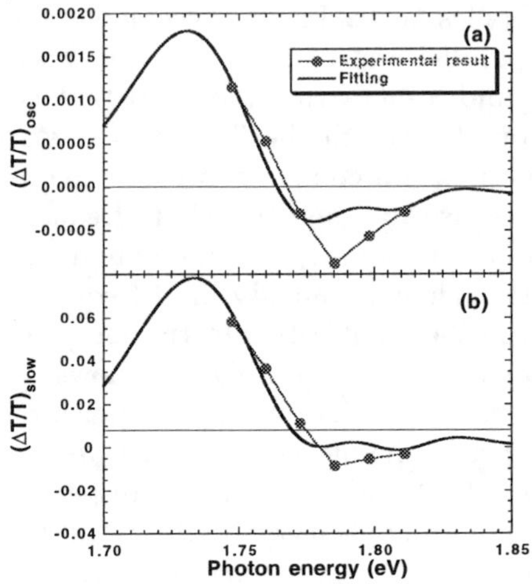

Figure 4. The results of the fitting for $(\Delta T/T)_{\rm osc}$ (a) and $(\Delta T/T)_{\rm slow}$ (b) using Eq. (6)

$(\Delta T/T)_{\rm osc}$, $(\Delta T/T)_{\rm slow}$ and $(\Delta T/T)_{\rm osc}$ correspond to the first and second terms in Eq. (6). By fitting the experimental results using $A(\omega)$ and $\frac{dA}{d\omega}$, $\delta A = (1.6 \pm 0.7) \times 10^{-2}$ and $\delta\omega = (1.4 \pm 0.2) \times 10^{-3}$eV are obtained. Figure 4 (a) and (b) show the results of the fitting for $(\Delta T/T)_{\rm osc}$ and $(\Delta T/T)_{\rm slow}$.

The value of $\delta\mu/\mu$ is derived from δA and has the value

$$\frac{\delta\mu}{\mu} = \delta A/2 = \frac{\delta(\mu^2)}{2\mu^2} \simeq 0.8 \pm 0.4\%. \tag{7}$$

On the other hand, the ratio of the frequency shift $\delta\omega$ to $2J$ is determined to be $\delta\omega/2J \simeq 0.8\%$.

3.6 Modulation of the Dipole-dipole Interaction

As is shown in Eq. (1), the modulation of J could be caused not only by μ but also by θ and a. The ratio of the modulation to $2J$ is given by

$$\frac{\delta(2J)}{2J} = \frac{2}{\mu}\delta\mu + \frac{3\sin\theta}{1-3\cos\theta}\delta\theta - \frac{3}{a}\delta a. \tag{8}$$

Since the amount of the modulation of the transition dipole moment calculated from the frequency modulation ($\delta\mu/\mu = 0.4\%$) is consistent with the calculated one from the amplitude modulation ($\delta\mu/\mu = 0.8 \pm 0.4\%$) within fitting error, we consider the modulation of μ to be the dominant origin of the frequency modulation, δJ.

3.7 Vibronic Coupling between Q- and B-band

In this subsection, we will discuss the origin of the modulation of the transition dipole moment ($\delta\mu$). As is discussed above, the transition to the Q-band is originally forbidden. The weak oscillator strength of the Q-band is caused by the mixing with the B-band due to configuration interaction. We consider that the modulation of the transition dipole moment is caused by the time-dependent re-mixing effect due to vibronic interaction between the Q- and B-bands, which is induced by the pump pulse through the ruffling mode. This effect is called intensity borrowing. We performed the following calculation to estimate the intensity borrowing effect from the intense B-band to the weak Q-band using $\delta\mu/\mu$ based on the vibronic coupling theory. Although the symmetry is important for the vibronic coupling mechanism, the symmetry of the ruffling mode is not clearly defined. The perturbed transition dipole moment between the Q-band and the ground state, $M_{\mathrm{GQ}} = \mu+\delta\mu$, is expressed as follows using vibronic coupling theory:

$$M_{\mathrm{GQ}} = M_{\mathrm{GQ}}^{0} + \frac{H_{\mathrm{vib}}}{E_{\mathrm{B}}^{0} - E_{\mathrm{Q}}^{0}} M_{\mathrm{GB}}^{0}. \tag{9}$$

Here $M_{\mathrm{GQ}}^{0} = \mu$, M_{GB}^{0}, E_{B}^{0}, E_{Q}^{0}, and H_{vib} denote the transition dipole moment between the ground state and the Q-band, the transition dipole moment between the ground state and the B-band, the transition energy to the B-band and the Q-band without the perturbation by the ruffling mode, and the perturbation to the Hamiltonian, respectively. The normalized value of the modulation of μ is expressed as

$$\begin{aligned} \frac{\delta\mu}{\mu} &= \frac{M_{\mathrm{GQ}} - M_{\mathrm{GQ}}^{0}}{M_{\mathrm{GQ}}^{0}} \\ &= \frac{H_{\mathrm{vib}}}{E_{\mathrm{B}}^{0} - E_{\mathrm{Q}}^{0}} \frac{M_{\mathrm{GB}}^{0}}{M_{\mathrm{GQ}}^{0}}. \end{aligned} \tag{10}$$

Using Eqs. (7) and (10), we can derive

$$H_{\mathrm{vib}} = (E_{\mathrm{B}}^{0} - E_{\mathrm{Q}}^{0}) \frac{M_{\mathrm{GQ}}^{0}}{M_{\mathrm{GB}}^{0}} \cdot \frac{\delta\mu}{\mu} \tag{11}$$

$E_{\mathrm{B}}^{0} - E_{\mathrm{Q}}^{0} \sim 0.93\mathrm{eV}$, and $M_{\mathrm{GB}}^{0}/M_{\mathrm{GQ}}^{0} \sim 2.6$ are estimated from the stationary absorption spectrum of the monomer and aggregate. By substituting these coefficients, $H_{\text{vib}} = 2.9\text{meV} = 23\text{cm}^{-1}$ is obtained, which is much less than the vibration frequency, 244cm^{-1}. As a consequence, the intensity borrowing from the B-band is estimated to be 0.3%. It should also be added that the intensity borrowing from the Q_y-band to the Q_x-band is negligible because the oscillator strength of the Q_y-band is much smaller than that of the B-band, and it is even smaller than that of the Q_x-band.

A similar interpretation has been reported in the stationary resonance Raman spectrum of metalloporphyrins. The resonance Raman spectrum with B- and Q-band excitations shows dominant contributions of the Albrecht A-term and the B-term scattering, respectively. The dominant contribution of the B-term in resonant Q-band excitation is also interpreted by the mixing of the Q- and B-bands.[31]

4 Conclusion

Coherent molecular vibration coupled to the Frenkel exciton in porphyrin J-aggregates is observed for the first time. The oscillation originates from molecular vibrations and is assigned to the ruffling mode. The in-phase oscillation in terms of $|\Delta T/T|$ is explained by the modulation of the transition dipole moment of the Q-band, which is due to the intensity borrowing mechanism from the intense B-band to the weak Q-band.

5 Acknowledgements

The authors would like to thank Drs. H. F. Hofmann and T. Fuji for enlightening discussions. The work is partly supported by Research for the Future of Japan Society for the Promotion of Science(JSPS-RFTF-97P-00101).

References

1. T. Kobayashi, *J-aggregates* (World Scientific, Singapore, 1996).
2. A. S. Davydov, *Theory of Molecular Excitons* (Plenum, 1971).
3. T. Kobayashi, *Mol. Cryst. Liq. Cryst.* **314,** 1 (1998).
4. E. E. Jelly, *Nature* **138,** 1009 (1936).
5. G. Sheibe, *Angew. Chem.* **49,** 563 (1936).
6. D. Barber *et al*, *J. Phys. Chem.* **95,** 4074 (1991).
7. O. Ohno *et al*, *J. Chem. Phys.* **99,** 4128 (1993).
8. J. M. Ribo *et al*, *J. Chem. Soc. Chem. Commun.* 681 (1994).
9. D. L. Akins *et al*, *J. Phys. Chem.* **98,** 3612 (1994).
10. N. Maiti *et al*, *J. Chem. Phys.* **99,** 17159 (1995).

11. N. C. Maiti *et al*, *J. Phys. Chem.* **99,** 10708 (1995).
12. D. L. Akins *et al*, *J. Phys. Chem.* **100,** 5420 (1996).
13. D. L. Akins *et al*, *J. Phys. Chem.* **100,** 14390 (1996).
14. I. Y.Chan and A. J. Hallock, *J. Chem. Phys.* **107,** 9297 (1997).
15. N. C. Maiti *et al*, *J. Phys. Chem.* **102,** 1528 (1998).
16. K. Misawa and T. Kobayashi, *J. Chem. Phys.* **110,** 5844 (1999).
17. G. A. Shick *et al*, *J. Phys. Chem.* **97,** 1339 (1993).
18. H. L. Fragnito *et al*, *Chem. Phys. Lett.* **160,** 101 (1989).
19. W. T. Pollard *et al*, *Chem. Phys. Lett.* **168,** 23 (1990).
20. W. T. Pollard *et al*, *J. Phys. Chem.* **96,** 6147 (1992).
21. G. Cerullo *et al*, *Phys. Rev. Lett.* **83,** 231 (1999).
22. A. Shirakawa *et al*, *App. Phys. Lett.* **74,** 2268 (1999).
23. T. Kobayashi and A. Shirakawa, *Appl. Phys.* to be published.
24. A. Vierheilig *et al*, *Chem. Phys. Lett.* **312,** 349 (1999).
25. J. Knoester, *Phys. Rev. A* **47,** 2083 (1993).
26. K. Minoshima *et al*, *Chem. Phys. Lett.* **218,** 67 (1994).
27. K. Misawa *et al*, *Appl. Phys. Lett.* **63,** 577 (1993).
28. H. Brito-Cruz *et al*, *IEEE J. Quantum Electron.* **24,** 261 (1988).
29. E. Tokunaga *et al*, *J. Opt. Soc. Am. B* **12,** 753 (1995).
30. H. Kano and T. Kobayashi, "Technical Digest of Quantum Electronics and Laser Science Conference 99", p10 (1999).
31. M. Gouterman *The Porphyrins*, volume III PART A. (Academic Press, 1979).

SUB-5FS REAL-TIME SPECTROSCOPY OF GEOMETRICAL RELAXATION PROCESSES IN POLYDIACETYLENES

TAKAYOSHI KOBAYASHI AND AKIRA SHIRAKAWA
Department of Physics, University of Tokyo7-3-1 Hongo, Bunkyo, Tokyo 113-0033, Japan
E-mail: kobayashi@phys.s.u-tokyo.ac.jp

Ultrafast dynamics of excitations in polydiacetylenes was studied with the shortest visible pulses with sub-5-fs duration. With this high time resolution it was clearly found that the relaxation from 1^1B_u free-excitons to 2^1A_g self-trapped excitons takes place with the time constant of about 70fs. The frequencies of C=C (23fs) and C-C (27fs) stretching modes are modulated with 145fs-period(C-C=C bending).

1 Introduction

Ultrafast dynamics of excitations in conjugated polymers such as polyacetylene and polydiacetylenes (PDA) has been attracting many scientists because they are model systems of quasi-one dimensional organic semiconductors with large optical nonlinearity. From our extensive studies[1] the initial kinetics after photo-excitation can be explained by a relaxation of a free exciton (FE) to a self-trapped exciton (STE) within 150fs. The recent achievement of sub-5-fs visible pulse generation based on a novel noncollinear optical parametric amplification (NOPA)[2] has enabled the real-time observation of the formation of an STE in a PDA for the first time. The stretching mode vibrations with longer oscillation periods than the pulse width are coupled with the excitoninc transition. The periods are 27 fs for C-C, 23 fs for C=C, and 16 fs for C≡C stretching modes. They drive the relaxation of the backbone structure from an acetylene(A)-like $(=CR-C \equiv C-CR=)_n$ to a butatriene(B)-like $(-CR=C=C=CR-)_n$ configuration.

2 Experimental

The pulse-front-matched NOPA generates transform-limited (TL) pulses centered at 650nm with a 250nm bandwidth 2. The width, energy, and repetition rate of the pulses are 4.7fs, 5μJ 1kHz, respectively. The PDA studied is a newly synthesized ladder-polymer. It is designed to increase the oscillator strength and optical nonlinearity. The formal (IUPAC) name of the compound is poly[5,7,17,19-tetracosatetraynylene-bis(butoxycarbonylmethyl)-carbamate], hereafter abbreviated as PDA4BCMU(8)), with two PDA-backbones linked with methylene chains 3. The transition energy to 11Bu FE is peaked at 1.96eV, which is classified to a typical blue-phase A-like PDA 1. Some of PDA s are classified as a red-phase B-like PDA, even though they do not have a butatriene-like configuration. The 370-nm thick film sample prepared by spin-coating on a glass substrate was pumped and probed with the visible sub-5fs pulses at a room temperature of 297K.

3 Results and discussion

The trace showing the delay time dependence of the normalized transmittance change is called a real-time spectrum hereafter. The real-time spectra at several photon energies are shown in Fig.1, which clearly shows a coherent molecular motion composed of several modes over the whole probe spectral range.

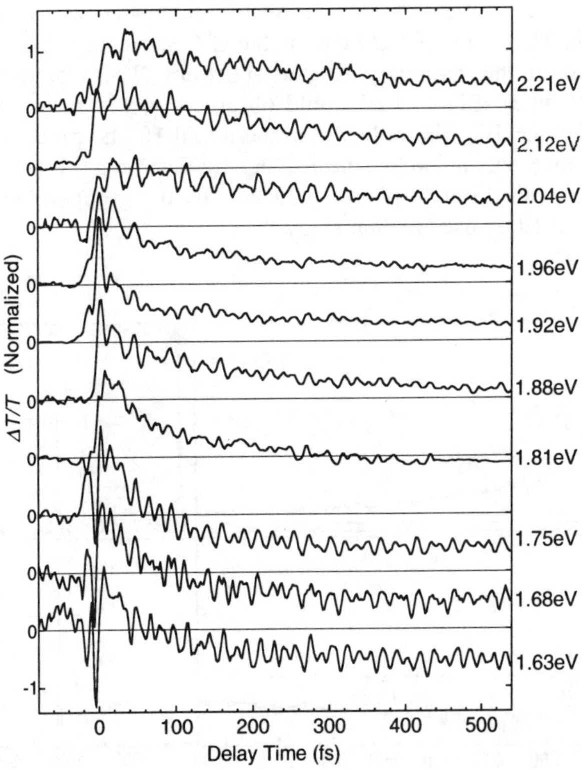

Figure 1. Delay time dependence of normalized transmittance change exihibiting coherent wavepacket motion. The figures on the right are the probe photon energies. The probe pulse intensity transmitted through the film sample of the PDA was measured by a combined system of a monochromater with 5-nm resolution, photomultiplier, and a lock-in amplifier.

Recent experimental and theoretical studies demonstrate the lowest singlet state in a blue-phase PDA to be a forbidden 2^1A_g state lying ~0.1 eV below the 1^1B_u FE state.[1] The fast 60fs decay observed around E_{probe}=1.96 eV is attributed to the internal conversion (IC) from a photoexcited 1^1B_u FE to a 2^1A_g FE. The long-lived oscillatory structure due to wavepacket appearing in the bleaching signal is thus clearly attributed to a ground-state motion driven by the impulsive stimulated Raman scattering. The stimulated emission from a 1^1B_u FE observed at $E_{probe}<1.8$eV is suppressed rapidly by the IC, followed by the relaxation to an STE which causes the excited-state absorption of the biexcitons→STE transition.[1]

The Fourier analysis of the real-time spectrum reveals two main modes of C=C(~1455 cm^{-1}) and C≡C(~2080cm^{-1}) stretching, in addition to weaker modes around 220, 700, and 1220 cm^{-1}, of which relative intensities are dependent on the probe wavelength. The mode of 1220 cm^{-1} is the C-C stretching. All of these modes are also observed in the stationary Raman spectrum. The vibration persists for longer than 1 ps(Fig.2). Bigot *et al.* could observed only a single mode of C=C stretching in a PDA-DCHD film only within the initial 100 fs[4] probably because the high frequency mode oscillation is smeared out by destructive interference among many components with different phases induced by the group-velocity dispersion in the non-TL pulses (9 fs) used in their study.

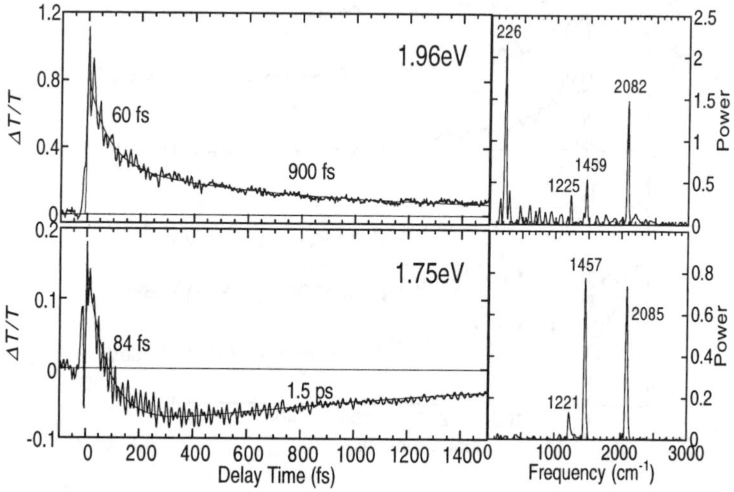

Figure 2. Transient data at the characteristic probe energies of 1.96 eV and 1.75 eV with an extended time-scale to .5 ps. The delay-time integrated Fourier-power-spectra of the corresponding oscillating components over 0 ps to 1.5 ps are shown on the right.

The net formation time of STE is determined to be about 80 fs and the lifetime is 1.5 ps. The wavepacket motion in an STE is transferred coherently from the 1^1B_u FE via IC, which clearly indicates that the coherent motions of the C=C and C≡C stretching along the polymer backbone act as the driving force of the geometrical relaxation to a localized B-like structure corresponding to the self-trapped exciton. The vibrational energy transfer among the modes coupled associated with the A→B isomerization and following thermalization of STE s are visualized by the frequency analysis of the the real-time spectra using the spectrograms.[5]

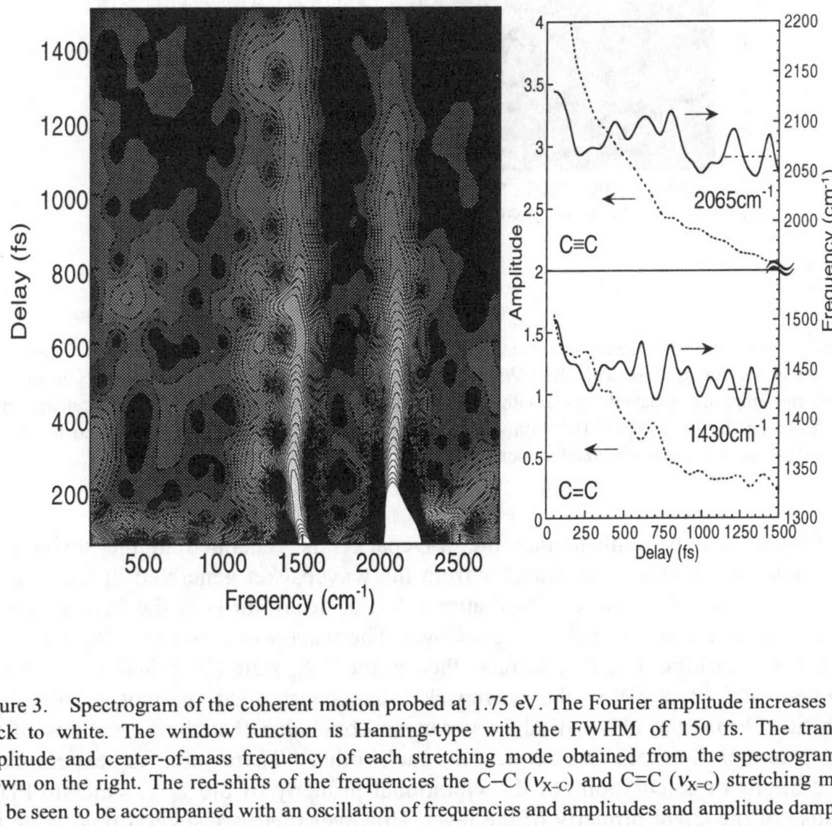

Figure 3. Spectrogram of the coherent motion probed at 1.75 eV. The Fourier amplitude increases from black to white. The window function is Hanning-type with the FWHM of 150 fs. The transient amplitude and center-of-mass frequency of each stretching mode obtained from the spectrogram are shown on the right. The red-shifts of the frequencies the C–C ($\nu_{X\text{-}C}$) and C=C ($\nu_{X=C}$) stretching modes can be seen to be accompanied with an oscillation of frequencies and amplitudes and amplitude damping, which visualize the vibrational energy exchange and dissipation, respectively, in the STE state. Each frequency of the thermal STE with the B-like backbone is smaller than that of the ground state by about $20cm^{-1}$.

The integrated Fourier analysis of the oscillations only offers information of the averaged vibrational frequency over the range of the wavepacket motion. For the study of the dynamic behavior of the geometrical relaxation, the time-frequency analysis was performed by means of a spectrogram.[5] The spectrograms clearly exhibited an interesting feature of mode-frequency modulations (Figs.3 and 4).

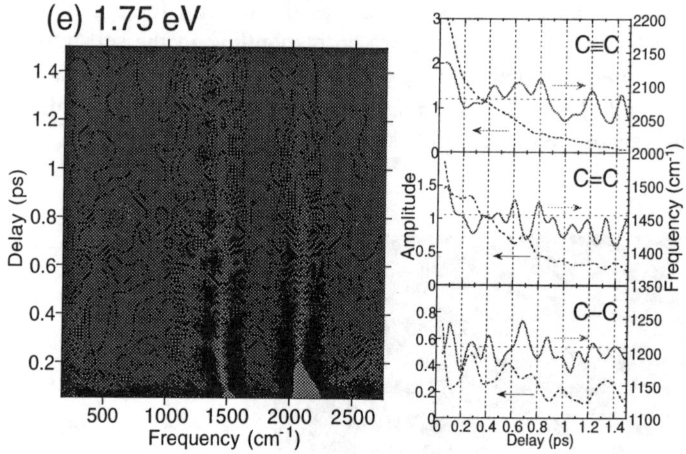

Figure 4. Spectrograms. The Fourier amplitude of the same value is plotted as contours. The window function is a Hanning-type with the FWHM of 150 fs. The averaged instantaneous frequency (solid curve) and amplitude (dashed curve) within the width of the window function of each stretching mode are shown on the right. The bandwidth for the integration is 200 cm^{-1} corresponding to the Fourier transform of the Hanning-type gatefunction.

Present results indicate that the wavepacket is generated in the self-trapped 2^1A_g state by the coherence transfer from the wavepacket generated in the 1^1B_u FE state. It is of interest that the oscillations do not decay through the IC. To explain this feature it can be considered as follows. The wavepacket in the 1^1B_u state leaks on both sides of the potential surface, then in the 2^1A_g state every leak is coherently summed, and after 80 fs the wavepacket is reconstructed without a substantial distortion. From the above finding it can be concluded that the curvatures of the two potential surfaces are equal to each other. There are no experimental observations or calculations of the vibrational property of the 2^1A_g states in PDAs because of the silent property of the level. However there are several reports on the vibrational characteristics of $C_{14}H_8$, showing that the mode-frequency differences of the three stretching modes are all smaller than 10 cm^{-1} between the 1^1B_u and 2^1A_g states.[6] After the IC, the wavepacket motion is kept for ~1 ps and decays via the nonradiative tunneling process to the ground state with the characteristic time constant of 1.5 ps.[1] This time constant is nearly universal for blue-phase PDAs.[1]

The fine structures were found to be well reproduced in the spectrograms probed at several photon energies. The time dependence of the transient amplitudes and center-of-mass frequencies at the two or three characteristic stretching modes is shown on the right side of each spectrogram. For E_{probe}<1.9 eV the oscillations show similar features: the frequencies of the C–C (ν_{X-C}) and C=C ($\nu_{X=C}$) stretching modes are modulated with the opposite phase to each other with the modulation frequency of ~230 cm^{-1}. On the other hand, the frequencies of the C≡C stretching mode show weaker modulations and seem to have less correlation with the other C–C and C=C stretching modes. These features can be explained by the butatrienic structure localized in one or a few units at most in the self-trapped state. The coherent nuclear motion coupled with the FE stays around this local deformation after self-trapping, and the above dynamic property is concentrated on the C–C and C=C stretching modes. It may be appropriate to say that the π–bond-orders of the C=C ($1 - \delta_0$) and C≡C ($2 - \delta_0$) bonds decrease to $1 - \delta$ and $2 - \delta$, respectively, where δ_0 and ($\delta_0 < \delta < 1$) are the π-bond-orders of the C–C bond in the ground state and equilibrium self-trapped 2^1A_g state.

This ~230cm^{-1} frequency is also observed in the coherent wavepacket motion, especially for E_{probe} > 1.9 eV. This mode can be assigned to a planar bending mode of C–C=C bonds. This lies in the typical frequency range of the C-C=C bending modes of olefines and alkynes. From recent *abinito* calculations, it has been shown that these bending modes have large Franck-Condon (FC) activities in molecules such as octatetraene (C_8H_{10}, 220–400 cm^{-1})[6] and diacetylene oligomers ($C_{14}H_8$, ~160 cm^{-1}).[7] However, as far as we know, the wavepacket motion has not been observed in any previous time-resolved spectroscopic studies of PDA s. The coupling feature was firstly and only observed by Nunzi *et al.* in a dynamic hole burning spectroscopy of a blue-phase PDA-4BCMU film.[8] They observed the creation of holes at both sides of the exciton energy with the separation of 290 cm^{-1}. The FC factor was estimated to be large enough from the hole depths, even though the corresponding Raman peak was quite weak. Our result is the first observation of the coherent wavepacket formation of this bending mode.

For E_{probe}<1.9 eV the Fourier peak around 230 cm^{-1} is weakened and disappears at 1.75 eV, even with the long-living oscillations. However, the large FC activity of this mode can create well-confined wavepackets both on the ground and excited state potential surfaces. If the FC factors of this mode between the self-trapped 2^1A_g state and the nB_u states are small, the PIA signal cannot contain a clear bending motion.

The frequency modulations of ν_{C-C} and $\nu_{X=C}$, on the other hand, clearly indicate the formation of this bending motion after the geometrical relaxation to the butatriene-like configuration. Because the equilibrium positions of the carbon atoms and the bond angle of C–C=C are different from those in the acetylenic configuration,[9] the coherent vibrations of the stretching modes and the C–C=C bending mode continue to exist until thermalization occurs among various vibrational modes. They include very low frequency modes of lattice (polymer chain) vibrations. The self-trapped state is not thermalized before the tunneling to the ground state, and these nuclear motions can be clearly observed for ~1.5 ps. In such non-equilibrium states the bending motion perturbs both force constants of the C–C and C=C stretching modes. Figure 5 illustrates the mechanism. For simplicity the case of a fully butatrienic structure is considered. If the bond angle is decreased, the bond-order in the C–C bond is increased while that in the C=C bond is decreased, which causes the corresponding frequencies to be up- and down-shifted, respectively. In the opposite case the changes are reversed, resulting in the out-of-phase modulations of ν_{X-C} and $\nu_{X=C}$. The out-of-phase property between each amplitude and the corresponding frequency can also be explained as follows. The potential energy curve is the intersection of the multi-dimensional potential energy surface along the corresponding configuration coordinate to the geometrical relaxation. The curvature of the potential curve of the stretching mode is modulated by the bending motion. It is steepened and the amplitude of the wavepacket motion is reduced, which is reflected by the decreased $\Delta T/T$ s, resulting in the smaller amplitude modulation.

The amounts of the modulated bond-length and bond-orders can be estimated from the frequency modulation. Lewis *et al.* introduced the empirical relation,[10]

$$r_{ij} = r_0 + \frac{B}{K_{ij}}, \tag{1}$$

where the stretching force constant K_{ij} between two atoms i and j is related to the bond-length r_{ij}. The parameters $r_0 = 0.106$ nm and $B = 1.65\times10^{-8}$ N were evaluated in the case of PDA-PTS (2,4-hexadiyne-1,6-diol). Batchelder *et al.* estimated K_{ij} s in the excited state as $K_{C-C} = 474$ N/m and $K_{C=C} = 481$ N/m.[11] Because the modulation depths of $\Delta\nu_{X\equiv C}$ and bending modes are small enough, the coupling between only the two stretching modes are considered, giving

$$\begin{pmatrix} \Delta\nu_{C-C} \\ \Delta\nu_{C=C} \end{pmatrix} = \begin{pmatrix} \dfrac{\partial\nu_{C-C}}{\partial K_{C-C}} & \dfrac{\partial\nu_{C-C}}{\partial K_{C=C}} \\ \dfrac{\partial\nu_{C=C}}{\partial K_{C-C}} & \dfrac{\partial\nu_{C=C}}{\partial K_{C=C}} \end{pmatrix} \begin{pmatrix} \Delta K_{C-C} \\ \Delta K_{C=C} \end{pmatrix}. \tag{2}$$

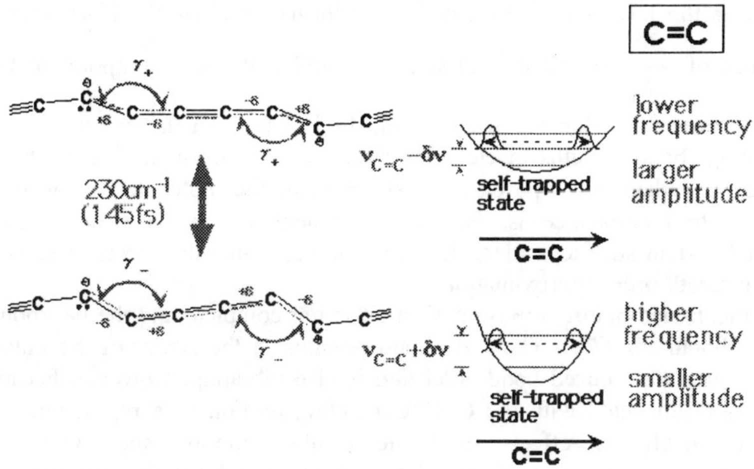

Figure 5. A model of frequency and amplitude modulations of C=C stretching vibration in PDA induced by C-C=C bending associated with the geometrical relaxation of free exciton photoexcited in the polymer.

The components of the matrix of differential coefficients $\frac{\partial v_j}{\partial K_{ij}}$ are given in Table 3 of ref. [12] for a PDA-PTS crystal in the ground state. There is no reported data of the excited state, which limits us to use these coefficients for the first calculation. The modulation depths $\Delta v_{C-C} \cong \Delta v_{C=C} = \sim 16\ \text{cm}^{-1}$ yields the modulated bond lengths of $\Delta r_{C-C} = \pm 0.0026$ nm and $\Delta r_{C=C} = \mp 0.0025$ nm from Eqs. (1) and (2). Using $r_{C-C} = r_{C=C} = 0.14$ nm in the excited state,[11] the relative modulations of r_{C-C} and $r_{C=C}$ to the equiribrium are ± 1.9 % and ∓ 1.8 %, respectively.

The modulations of the π bond orders of the single and double bonds are estimated by using the simple relationship between the mode frequency and bond order derived by Baughman *et al.*[13] In the case using Dewar s relationship $dv_{X=C}/dK = 512.4\ \text{cm}^{-1}$, assuming that it is also valid for the 2^1A_g state of a PDA-4BCMU4A(8), the modulation depth $\Delta\delta$ is given as ± 0.031 from $\Delta v_{C=C}$. In the case of $r_{C-C} = r_{C=C} = 0.14$ nm with $\delta = 0.41$ being used[14], the relative modulation is estimated as 7.6 %.

These estimates are based on several assumptions. First, we use the values of PDA-PTS investigated in detail because there are no such experiments for PDA-4BCMU4A(8). Second, in ref. 11 K_{ij} and r_{ij} in the excited state were estimated by measuring resonance Raman excitation profiles, which only reflects the molecular structure in the 1^1B_u state and includes no information about the 2^1A_g state. Third, the values of $\frac{\partial \nu_j}{\partial K_{ij}}$ in the ground state are used with the assumption of Eq. (2). Finally, Dewar s relationship used in ref. 14 is based on the assumption that in determining bond lengths π-election resonance is much less important than hybridization when only one classical structure of the molecule can be written.[14] The π-electron resonance is, however, non-negligible in a strongly electron-correlated system such as a PDA. It should be noted that the present analysis only gives the zeroth-order approximation.

As mentioned before, the isomerization is not complete and the backbone may be only butatriene-like . The C≡C bond remains in the center of the butatriene-like unit with the reduced bond-order and is also vibrating. However, because the bond is not connected with the C–C=C bending portion of a repeat unit, it may experience negligible effects from the bending motion and evolve almost independently. The amount of modulation of the bond lengths and bond orders mentioned above may be overestimated because of the assumption of a completely delocalized structure ($r_{C-C}=r_{C=C}$). In fact it is difficult to determine the amounts of the bond orders in the 2^1A_g state. Even though there is some ambiguities in the above estimated amounts of modulation depth, the present results are the first observation of coupling between normal modes via the vibration of other modes. These phenomena are expected to be widely observed in such a photoexcited non-equilibrium system in such a short time scale. The present results will offer valuable information about the geometrical relaxation and encourage theoretical investigations of the electron-phonon and electron-electron interactions in non-degenerate conjugated polymers. The experimental technique of real-time spectroscopy with sub-5 fs pulses can reveal the structure of intermediates or even in transition states in the dynamical processes from the dynamic molecular vibrational spectra obtained by a windowed Fourier transform of the real-time specra. It offers a powerful method to clarify the mechanism of chemical reactions, such as isomerization and tautmerization.

Acknowledgement

The authors would like to express their thanks to Prof. H. Nakanishi and Mr. H. Matsuzawa for providing the polymer sample. They thank Prof. H. Okamoto and Dr. Takasu for helpimg the Raman spectra measurement of the PDA sample. They are also grateful for useful discussions with Profs. Y. Toyozawa, M. Yoshizawa, and K. Misawa. This work was partly supported by Research for the Future of Japan Society for the Promotion of Science(JSPS-RFTF-97P-00101).

References

1 E.g, T. Kobayashi, M. Yoshizawa, U. Stamm, M. Taiji, and M. Hasegawa, J. Opt. Soc. Am. B. 7, 1558 (1990); M. Yoshizawa, Y. Hattori, and T. Kobayashi,

Phys. Rev. B **123**, 121 (1994); T. Kobayashi, M. Yasuda, S. Okada,H. Matsuda, H. Nakanishi, Chem. Phys. Lett. **267**, 472 (1997).

2 A. Shirakawa, I. Sakane, and T. Kobayashi, Opt. Lett. **23**, 1292 (1998); A. Shirakawa, I. Sakane, M. Takasaka, and T. Kobayashi, Appl. Phys. Lett. **74**, 2268 (1999).

3 H. Matsuzawa, S. Okada, H. Matsuda, and H. Nakanishi, Chem. Lett., 1105 (1997).

4 J. Y. Bigot, T. A. Pham, and T. Barisien, Chem. Phys. Lett. **259**, 469 (1996).

5 M. J. J. Vrakking, D. M. Villeneuve, and A. Stolow, Phys. Rev. A **54**, 37 (1996)

6 F. Zerbetto and M. Z. Zgierski, J. Chem. Phys. **101**, 1842 (1994).

7 F. Zerbetto, J. Phys. Chem. **98**, 13157 (1994).

8 J. M. Nunzi, C. Hirlimann, J. F. Morhange, Chem. Phys. Lett. **221**, 199 (1994).

9 A. Karpfen, J. Phys. C **13**, 5673 (1980).

10 W. F. Lewis and D. N. Batchelder, Chem. Phys. Lett. **60**, 232 (1979).

11 D. N. Batchelder and D. Bloor, J. Phys. C **15**, 3005 (1982).

12 A. C. Cottle, W. F. Lewis, D. N. Batchelder, J. Phys. C **11**, 605 (1978).

13 R. H. Baughman, J. D. Witt, K. C. Yee, J. Chem. Phys. **60**, 4755 (1974).

14 M. J. S. Dewar, Hyperconjugation (Ronald Press, New York, 1962), p. 48.

LOCALIZATION IN DISSIPATIVE TUNNELING SYSTEMS DRIVEN BY MONOCHROMATIC LIGHT

NANCY MAKRI

Department of Chemistry, University of Illinois, 601 S. Goodwin Avenue, Urbana, Illinois 61801, USA
E-mail: nancy@makri.scs.uiuc.edu

This lecture reviews recent theoretical work on the role of continuous-wave radiation in extending the lifetime of localized states in dissipative two-state systems. Through a combination of path integral methods and semiclassical arguments it is shown that strong fields can stabilize localized states over long time intervals. At moderate dissipation strength this phenomenon is quite robust and insensitive to details of the phonon spectrum, temperature and driving frequency, depending primarily on the intensity of the applied field.

1 Introduction

Much of current research at the interface of physics, chemistry and engineering is concerned with the possibility of controlling elementary processes in systems of nanoscale dimensions, with the goal of constructing ultrasmall devices with well-defined function. Combination of techniques based on nanolithography and scanning tunneling microscopy have enabled the construction of structures such as quantum dots or wells with shape resolution on the atomic level, and the manufacturing of nanotubes or molecular wires with intriguing properties has become routine. It is thus of great interest to devise schemes of reversibly controlling the carrier dynamics in these structures.

The present lecture focuses on control schemes pertaining to generic tunneling systems in contact with a phonon environment that plays the role of a "reservoir", leading to quenching of the tunneling oscillations. The resulting behaviors are typical of many diverse phenomena in molecular or electronic nanostructures. For instance, damped exciton tunneling has been predicted and observed directly in double quantum wells composed of alternating layers of two semiconductor materials.[1,2] In these experiments the two wells are brought into resonance via an external dc field. Electron-hole excitation can then launch a localized state, which executes observable intrawell tunneling oscillations. Electron-phonon and carrier-carrier interactions tend to quench coherence, and the system reaches thermal equilibrium typically after a few oscillations. Of primary concern to the dynamics are the dissipative effects induced by these interactions,[3-5] but carrier screening is known to play an important role at higher excitations.[6,7] It is of interest to manipulate this dynamics in two distinct ways: (a) to suppress tunneling, thereby extending indefinitely the initial state lifetime, or (b) to induce coherent intrawell oscillations that persist over time lengths that are long compared to the time required for thermalization. This lecture focuses on thc first of these questions.

Modeling the dissipative effects of phonon coupling in driven dissipative two-level systems has been the subject of numerous analytical studies,[8-15] many of which employ treatments based on extensions of the noninteracting blip approximation pioneered for time-independent systems by Leggett and coworkers.[16] The present work employs a numerical methodology developed by Makri and coworkers.[17-23] In this, the dissipative effects of the environment are introduced by coupling the tunneling subsystem to a bath of harmonic degrees of freedom whose frequencies and couplings are specified collectively via a spectral density function. The evolution of the composite Hamiltonian, including an external radiation field, is expressed in terms of the path integral representation of quantum mechanics, and the integrations with respect to the degrees of freedom of the environment are performed analytically. This leads to a path integral for the driven tunneling subsystem augmented by an influence functional that incorporates the dissipative effects of the medium. The influence functional introduces nonlocal interactions in time in a way similar to the memory kernel in the classical generalized Langevin equation. Due to dephasing, the range of this temporal nonlocality is finite. As a consequence, the path integral can be cast in an iterative form, which involves successive multiplication of the initial state density by a multi-time propagator matrix, a procedure that is easily carried out numerically. The scheme can be iterated indefinitely, yielding the exact quantum dynamics of the system over very long periods.

The iterative path integral methodology is reviewed in section II. Section III discusses the results of applying continuous wave electromagnetic fields to dissipative two-level systems with parameters typical of double quantum wells. The numerical results are also analyzed with the aid of simple semiclassical arguments. Finally, some concluding remarks appear in the last section.

2 Model systems and path integral methodology

We focus on the dynamics of localized states in a two-level system coupled to a harmonic dissipative bath and driven by a periodic force. The model is described by the following Hamiltonian:

$$H = -\hbar\Omega\sigma_x + \varepsilon\sigma_z + \sum_j \frac{p_j^2}{2m_j} + \frac{1}{2}m_j\omega_j^2\left(x_j + \frac{c_j\sigma_z}{m_j\omega_j^2}\right)^2 + V(t)\sigma_z \tag{1}$$

Here σ_x and σ_z are the 2×2 Pauli spin matrices, $2\hbar\Omega$ is the splitting of the bare TLS tunneling doublet, $\{x_j\}$ are the coordinates of the harmonic bath degrees of freedom responsible for dissipation and $V(t)$ is the external driving field applied to the TLS which represents in the semiclassical limit its interaction with radiation.

Time evolution can be expressed in terms of the propagator $U(t'',t')$ for the time-dependent Hamiltonian of Eq. (1). Since one is not interested in recording the detailed dynamics of each bath degree of freedom but only the overall influence of the environment on the TLS state, we integrate out the bath and focus on the evolution of the reduced density matrix

$$\tilde{\rho}(t) = \mathrm{Tr}_b[U(t,0)\rho(0)U^{-1}(t,0)] \tag{2}$$

where Tr_b denotes the trace with respect to the harmonic bath.

Below we proceed to express the propagators in the form of a discretized path integral which is to be evaluated numerically. It is thus advantageous to use a short time propagator that is accurate over as long a time step as possible. A successful strategy is to identify a meaningful low-dimensional reference system and partition the Hamiltonian accordingly.[24] In the present case we choose a time dependent reference that includes the TLS Hamiltonian along with the driving term

$$H_0(t) = -\hbar\Omega\sigma_x + \varepsilon\sigma_z + V(t)\sigma_z \tag{3}$$

The remaining terms in the Hamiltonian of Eq. (1) correspond to harmonic oscillators displaced by an amount that depends on the TLS site. We employ a symmetric splitting of the time evolution operator,

$$U(t+\Delta t,t) \approx \exp\left(-\frac{i}{\hbar}(H-H_0)\frac{\Delta t}{2}\right)U_0(t+\Delta t,t)\exp\left(-\frac{i}{\hbar}(H-H_0)\frac{\Delta t}{2}\right), \quad (4)$$

where U_0 is the evolution operator that corresponds to H_0, and assume for simplicity that the system and bath are uncoupled at $t=0$ such that the initial density matrix can be written as a product of system and bath density matrices,

$$\rho(0) = \tilde{\rho}(0)\rho_b(0). \quad (5)$$

The bath is chosen to be at thermal equilibrium, i.e.,

$$\rho_b(0) = \frac{e^{-\beta H_b}}{\mathrm{Tr}\, e^{-\beta H_b}} \quad (6)$$

where $\beta = 1/k_B T$. After integrating out the path integral variables corresponding to the bath oscillators, the reduced density matrix is brought in the form:[25]

$$\begin{aligned}\langle\sigma''|\tilde{\rho}(t)|\sigma'\rangle = &\sum_{\sigma_0^+=\pm1}\sum_{\sigma_1^+=\pm1}\cdots\sum_{\sigma_{N-1}^+=\pm1}\sum_{\sigma_0^-=\pm1}\sum_{\sigma_1^-=\pm1}\cdots\sum_{\sigma_{N-1}^-=\pm1}\\ &\times\langle\sigma''|U_0(t,t-\Delta t)|\sigma_{N-1}^+\rangle\cdots\langle\sigma_1^+|U_0(\Delta t,0)|\sigma_0^+\rangle\\ &\times\langle\sigma_0^+|\rho_0(0)|\sigma_0^-\rangle\langle\sigma_0^-|U_0^{-1}(\Delta t,0)|\sigma_1^-\rangle\cdots\\ &\times\langle\sigma_{N-1}^-|U_0^{-1}(t,t-\Delta t)|\sigma'\rangle\\ &\times F(\sigma_0^+,\sigma_1^+,\cdots\sigma_{N-1}^+,\sigma'',\sigma_0^-,\sigma_1^-,\cdots\sigma_{N-1}^-,\sigma';\Delta t).\end{aligned} \quad (7)$$

Here $\sigma=\pm1$ is a discrete variable indicating the TLS state in the left-right representation. The superscripts $\pm$ refer to forward and backward paths that evolve in the positive and negative time directions and F is the influence functional arising from the coupling to the bath. If the latter is described by a canonical ensemble at temperature $T = 1/k_B\beta$, the influence functional has the form[26]

$$F = \exp\left(-\frac{1}{\hbar}\sum_{k=0}^{N}\sum_{k'=0}^{k}(\sigma_k^+ - \sigma_k^-)(\eta_{kk'}\sigma_{k'}^+ - \eta_{kk'}^*\sigma_{k'}^-)\right) \quad (8)$$

The coefficients $\eta_{kk'}$ constitute the discrete analog of the bath response function (two-time kernel) $\alpha(t-t')$ and have been given explicitly in earlier reviews.[20] It can be shown that these coefficients depend only on collective characteristics of the environment as specified by the spectral density function,

$$J(\omega)=\frac{\pi}{2}\sum_j \frac{c_j^2}{m_j\omega_j}\delta(\omega-\omega_j) \tag{9}$$

Eq. (7) contains nonlocal terms through the double sum in (8), which may be thought of as the quantum mechanical generalization of medium-induced "memory" entering the generalized Langevin equation.

Evaluation of the reduced density matrix via a global summation over all variables appearing in (7) is feasible only for short times, i.e., $N\leq 10$ or so. To access longer times we have devised an iterative procedure. The possibility of iterative evaluation of the path integral expression, Eq. (7), arises from the fact that the bath response function characterizing dissipative environments drops off rapidly with time. This feature is a consequence of dephasing in media with broad spectra and implies that "long memory" terms make negligible contribution to the dynamics. Retaining terms with $|k-k'|\leq \Delta k_{\max}$, where $\Delta k_{\max}\Delta t$ is the effective memory length, leads to quasi-Markovian dynamics for a higher dimensional quantity and an iterative scheme. Specifically, an augmented reduced density tensor **R** of rank $\Delta k_{\max}$ can be constructed whose evolution is given by the iterative procedure[18,19]

$$\begin{aligned} R(\sigma_{k+1}^{\pm},\ldots,\sigma_{k+\Delta k_{\max}}^{\pm};(k+1)\Delta t) = \sum_{\sigma_k^{\pm}=\pm 1} & \Lambda(\sigma_k^{\pm},\ldots,\sigma_{k+\Delta k_{\max}}^{\pm};k\Delta t) \\ & \times R(\sigma_k^{\pm},\ldots,\sigma_{k+\Delta k_{\max}-1}^{\pm};k\Delta t) \end{aligned} \tag{10}$$

where Λ is an appropriate time-dependent propagator tensor. The path integral interactions included in a single iteration of this scheme are indicated in the diagram of Fig. 1.

Projection of **R** yields the reduced density matrix $\tilde{\rho}$ of the TLS:

$$\tilde{\rho}(\sigma_k^{\pm};k\Delta t)\equiv\left\langle\sigma_k^{+}\left|\tilde{\rho}(k\Delta t)\right|\sigma_k^{-}\right\rangle = R(\sigma_k^{\pm},\sigma_{k+1}^{\pm}=\cdots=0;k\Delta t) \tag{11}$$

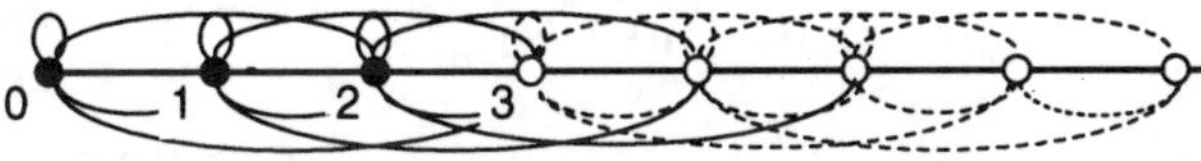

Figure 1. Diagrammatic representation of the path integral interactions in a case where the memory length is equal to three time steps. The loops indicate terms in the exponent of the influence functional. Solid and dashed loops indicate the decomposition of the path integral into an iterative procedure.

The above procedure can be iterated to yield the reduced density matrix over long propagation intervals. Stability is guaranteed, as the tensor multiplication scheme preserves the trace of the reduced density matrix.

The iterative procedure outlined above avoids global evaluation of the multi-dimensional discretized path integral. It is easily generalizable to systems described by continuous potentials via the use of discrete variable representations.[27] For a TLS, propagation of the augmented reduced density tensor requires storage of $2^{(2\Delta k_{\max}+1)}$ complex numbers. This is an easy task for $\Delta k_{\max}$ up to about 10. The above version of the tensor multiplication scheme is not practical if the medium-induced memory spans many more time steps. A filtering procedure that selects statistically significant path segments reduces the required propagator storage and makes the scheme practical in cases of long memory.[21,22]

Finally, it is worth pointing out that the formulation of an iterative scheme is not specific to harmonic baths but can be extended to arbitrary environments as long as there is an adequate number of degrees of freedom to restrict the range of memory interactions.[28] Since the influence functional from arbitrary media is not available analytically, it is necessary to evaluate it numerically, and this has recently become feasible via forward-backward semiclassical approximations.[29,30]

3 Driven dissipative two-level systems

Below we specialize to dissipative TLSs driven by continuous wave fields of the type

$$V(t) = V_0 \cos \omega_0 t \tag{12}$$

The amplitude of the driving term is chosen in the range $V_0 \le 30\hbar\Omega$ which (for typical values of the system dipole moment) corresponds to laser fields of intermediate to large strength. The TLS is initially in the +1 (right-localized) state.

The calculations reported in this section are performed with the so-called Ohmic spectral density described by the equation

$$J_1(\omega) = m_0 \gamma \ \omega \ e^{-\omega / \omega_c} \tag{13}$$

Here m_0 is the effective mass of the tunneling system, γ is the friction coefficient and ω_c is a cutoff which ensures that the phonon spectrum of the environment decays to zero at high frequencies. We choose $\omega_c = 20\Omega$; this magnitude of the cutoff frequency is typical of semiconductor double quantum well structures where $\Omega \sim 10-50$ cm^{-1} while the crystal Debye frequency is a few hundred cm^{-1}. The strength of the dissipation is commonly characterized by the dimensionless Kondo parameter $\alpha = 2m_0\gamma / \pi\hbar$.[16]

In the absence of coupling to a bath, the most pronounced effect of driving on symmetric TLSs is the suppression of tunneling along a one-dimensional manifold in the V_0, ω_0 parameter space[31] referred to as "optimal localization condition". This phenomenon is related to a degeneracy in the driven TLS quasienergy spectrum generally when $2V_0 / \hbar\omega_0 = z_i$, where z_i are nodes of the zero-order Bessel function, and leads to rigorous survival of localized states at times which are multiples of the field period. Interaction with a dissipative bath offers the possibility of delocalization via destruction of phase coherence, which generally leads to a time-dependent steady state whose time-average corresponds (for a symmetric TLS) to Boltzmann equilibrium.

Fig. 2 shows the average position of a symmetric dissipative TLS driven by a monochromatic field of strength $V_0 = 20\hbar\Omega$ at a temperature $k_B T = 10\hbar\Omega$.[32] In agreement with earlier work,[9] the decay of a localized state is seen to be practically exponential (with superposed small amplitude oscillations) at intermediate to high temperature and/or sufficiently strong friction, at least as long as the driving frequency is sufficiently high with respect to the tunneling splitting.

Analogous, though more complex, behaviors are observed in driven asymmetric TLSs.[33] When a driving field is introduced to the bare asymmetric TLS, there is competition between the field and the bias for control of the system dynamics.

In the absence of TLS-bath coupling generic driving again produces frequency-modulated tunneling oscillations, but the amplitude of these oscillations is diminished with increasing bias. As expected, except in cases of resonant driving (asymmetry equal to a multiple of the driving frequency, $2\varepsilon = m\hbar\omega_0$) large asymmetry ($\varepsilon >> \hbar\Omega$) results again in localization. The optimal localization condition is now modified to[34]

$$J_m(2V_0 / \hbar\omega_0) = 0 \tag{14}$$

where J_m is the Bessel function of order m. These behaviors persist in a weakly dissipative environment. Finally, if the driven asymmetric TLS is coupled to a dissipative bath, one observes the same trend in the dynamics as in the symmetric case. Again, as the dissipative strength of the bath increases, the results of differing driving fields become nearly identical.

In order to summarize these results Makri and coworkers have calculated the decay rate of $\langle\sigma_z(t)\rangle$ in the regime where the TLS position exhibits on average exponential decay to a steady state value. For this purpose the data from various simulations are fit to the form[32,33,35]

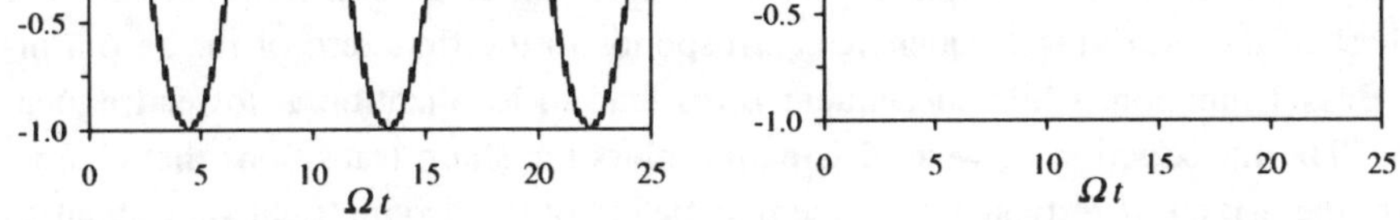

Figure 2. Path integral results for the expectation value $\langle\sigma_z(t)\rangle$ as a function of time at a high temperature $\hbar\Omega\beta = 0.1$ for a symmetric TLS coupled to an Ohmic bath with $\alpha = 0$, 0.16 and 0.64. (a) Generic driving field characterized by $V_0 = 20\hbar\Omega$, $\omega_0 = 12\Omega$. (b) Optimal localization condition, $V_0 = 20\hbar\Omega$, $\omega_0 = 16.714\Omega$. In all cases the oscillatory line corresponds to the dissipationless case.

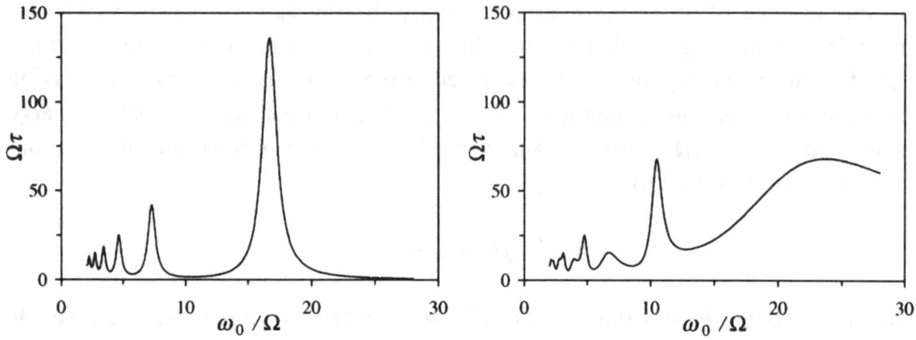

Figure 3. Path integral calculations of the lifetime τ of localized states in the weak dissipation regime $(\alpha = 0.025)$ as a function of field driving frequency ω_0 for the parameters $V_0 = 20\hbar\Omega$ and $\hbar\Omega\beta = 0.1$. (a) $\varepsilon = 0$. (b) $\varepsilon = 5\hbar\Omega$. The relatively sharp peak near $\omega_0 = 10\Omega$ is an example of an optimum localization condition for a driven asymmetric TLS.

$$\langle\sigma_z(t)\rangle_{\text{av}} = \left[\langle\sigma_z(0)\rangle - \langle\sigma_z\rangle_{\infty,\text{av}}\right]\exp(-kt) + \langle\sigma_z\rangle_{\infty,\text{av}} \tag{15}$$

where k is the decay rate and $\langle\sigma_z\rangle_{\infty,\text{av}}$ the asymptotic time average of the TLS coordinate.

The TLS lifetime in weakly dissipative environments is diplayed in Fig. 3 for several values of the asymmetry parameter ε. For a given driving strength, the isolated symmetric TLS displays perfect localization at several frequencies. The highest of these driving frequencies corresponds to the first zero of the zeroth order Bessel function, while subsequent zeros lead to localization at lower frequencies. The introduction of weak dissipation blurs the sharp transitions that characterize the driven TLS dynamics as the parameters of the driving field vary about an optimal localization condition. Under such conditions, the driven symmetric TLS exhibits a finite lifetime whose maximum corresponds to a frequency determined by the first zero of the Bessel function. Nearby frequencies also give rise to extended survival of initially localized states.[32] Localization at other frequencies satisfying Eq. (14) is less robust. The peaks in the TLS lifetime become progressively smaller as $2V_0/\hbar\omega_0$ increases and their spacing and width become smaller. As the dissipation strength increases these peaks begin to overlap. Asymmetric two-level systems exhibit similar trends. Away from resonances, i.e., if 2ε is sufficiently different from $m\hbar\omega_0$ where m is an integer, localized states are preserved indefinitely. With weak dissipation the decay is slow for off-resonant driving in the absence of dissipation. Near-resonant driving leads to different behaviors, depending on the field intensity. The path integral calculations show that increasing the asymmetry generally broadens the lifetime peaks seen in the driven symmetric dissipative TLS. However as the asymmetry grows, certain fairly sharp peaks emerge. These sharp antiresonance peaks correspond to localization conditions that are modified due to the presence of bias.

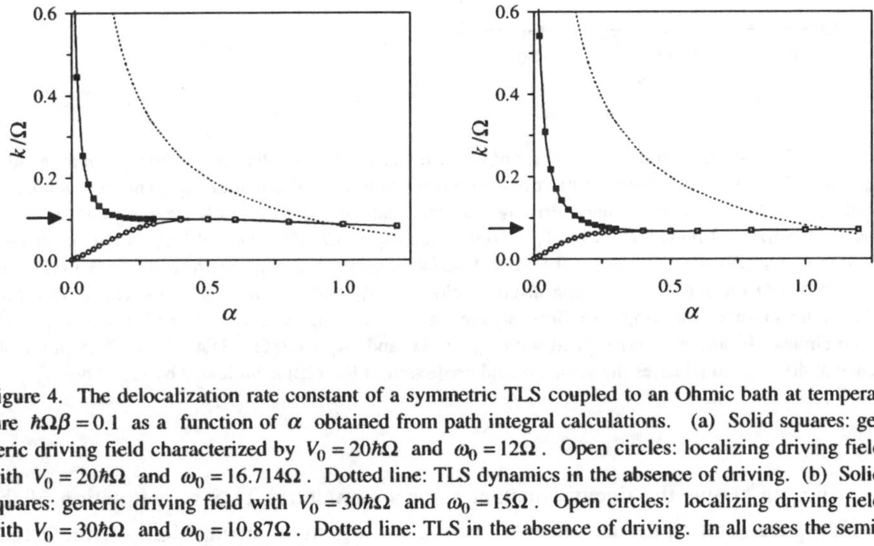

Figure 4. The delocalization rate constant of a symmetric TLS coupled to an Ohmic bath at temperature $\hbar\Omega\beta = 0.1$ as a function of α obtained from path integral calculations. (a) Solid squares: generic driving field characterized by $V_0 = 20\hbar\Omega$ and $\omega_0 = 12\Omega$. Open circles: localizing driving field with $V_0 = 20\hbar\Omega$ and $\omega_0 = 16.714\Omega$. Dotted line: TLS dynamics in the absence of driving. (b) Solid squares: generic driving field with $V_0 = 30\hbar\Omega$ and $\omega_0 = 15\Omega$. Open circles: localizing driving field with $V_0 = 30\hbar\Omega$ and $\omega_0 = 10.87\Omega$. Dotted line: TLS in the absence of driving. In all cases the semiclassical prediction is indicated by the arrow.

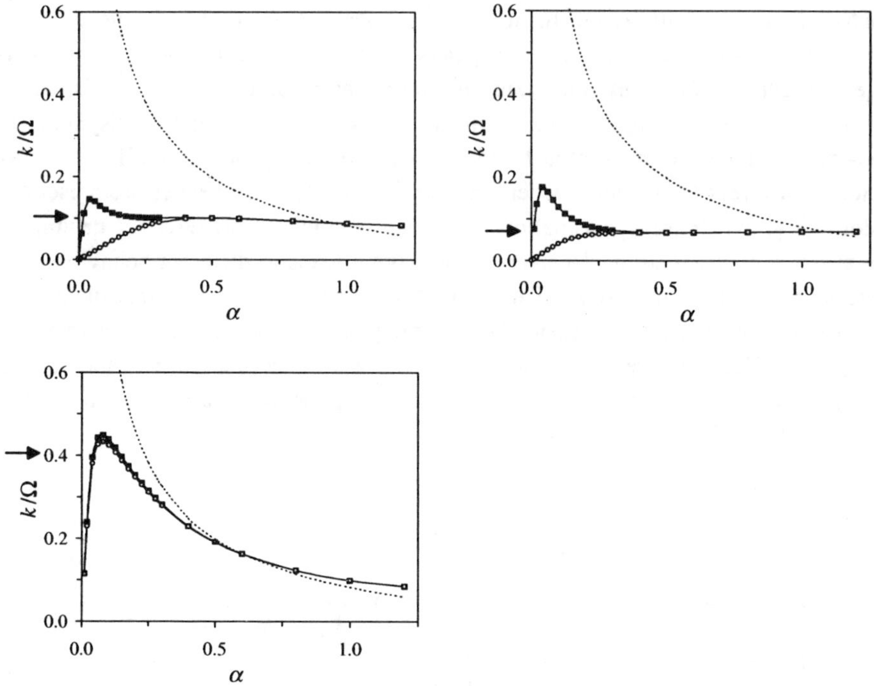

Figure 5. The delocalization rate constant of an asymmetric TLS coupled to an Ohmic bath at the temperature $\hbar\Omega\beta = 0.1$ as a function of α obtained from path integral calculations. The case $\varepsilon = \hbar\Omega$ is plotted. (a) Solid squares: generic driving field characterized by $V_0 = 20\hbar\Omega$ and $\omega_0 = 12\Omega$. Open circles: localizing driving field with $V_0 = 20\hbar\Omega$ and $\omega_0 = 16.714\Omega$. Dotted line: NIBA result for a TLS in the absence of driving. (b) Solid squares: generic driving field with $V_0 = 30\hbar\Omega$ and $\omega_0 = 15\Omega$. Open circles: localizing driving field with $V_0 = 30\hbar\Omega$ and $\omega_0 = 10.87\Omega$. Dotted line: TLS in the absence of driving. (c) Solid squares: generic driving field with $V_0 = 5\hbar\Omega$ and $\omega_0 = 6\Omega$. Open circles: localizing driving field with $V_0 = 5\hbar\Omega$ and $\omega_0 = 4.28\Omega$. Dotted line: TLS in the absence of driving. In all cases the semiclassical prediction of Eq. (22) is indicated by the arrow.

Fig. 4a shows the decay constant k for a symmetric TLS as a function of the Kondo parameter in the absence of driving for a generic strong field characterized by $V_0 = 20\hbar\Omega$, $\omega_0 = 12\Omega$ and under the "optimal localization" condition $V_0 = 20\hbar\Omega, \omega_0 = 16.714\Omega$ at $\hbar\Omega\beta = 0.1$. It is seen that both driving conditions extend significantly the lifetime of a localized state. While the decay rate under optimal driving first increases as a function of friction, it soon reaches a plateau in which the lifetime becomes friction-independent. By contrast, driving with generic fields leads to a rate that drops off initially while settling to a constant value once dissipation is sufficiently strong. Most intriguing is the observation that the delocalization rate displays *the same plateau* that spans the friction range $0.3 < \alpha < 0.9$ in both systems driven by the same intensity field, irrespective of the precise values of the driving frequencies and in spite of the fact that the dissipationless dynamics is vastly different in these two cases. At very large values of the dissipation parameter the delocalization rate begins to drop off in both cases.

The rate plateau observed at high temperatures is even more pronounced in Fig. 4b, which displays results with a stronger driving field of amplitude $V_0 = 30\hbar\Omega$ at $k_B T = 10\hbar\Omega$. The frequency of the generic driving field is $\omega_0 = 15\Omega$, while the choice $\omega_0 = 10.87\Omega$ is the optimal localization condition. Again, the rate curves merge as soon as the friction strength becomes considerable, but here the plateau extends up to or even beyond $\alpha \approx 1.5$.

The onset of the constant rate regime occurs at stronger friction and the rate plateau is overall less pronounced at lower temperatures. This is a consequence of more persisting phase relations at lower temperature, which favor state-specific behaviors. In addition, it is seen that driving can increase the decay rate at strong friction.

Similarly, Fig. 5 presents the results of path integral calculations for a driven asymmetric TLS coupled to a dissipative bath. Once again, the plateau value of the rate constant is generally smaller than that of the undriven dissipative TLS.

Recent work in Makri's group has presented a qualitative account of the behaviors observed in driven dissipative tunneling systems.[32,33,35] It is useful to resort to the quantized representation of the radiation field in which the Hamiltonian of Eq. (1) is time-independent:

$$\begin{aligned} H = & -\hbar\Omega\sigma_x + \varepsilon\sigma_z \\ & + \sum_j \frac{p_j^2}{2m_j} + \frac{1}{2}m_j\omega_j^2\left(x_j + \frac{c_j\sigma_z}{m_j\omega_j^2}\right)^2 + (a^+a + \tfrac{1}{2})\hbar\omega_0 + C\sigma_z(a + a^+) \end{aligned} \tag{16}$$

Here a and a^+ represent field annihilation and creation operators, respectively, and C is a measure of the field-TLS interaction strength. In order for Eq. (16) to be equivalent to (1) in the semiclassical limit, the coupling constant C must satisfy

$$C = \frac{V_0}{2\sqrt{n+\frac{1}{2}}} \tag{17}$$

where n is a quantum number that specifies the number of photons. In the semiclassical limit where the effects of the radiation field are equivalent to those of a time-dependent driving term, $n >> 1$.

In the absence of a bath, Eq. (16) represents two parabolic "diabatic" potential surfaces which are coupled via the constant term $\hbar\Omega$ and which intersect at the point

$$q_c = -\frac{\varepsilon}{V_0}\sqrt{\frac{\hbar(2n+1)}{\omega_0}}$$

and are coupled via the term $\hbar\Omega$. The minima of these curves are located at

$$q_{1,2} = \mp\frac{V_0}{\omega_0\sqrt{\hbar\omega_0(2n+1)}} \quad .$$

In the semiclassical limit $n >> 1$ one sees that $|q_c| >> |q_{1,2}|$ for any nonzero value of ε. These potential curves are sketched in Fig. 6 for a symmetric and an asymmetric TLS. In this manner the zero order Hamiltonian has been mapped onto a curve-crossing problem.

In the context of system dynamics one can treat the field coordinate q as a "reaction coordinate," since transitions between left and right states of the TLS correspond to hopping events between the two parabolic potential curves. Thus we may refer to these curves as diabatic potential surfaces for the TLS. Including the coupling terms $\hbar\Omega$, this set of diabatic curves can be diagonalized to produce a set of adiabatic potential curves (also shown in Fig. 6). Transitions between diabatic curves are known to occur primarily in the vicinity of the crossing point. It is important to note that in this case the reaction coordinate must be assumed to maintain constant amplitude for the driving field; this is indeed the case if the field quantum number approaches infinity. As a consequence, q does not go to equilibrium but rather approaches a steady state value.

The effects of resonant driving on the bare TLS dynamics can be understood by applying degenerate perturbation theory to the quantized field Hamiltonian. This physically appealing approach shows that localization in the case of a symmetric TLS is observed when Eq. (14) with $m=0$ is satisfied.[36,37] It is easy to extend that treatment to asymmetric systems. In this case, the parabolic diabatic surfaces are shifted in energy from one another by an amount 2ε. In the absence of dissipation one can induce transitions between these diabatic surfaces by tuning the frequency of the driving field such that 2ε is an integer multiple of $\hbar\omega_0$. The tunneling splitting of a degenerate doublet is proportional to the Franck-Condon overlap between the corresponding eigenstates of the left and right potential curves. In the limit $n \to \infty$ and if $2\varepsilon \approx m\hbar\omega_0$ this overlap can be shown to be proportional to $J_m(2V_0/\hbar\omega_0)$. It follows that the eigenstates of Eq. (16) are degenerate and thus localization is preserved if the driving field satisfies Eq. (14). Thus setting $2V_0/\hbar\omega_0$ to a zero of J_m leads to localization, despite the fact that the driving field satisfies the resonance condition $2\varepsilon/\hbar\omega_0 = m$ where m is an integer. Stated another way, in order to induce transitions between states in the driven asymmetric TLS the driving frequency must be tuned to the asymmetry - but having fixed the frequency of driving, one can still create a localizing field by tuning appropriately the amplitude of the driving field.

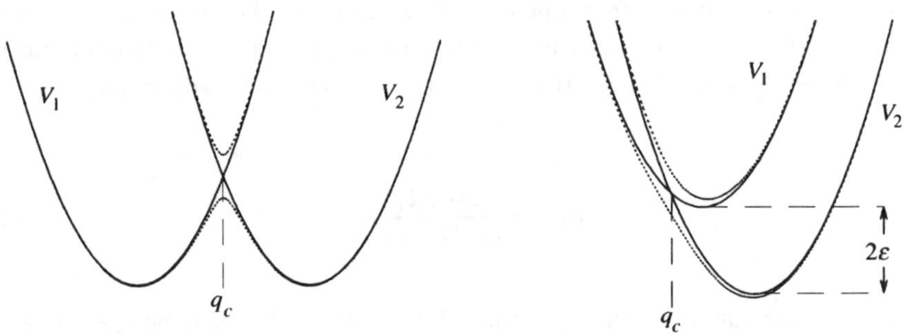

Figure 6. Parabolic diabatic potential surfaces for a driven TLS in the space of the quantized field oscillator coordinate q. The adiabatic curves are shown as dotted lines. The diabatic and adiabatic surfaces differ significantly only in the vicinity of the crossing, the Landau-Zener region. (a) Symmetric TLS. (b) Asymmetric TLS.

The remaining terms in Eq. (16) couple this two-surface system to a dissipative harmonic environment. This way the driven dissipative TLS is mapped on a time-independent dissipative multi-level curve-crossing problem, although the system-bath coupling differs from that employed in conventional system-bath models. As will become clear below, the details of the system-bath interaction are generally unimportant except at weak friction where dissipative processes are slow and mode specificity prevails.

By virtue of the quantized field analog, transitions between left and right states of the driven TLS correspond to hopping between diabatic curves in the time-independent quantized field Hamiltonian. Under conditions that lead to exponential decay, the situation is similar to that in electron transfer reactions with the field position q playing the role of a "reaction coordinate". However, there is an important difference between canonical reactions and the problem at hand: in order to give rise to constant amplitude driving, the field oscillator should maintain essentially constant energy. This is indeed the case if the field oscillator is in a state of very large quantum number, a necessary condition for validity of the quantum-classical correspondence. Thus, the "reaction coordinate" does not reach thermal equilibrium in this case. Instead, a steady state is reached at long time, in which the populations of the two diabatic curves become equal on average and the average TLS position is zero.

In the absence of dissipation, a standard procedure for analyzing curve-crossing problems is to use Landau-Zener theory[38-40]. This theory provides us with an expression describing the probability $P_{\rm LZ}$ that the system, having entered the region of the crossing from one diabatic surface, will hop to the other diabatic surface and finally exit the crossing region. The hopping probability (and thus the nature of the dynamics) is determined by the Landau-Zener "adiabaticity parameter"

$$\delta_{\rm LZ} = \frac{2\pi(\hbar\Omega)^2}{\hbar v_c \left|\lambda_2 - \lambda_1\right|} \,. \tag{18}$$

Here, v_c is the classical velocity (assumed constant) with which the system moves through the crossing region and λ_1 and λ_2 are respectively the slopes of V_1 and V_2 at the crossing point. When $\delta_{\rm LZ} > 1$ the motion tends to follow the adiabatic curves; in the opposite limit the dynamics is nonadiabatic. Using the energy conservation condition for the field at the crossing point and dropping terms that are negligible in the limit $n \to \infty$ leads to the following result for the classical velocity for $\varepsilon_0 < V_0$:

$$v_c = \sqrt{(2n+1)}\left(1-\frac{\varepsilon^2}{V_0^2}\right)^{1/2}. \tag{19}$$

A straightforward calculation yields the slopes at the crossing point, such that

$$|\lambda_2 - \lambda_1| = 2V_0\sqrt{\frac{\omega_0}{\hbar(2n+1)}}. \tag{20}$$

Insertion of this in Eq. (18) leads, in the limit of $n >> 1$, to the result

$$\delta_{\rm LZ} = \frac{\pi\hbar\Omega^2}{V_0\omega_0\sqrt{1-\varepsilon^2/V_0^2}}.$$

This expression shows that the hopping between electric field curves (and thus the dynamics of right-left transitions) is primarily nonadiabatic if $V_0 >> \hbar\Omega$. The Landau-Zener hopping probability

$$P_{\rm LZ} = 1 - e^{-\delta_{\rm LZ}}$$

is determined as

$$P_{\rm LZ} = 1 - \exp\left(\frac{-\pi\hbar\Omega^2}{V_0\omega_0\sqrt{1-\varepsilon^2/V_0^2}}\right)$$

Although the quantized field coordinate is not directly coupled to the bath in the present case, the mixing of the diabatic curves in the Landau-Zener region dilutes the TLS-bath coupling over the region where nonadiabatic transitions are most probable. Frauenfelder and Wolynes[41] argue that if the curve-crossing system is allowed to interact with a dissipative environment, the system dynamics at moderate dissipation and high temperature displays an exponential decay with a rate constant that does not change over a broad range of friction. This occurs when the dissipation is strong enough to destroy the quantum mechanical phase memory of the system on a time scale smaller than the time that the system spends in the crossing region, but at the same time the dissipation is not yet so strong that the motion of the system is diffusive on the length scale of the crossing region. These arguments are in line with the results obtained earlier by Tembe, Friedman and Newton.[42] In this regime the forward hopping rate is given semiclassically by the frequency π/ω_0 of passing through the curve crossing point times the Landau-Zener hopping probability. In the limit $n >> 1$ the forward and backward rates are equal and thus the decay constant becomes in the semiclassical approximation

$$k_{SC} = 2\frac{\omega_0}{\pi} P_{LZ} \tag{21}$$

Use of this expression leads to the final result for the semiclassical delocalization rate of the driven dissipative TLS:

$$k_{SC} = \frac{2\omega_0}{\pi}\left(1 - \exp\left(\frac{-\pi\hbar\Omega^2}{V_0\omega_0\sqrt{1-\varepsilon^2/V_0^2}}\right)\right) \approx \frac{2\hbar\Omega^2}{V_0\sqrt{1-\varepsilon^2/V_0^2}} \tag{22}$$

This result shows that at high temperature and intermediate friction the TLS decay dynamics are independent of the driving frequency and the specifics of the environment (the temperature as well as the strength and mechanism of dissipation). Eq. (22) is valid in the regime of strong coupling between the driving field and the TLS. As the field amplitude becomes smaller the plateau in the rate shifts to lower friction and extends over a smaller range of dissipation strengths. In the limit $V_0 \to 0$, the dynamics reverts to that of the undriven dissipative TLS. The same applies also to driving fields of very high frequency, $\omega_0 >> \Omega$, whose effects average to zero without affecting the TLS dynamics.

4 Summary and outlook

In spite of their apparent simplicity, driven dissipative TLSs exhibit very rich dynamics. In the regime most relevant to technological applications, that of high temperature and intermediate dissipation strength, initially localized exciton states in TLSs interacting with monochromatic radiation decay essentially exponentially with a time constant that is independent of material properties and external conditions other than the field intensity. In addition, the decay is generally slow compared to that in the absence of dissipation and the lifetime can be controlled by adjusting the field strength. This robust behavior makes such systems ideal candidates for reversible information storage and signal transmission.

Whether these effects persist in systems with multiple states, in particular in the presence of strongly anharmonic environments as in the case of organic materials or biological systems, is a very interesting question that awaits future research.

Acknowledgment

This material is based upon work supported by the National Science Foundation under grants No. NSFCHE 93-13603, NSF CHE 93-57102 NYI and NSF DMR 96-12194.

References

1 S. Luryi, *Solid State Communications* **65,** 787-789 (1988).
2 K. Leo, J. Shah, E. O. Goebel, T. C. Damen, S. Schmitt-Rink, W. Schaefer, and K. Koehler, *Phys. Rev. Lett.* **66,** 201 (1991).
3 M. Lindberg and S. W. Koch, *Phys. Rev. B* **38,** 3342 (1988).
4 W. Pötz and J. Zhang, *Phys. Rev. B* **45,** 11496-11510 (1992).
5 T. Kuhn and F. Rossi, *Phys. Rev. B* **46,** 7496 (1992).
6 P. Resibois, *Phys. Fluids* **6,** 817 (1963).
7 U. Hohenester and W. Pötz, *Phys. Rev. B* **56,** 13177-13189 (1997).
8 T. Dittrich, B. Oelschlagel, and P. Hänggi, *Europhys. Lett.* **22,** 5-10 (1993).

9 M. Grifoni, M. Sassetti, J. Stockburger, and U. Weiss, *Phys. Rev. E* **48,** 3497-3509 (1993).
10 F. Grossmann, T. Dittrich, P. Jung, and P. Hänggi, *J. Stat. Phys.* **70,** 229-245 (1993).
11 M. Grifoni, M. Sassetti, P. Hänggi, and U. Weiss, *Phys. Rev. E* **52,** 3596-3607 (1995).
12 M. Grifoni and P. Hänggi, *Phys. Reports* **304,** 229-358 (1998).
13 Y. Dakhnovskii, *Phys. Rev. B* **49,** 4649-4657 (1994).
14 Y. Dakhnovskii and R. D. Coalson, *J. Chem. Phys.* **103,** 2908-2916 (1995).
15 D. G. Evans, R. D. Coalson, H. Kim, and Y. Dakhnovskii, *Phys. Rev. Lett.* (1996).
16 A. J. Leggett, S. Chakravarty, A. T. Dorsey, M. P. A. Fisher, A. Garg, and M. Zwerger, *Rev. Mod. Phys.* **59,** 1-85 (1987).
17 D. E. Makarov and N. Makri, *Chem. Phys. Lett.* **221,** 482-491 (1994).
18 N. Makri and D. E. Makarov, *J. Chem. Phys.* **102,** 4600-4610 (1995).
19 N. Makri and D. E. Makarov, *J. Chem. Phys.* **102,** 4611-4618 (1995).
20 N. Makri, *J. Math. Phys.* **36,** 2430-2456 (1995).
21 E. Sim and N. Makri, *Chem. Phys. Lett.* **249,** 224-230 (1996).
22 E. Sim and N. Makri, *Comp. Phys. Commun.* **99,** 335-354 (1997).
23 Makri, *J. Phys. Chem.* **102,** 4414-4427 (1998).
24 N. Makri, *Chem. Phys. Lett.* **193,** 435-444 (1992).
25 R. P. Feynman and A. R. Hibbs, *Quantum Mechanics and Path Integrals* (McGraw-Hill, New York, 1965).
26 R. P. Feynman and J. F. L. Vernon, *Ann. Phys.* **24,** 118-173 (1963).
27 M. Topaler and N. Makri, *Chem. Phys. Lett.* **210,** 448 (1993).
28 N. Makri, *J. Chem. Phys.* **111,** 6164-6167 (1999).
29 N. Makri and K. Thompson, *Chem. Phys. Lett.* **291,** 101-109 (1998).
30 K. Thompson and N. Makri, *J. Chem. Phys.* **110,** 1343-1353 (1999).
31 F. Grossmann, T. Dittrich, P. Jung, and P. Hänggi, *Phys. Rev. Lett.* **67,** 516-19 (1991).
32 N. Makri, *J. Chem. Phys.* **106,** 2286-2297 (1997).
33 G. Taft and N. Makri, *J. Phys. B* **31,** 209-226 (1998).
34 M. Grifoni, M. Sassetti, P. Hänggi, and U. Weiss, *Phys. Rev. E* **52,** 3596-3607 (1995).
35 N. Makri and L. Wei, *Phys. Rev. E* **55,** 2475-2478 (1997).

36 J. Plata and J. M. Gomez Llorente, *Phys. Rev. A* **48,** 782 (1993).
37 D. E. Makarov, *Phys. Rev. E* **48,** R4146 (1993).
38 L. D. Landau, Z. *Sowjun.* **2,** 46 (1932).
39 C. Zener, *Proc. R. Soc. A* **137,** 696-703 (1932).
40 E. Stueckelberg, *Helv. Phys. Acta* **5,** 369 (1932).
41 H. Frauenfelder and P. G. Wolynes, *Science* **228,** 337 (1985).
42 B. L. Tembe, H. L. Friedman, and M. D. Newton, *J. Chem. Phys.* **76,** 1490-1507 (1982).

EXCITED- AND GROUND-STATE WAVE PACKET DYNAMICS IN ORGANIC MATERIALS INDUCED BY FEMTOSECOND CHIRPED PULSES

KAZUHIKO MISAWA
Department of Applied Physics, Tokyo University of Agriculture and Technology,
CREST, Japan Science and Technology Corporation
2-24-16 Naka-cho, Koganei, Tokyo 184-8588, Japan
E-mail: kmisawa@cc.tuat.ac.jp

TAKAYOSHI KOBAYASHI
Department of Physics, University of Tokyo,
7-3-1 Hongo, Bunkyo-ku, Tokyo 113, Japan
E-mail: kobayashi@phys.s.u-tokyo.ac.jp

Wave packet motion is observed to be dependent on chirp direction and rate of excitation by femtosecond pulses in a cyanine dye molecule and a halogen-bridged metal complex. Strong reduction in excited state population is efficient for negatively chirped pulses in the cyanine dye molecule, which is explained in terms of a pump-dump process. Vibrational oscillation for positively chirped excitation is retarded by $\pi/4$ with respect to negatively chirped excitation in the halogen-bridged metal complex, which originates from the difference between excited and ground state wave packets.

1 Introduction

Recent progress in femtosecond pulse generation has enabled us to observe real-time dynamics of nuclear motions in molecular systems.[1] Femtosecond pulses of less than 10 fs are easily available nowadays. Even sub-5 fs pulses in the visible[2] are applied to real-time probing of structural changes in large organic molecules. In particular, femtosecond wave packet dynamics has been recently studied to realize the possibility of controlling molecular dynamics using tailored ultrashort pulses.[3] For instance, the products of photoinduced chemical reactions can be selected. Selective dissociation of NaI with chirped pulses was recently demonstrated by controlling the wave packet dynamics.[4] Ref. 4 reported that negatively chirped pulses brings larger yield of Na^*.

One of our purposes is to realize such selective chemistry in the condensed phase. In the present paper, we show experimental results on the vibrational wave packet in a cyanine molecule and a halogen-bridged metal complex. Especially, different properties of an excited wave packet with negatively and positively chirped pulses will be discussed.

2 Experimental

A variable chirp laser was developed to observe chirp dependent dynamics of organic molecules. The excitation and probe pulses are both derived from a Ti:sapphire multi-pass amplifier. The positive dispersion inside the amplifier is compensated with a pair of prisms. The chirp of the pulses can be controlled by changing the amount of the dispersion in the prism compressor. The first prism is put on a computer controlled translation stage to change the insertion distance in the prism. The autocorrelation trace of the output pulses is monitored with a modified Michelson interferometer, and the second order dispersion is compensated so as to minimize the pulse duration via a feedback loop.

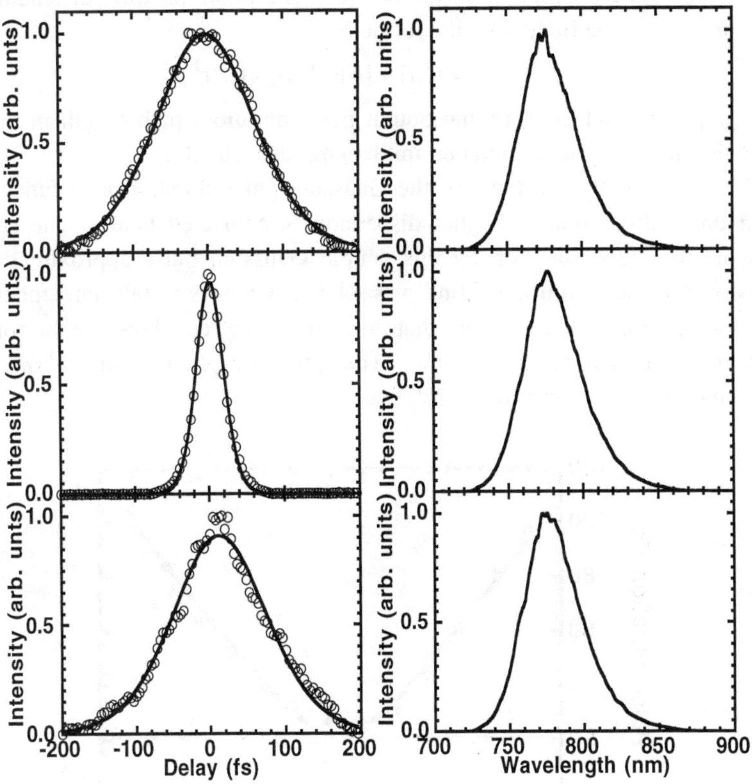

Figure 1. Autocorrelation traces (left) and spectra (right) of negatively chirped, transform limited, and positively-chirped pulses. The second order dispersions are –750 fs^2, 0fs^2, and 750fs^2, respectively, from the top to bottom.

The shortest available duration of the output after the compressor is 21 fs. The dispersion is automatically adjusted to obtain the desired value of the chirp. Figure 1 shows the autocorrelation traces and spectra of the output pulses with different chirps. The middle panel shows the case of minimum duration, which we call transform limited (TL). In the top and bottom panels, the correlation traces look similar to each other, but the chirp direction is opposite, i.e. the former is negatively chirped (NC) while the latter is positively chirped (PC). Note here that the spectral profiles do not change by chirp introduction and its direction. The pulse energy, pulse duration, and the spectral profile are all the same except the chirp direction.

The pulse duration obtained from the autocorrelation trace is plotted as a function of the prism position in Fig. 2. The prism insertion distance is proportional to the path length in the prism. The prism position dependence of the pulse duration can be fitted with the formula

$$t_{\text{out}} = t_0\sqrt{1+\left(4\ln 2\cdot\beta(\lambda)L/t_0^2\right)^2}\,. \tag{1}$$

Here, t_{out} , L, t_0, and $\beta(\lambda)$ are the output pulse duration, path length in quartz, FT-limited duration, and dispersion per unit length, respectively.

This is the analytical form of the Gaussian pulse duration as a function of the second order dispersion.[5] Higher dispersion is neglected here. The solid curve represents the fitted function by this simple Gaussian pulse approximation. The dispersion obtained from the fitting was 44 fs^2, which is in fair agreement with the real value of 36fs^2. We consider that the dominant effect shown in the following is due to the second order dispersion. The quadratic phase term, $\Phi''(\omega)=\beta(\omega)L$, can be changed in the range of ± 750 fs^2.

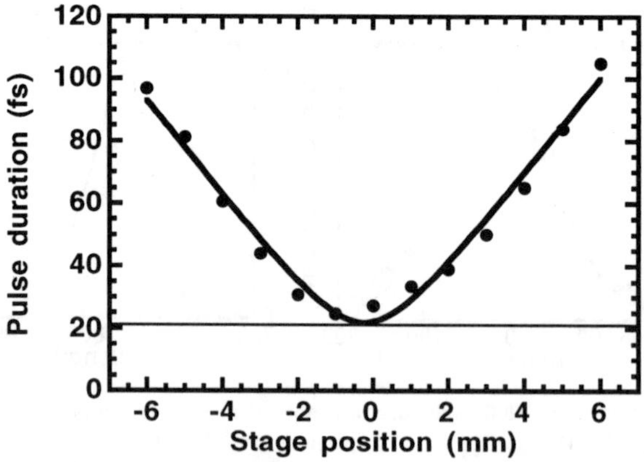

Figure 2. Prism position dependence of pulse duration obtained from the autocorrelation trace.

The dye molecule studied here is 3-3'-diethylthiatricarbocyanine iodide (DTTCI). An ethanol solution of 2×10^{-4} M DTTCI is circulated in a 1-mm thick quartz cell. The material dispersion is precompensated in the prism compressor. The incident light is resonant with the lowest edge of the S0→S1 transition of DTTCI. A mixed-valence metal complex is $[Pt(en)^2]\ [Pt(en)_2Br_2](ClO_4)_2$ (Pt-Br). A single crystal of Pt-Br used in this experiment was synthesized by the method reported previously.[6] The incident light is resonant with the lower edge of the excitonic transition of Pt-Br.

3 Pump-dump process in a cyanine molecule

3.1 Basic transients

The right panel of Figure 3 shows the time dependence of the transmittance change measured with the TL pulses at the excitation photon density of 4.5×10^{16} photons/cm^2 at 800 nm. The long-lived signal associated with an oscillatory structure was observed. The oscillation corresponds to the vibrational mode of 160 ± 10 cm^{-1}. The left panel shows the Fourier spectrum of the transient data. This mode has been also observed in a photon echo experiment.[7]

In order to discuss the excited state dynamics in more detail, we measured the transient transmittance change spectrum. Figure 4 shows the transient spectrum measured with transform-limited pulses. The left panel is a three dimensional bird's eye view plot, and the right is a contour plot. A prominent oscillatory structure can be seen in the contour plot. As the probe wavelength becomes longer, the oscillation starts to split, and the finally the oscillation peak appears twice during the period. This splitting is more remarkable in the temporal profile at each probe

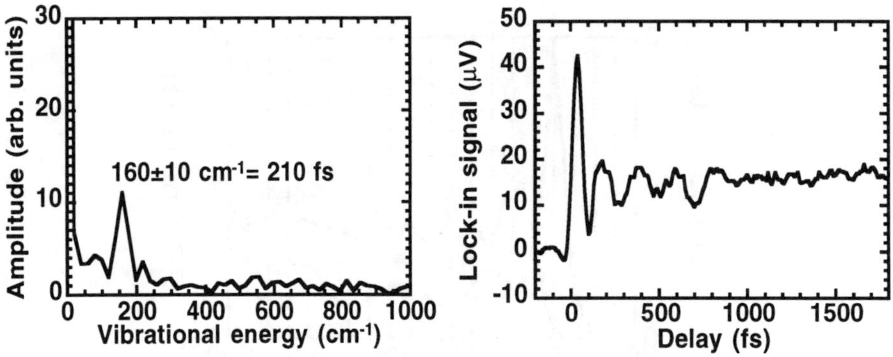

Figure 3. Fourier spectrum (left) of the time dependence of the transmittance change (right).

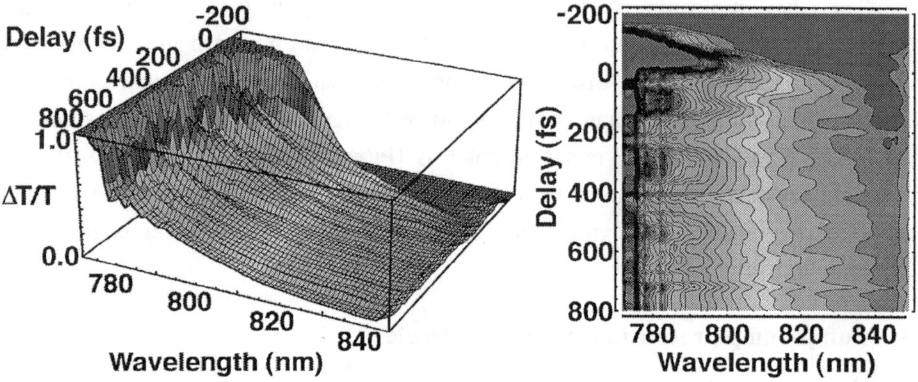

Figure 4. Three dimensional bird's eye view plot (left) and contour plot (right) of the transient spectrum measured with transform-limited pulses.

wavelength as shown in Fig. 5. At 1.57eV, which is the probe peak photon energy, the oscillation period is 210 fs. The period at 1.54eV, corresponding to the peak photon energy of the luminescence intensity, is half of that at 1.57 eV.

Figure 6 shows the potential energy surface estimated from the absorption and emission spectra. During one period of the molecular vibration, the wave packet pass through the turning point once in a single round trip. On the other hand, the wave packet passes the potential minimum twice in the trip. This is the reason why the oscillation period changes with the probe photon energy. The potential energy diagram indicates that the signal probed at 1.57 eV corresponds to the excited state wave packet at one of the turning points, while that at 1.54 eV corresponds to the potential minimum in the excited state. This result tells us that we mainly observe the wave packet dynamics in the excited state, not in the ground state. Note that

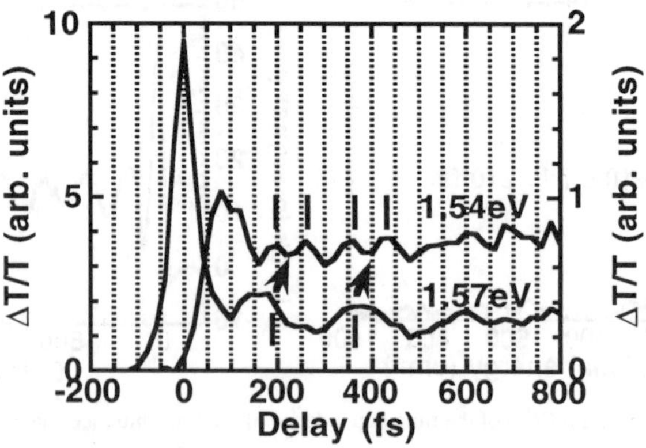

Figure 5. Time dependence of the transmittance change at 1.54 and 1.57 eV.

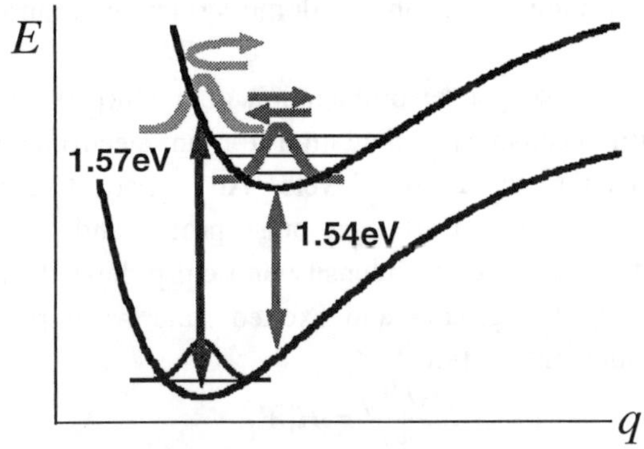

Figure 6. Potential diagram estimated from the absorption and emission spectra.

the ground state wave packet should show the full period oscillation at 1.54 eV and the half period at 1.57 eV. The similar doublet structure due to the wave packet round trip through the probe region has been observed in a laser-induced fluorescence experiment for NaI molecules.[8] The peak separation between the two maxima is also dependent on the probe wavelength. The explanation of this splitting is established to explain inward and outward bound propagation of the wave packet.

3.2 Chirp-dependent dynamics

Figure 7 shows the time dependence of the transmittance change induced by several different chirp rates. The signal intensity is considered to be proportional to the population difference between the excited and ground states. The amplitude of the damped oscillation is substantially decreased and enhanced in the cases of negatively and positively chirped excitations, respectively. At $-500fs^2$, excited state population is almost completely cancelled, but some oscillatory structure can still be seen. This should be a contribution from the ground-state wave packet, which has been reported to give a larger contribution with negatively-chirped pulses.[9] This cancellation can be explained in terms of a pump-dump process.[10]

At higher excitation intensity, both absorption and stimulated emission take place. The motion of the excited wave packet results in the dynamic Stokes shift, where the population moves from higher to lower energy. Negatively chirped pulses, in which the blue part comes earlier and the red part later, can follow this motion. This dynamical resonance enhances the stimulated emission, and as a result, excited state population is dumped. Actually, we observed that the cancellation of the long-lived signal disappears in a weak excitation regime. On the contrary, positively chirped pulses cannot induce the above dumping process, and pure absorption becomes dominant.

To understand the wave packet propagation after chirped pulse excitation, we performed a quantum mechanical calculation for an electronic two-level system associated with molecular vibrational levels. An unperturbed electric field, an arbitrary potential surface, and an arbitrary pulse profile and phase are used in the calculation in order to reproduce the intensity and chirp dependent dynamics.. The temporal propagation of the ground and excited state wave packets satisfies the time-dependent Schrödinger equation,

$$i\hbar\frac{\partial|\Psi\rangle}{\partial t} = H|\Psi\rangle \tag{2}$$

The Hamiltonian H is expressed in the following form

$$\begin{aligned} H &= H_{\text{ele}} + H_{\text{vib}} + H_{\text{int}} \\ &= |1\rangle H_{\text{vib}}^{(1)}\langle 1| + |2\rangle\left(H_{\text{vib}}^{(2)} + \hbar\omega_0\right)\langle 2| - \left(|1\rangle\mu E(t)\langle 2| + |2\rangle\mu^{*}E(t)\langle 1|\right). \end{aligned} \tag{3}$$

Substituting the total wave function $|\Psi\rangle = |\psi_1(t)\rangle|1\rangle + |\psi_2(t)\rangle|2\rangle$ into the Schrödinger equation, a coupled set of partial differential equations for the vibrational wave functions are deduced as follows:

$$i\hbar\frac{\partial|\psi_1\rangle}{\partial t} = H_{\text{vib}}^{(1)}|\psi_1\rangle - \mu E(t)|\psi_2\rangle \tag{4a}$$

$$i\hbar\frac{\partial|\psi_2\rangle}{\partial t} = \left(H_{\text{vib}}^{(2)} + \hbar\omega_0\right)|\psi_2\rangle - \mu E(t)|\psi_1\rangle. \tag{4b}$$

Here $|\psi_1(t)\rangle$ and $|\psi_2(t)\rangle$ are the vibronic wave functions for these electronic states. We assume these electronic states couple to a single vibrational mode along a one-dimensional normal coordinate. The transition dipole moment μ is assumed to be independent of the molecular displacement.

The integration of these equations by a small temporal step Δt is given by

$$\begin{bmatrix} |\psi_1(t+\Delta t)\rangle \\ |\psi_2(t+\Delta t)\rangle \end{bmatrix} = \exp\left[-\frac{i\hat{H}\Delta t}{\hbar}\right]\begin{bmatrix} |\psi_1(t)\rangle \\ |\psi_2(t)\rangle \end{bmatrix}. \tag{5}$$

The Hamiltonian is decomposed into kinetic, potential, and interaction operators, which are expanded according to the split-operator scheme.[11]

In the present calculation, both population relaxation and dephasing are neglected. Recent calculation on vibrational coherence induced by chirped pulses includes relaxation processes in the calculation.[12] The intra-pulse, pump-dump process finishes within the pulse duration, which is 20 fs for TL pulses and as long as 80 fs for chirped pulses. Within such a short duration, any relaxation process is not considered to be significant.

The temporal profile of the electric field $E(t)$ is given by

$$E(t) = \mathrm{Re}\left[E_0\sqrt{\frac{1}{2\pi(1-iC)t_0^2}}\exp\left[-\frac{(1+iC)t^2}{(1+C^2)t_0^2}\right]\exp[-i\omega t]\right], \tag{6}$$

where ω, t_0, and E_0 are the central angular frequency, the temporal duration of Fourier-transform limited pulse, and the electric field amplitude, respectively. The spectrum is unchanged, and second-order dispersion is introduced. The a chirp parameter, C, is directly related with the second-order dispersion as $C = \Phi''(\omega)/t_0^2$. It represents broadening of a Gaussian pulse due to the linear chirp. The pulse duration of the linearly chirped pulse with a parameter C becomes $t_0' = \sqrt{1+C^2}\,t_0$. The full width at half maximum (FWHM) of the pulse is given by $t_{\mathrm{FWHM}} = 2\sqrt{\ln 2}\,t_0$.

Figure 8 shows the simulated temporal propagation of excited- and ground-state wave packets for the shifted Morse potential. The solid and gray curves represent the squared amplitudes of the ground- and excited-state wave functions, $|\psi_2\rangle$ and $|\psi_1\rangle$, respectively. The excited pulse duration is set to be 20fs, and the chirped parameters ±2 are used for negatively and positively chirped cases. Note that the spectral profile $|E(\omega)|^2$ is always the same, and the temporal profile $|E(t)|^2$ depends on C^2. Hence there are no differences in the spectrum, pulse energy, and temporal duration between two cases of the equal C^2. However, the remaining population of the two-level system after termination of the pulse is different for the negative and positive chirps. This calculation explains the experimental result of the chirp-dependent excitation. The PC excitation efficiently produces the excited-state population, while the NC pulse excites and dumps the population by the leading and trailing edges, respectively.

The calculation present a quantum mechanical picture of the pump-dump process. It is worth mentioning the spatial extension of the excited wave packets. For the NC case, the spatial distribution is narrower than that for the PC case at 0 fs. The pulse chirp compensates the dispersion among the frequency components of the wave packet. The overlap integral between the excited- and ground-state wave packets is large, and therefore the trailing edge interacts with the narrowly distributed wave packet more efficiently in the NC case. As a result, the pump-dump process becomes efficient in this case. From this figure the ground state wave packet is found to be driven more efficiently for the NC excitation. Although the excited state population is almost completely cancelled at $-500\mathrm{fs}^2$, still some oscillatory structure can be seen. A larger contribution from the ground-state wave

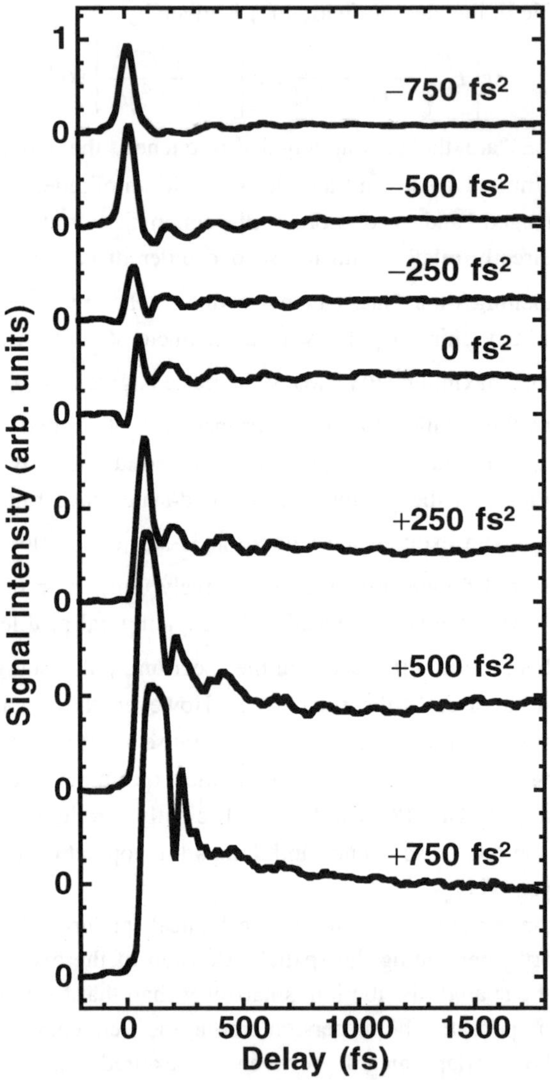

Figure 7. Time dependence of the transmittance change induced by different chirp rates in the range of $\pm750\ fs^2$.

packet has been reported with the NC pulses[9] than in the opposite case as discussed below.

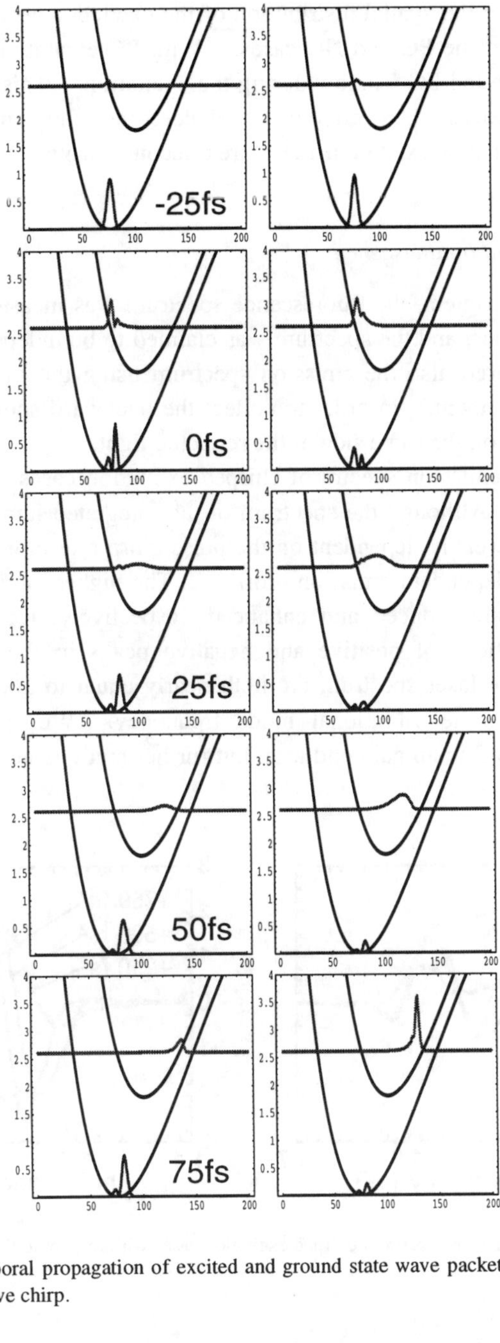

Fig. 8. Simulated temporal propagation of excited and ground state wave packets, (left column) negative chirp and (right) positive chirp.

The difference in the spatial distribution of the excited-state wave packet can be seen at 0 fs between the PC and NC cases. With PC excitation, the wave packet rapidly escapes from the FC window during the excitation. This broadening makes the pump-dump process less effective. Hence, the chirp dependence of the population at a particular excitation pulse area becomes asymmetric with respect to the chirp direction.

3.3 Chirp-dependent luminescence

In the previous experiment, the fluorescence spectrum was measured to monitor the excited-state dynamics, and the spectrum was claimed to be independent of the pulse chirp.[10] We measured also the emission spectrum using the same setup as in the pump-probe measurement. In order to collect the backward spontaneous emission, the detector was put on the same side of the reflected light.

The difference emission spectra of chirped excitation cases for the TL case are depicted in Fig. 9. Although the spectrum of the integrated luminescence intensity was reported to be weakly dependent on the pulse chirp rate in the previous paper,[10] we observed chirp-dependent emission profiles. The higher- and lower-sides of the emission intensity are reduced and enhanced, respectively, for both PC and NC cases. The separations of positive and negative peaks in the difference spectra from the peak of the laser spectrum are both nearly equal to 160 cm^{-1}, which is in fair agreement with the vibrational mode frequency. We speculate that these structures are due to Raman gain and loss, but further study is needed to confirm the mechanism.

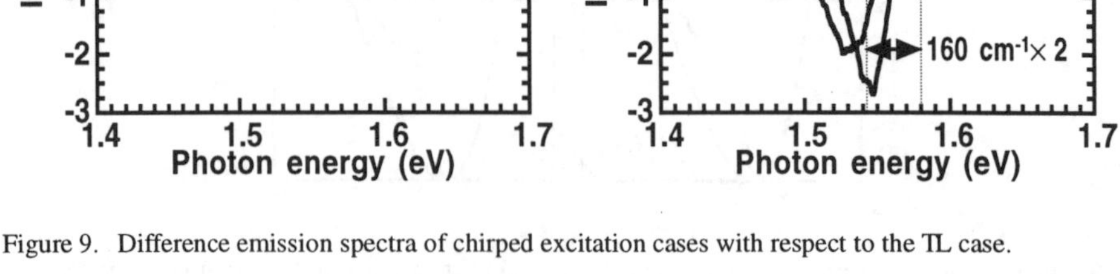

Figure 9. Difference emission spectra of chirped excitation cases with respect to the TL case.

4 Oscillation phase shift in a halogen-bridged mixed-valence metal complex

4.1 Sample properties

Halogen-bridged mixed-valence metal complexes are composed of one-dimensional chains of a transition metal ion such as Pt, Pd, and Ni and a halogen ion like Br, Cl, and I.[6] Each metal ion is surrounded by ligands such as ethylenediamine, and, as a result, each chain is isolated from each other. The optical property of these materials is considered to be quasi one-dimensional.[13] One of the most important features of halogen-bridged metal complexes is their strong electron-phonon coupling.[14] The vibrational mode strongly coupled to the electronic states is symmetric vibration of the halogen ions. Many overtones are observed in the resonance Raman spectrum, and a large Stokes shift is also observed.

Assume a shifted harmonic oscillator with an equal mode frequency for both the ground and excited states. Typically, the peak photon energy of the photoluminescence spectrum is less than half, or even close to a third of the absorption peak energy.[15] In Pt-Br, the photon energy at the absorption peak is 1.95 eV, while the emission peak is at 0.76 eV. The vibrational mode frequency obtained from Raman scattering corresponds to 21.7 meV. The Franck-Condon transition in Pt-Br is resonant with the twenty fifth vibrational level. The detuning from the FC transition of the incident light is 0.38 eV, which corresponds to the seventh vibrational level.

4.2 Excited- and ground-state wave packets

Figure 10 shows the time dependence of the transmittance change measured at the excitation power density of 3×10^{16} photons/cm^2 with positively chirped 60-fs pulses. The pronounced oscillatory structure is observed, which corresponds to a vibrational mode frequency of 170 ± 3 cm^{-1}. The upper panel shows the Fourier spectrum of the transient data after removing the low frequency component. This mode is already assigned as the Pd-Br symmetric stretching vibrational mode.[14] The main decay profile can be fitted with a sum of exponential rise and decay components, which correspond to the formation and relaxation due to thermalization of the self-trapped exciton (STE), respectively. A long-lived component corresponds to a thermal STE. The time constants of the rise and decay are determined to be 130 ± 40 and 670 ± 20 fs, respectively. The relaxation pathways of the excitons in these metal complexes in a femtosecond regime have been discussed elsewhere.[16]

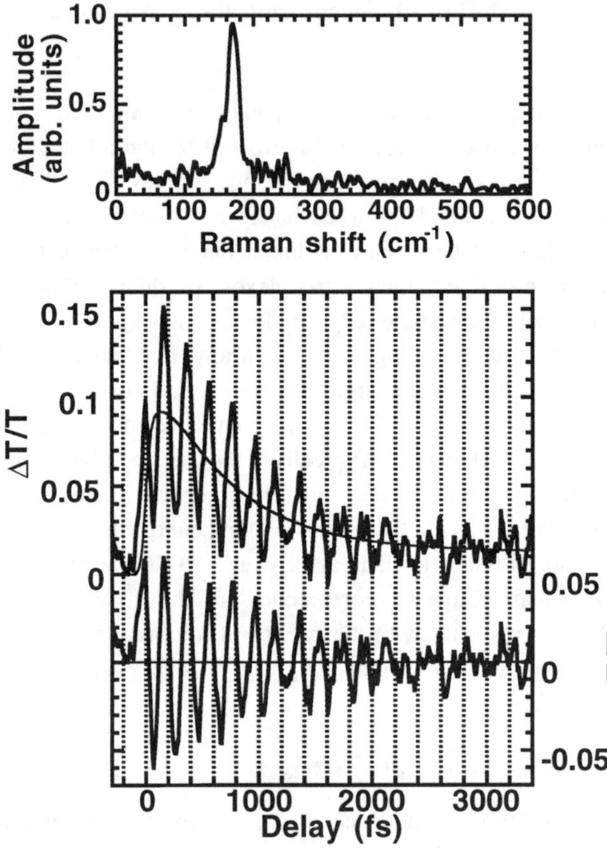

Figure 10. Time dependence of the transmittance change (lower) and Fourier spectrum (upper) of the transient data after removing the low frequency component.

In the present paper we focus on the dynamics of the vibrational wave packet. This oscillatory structure due to the vibrational wave-packet motion is sensitive to the chirp direction of the excitation pulses. The transient transmittance change with the PC and NC excitations are shown in Fig. 11. The lower panels show the Fourier spectra of the transient data. The center frequencies in both cases are equally 170 cm^{-1}, but the Fourier amplitude is smaller in the negative-chirp case than the positive one.

The most remarkable difference is the phase of the oscillation due to the vibrational wave packet. Figure 12 shows magnified traces of the transient data. The oscillation induced by the NC pulses leads by 49 fs with respect to the positive chirp case by almost exactly a quarter of the period of 195 fs. The bottom curves show sine and cosine functions of the period. Hence the data shows that the negatively chirped pulses excite the sine phase oscillation, while the positively chirped pulse induces cosine oscillation. The frequency of the STE is expected to be lower than that of the ground state, but the frequency difference is not evident in the present experiment.

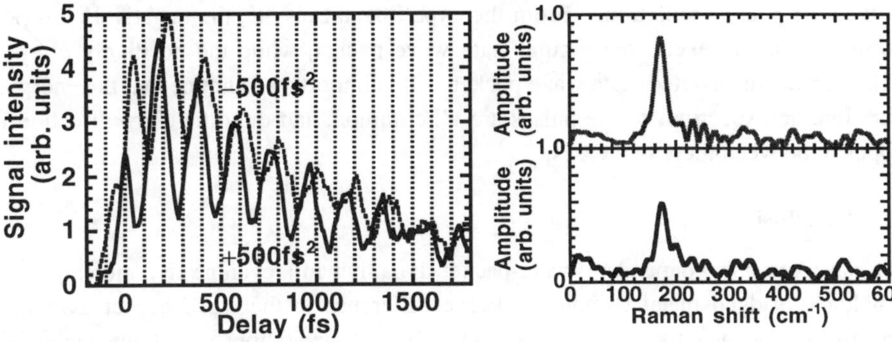

Figure 11. Transient transmittance change with the PC and NC excitations (right) and the Fourier spectra (left) of the transient data.

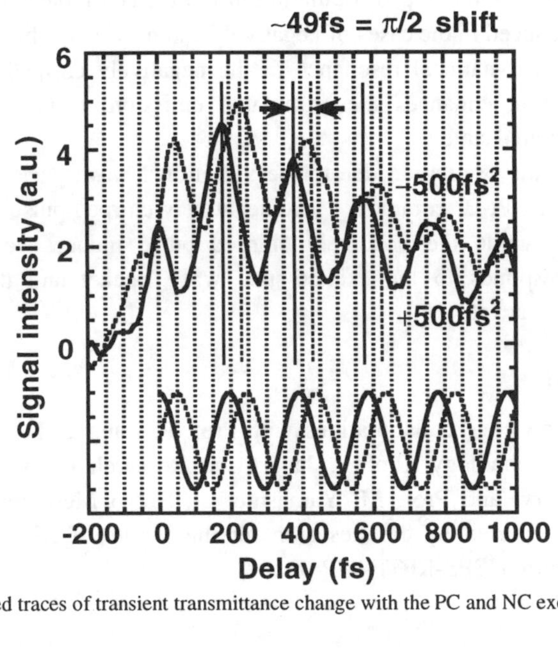

Figure 12. Magnified traces of transient transmittance change with the PC and NC excitations.

The potential energy diagram is a key to the explanation of this phase shift. The vibrational wave packet is possibly generated on both the ground and excited potential energy surfaces. The FC transition induces the wave packet just after excitation at one of the turning points on the excited state potential energy surface. Therefore, the excited-state wave packet shows cosine phase oscillation. On the other hand, the ground-state wave packet is not initially displaced from the potential minimum, and it is accelerated by the impulsive stimulated Raman process resulting in the sine-phase oscillation. From the experimental results, the negatively chirped pulses efficiently excite the ground-state wave packet, while the positively chirped pulses create the excited-state wave packet. It is important to notice that the energy, duration, and spectrum of the pulses are all common, and only the chirp direction is opposite between these two cases.

5 Conclusions

We observed chirp-dependent wave-packet dynamics in a cyanine dye molecule and a halogen-bridged metal complex. Large difference in the wave packet dynamics are found for negatively and positively chirped excitations for both materials, although the energy, duration, and spectrum of the pulses are all common and only the chirp direction is opposite.

The amplitude of the damped oscillation in the DTTCI molecule is substantially decreased and enhanced in the cases of negatively and positively chirped excitations, respectively. The dynamical resonance of the negatively chirped pulses with the wave packet motion enhances the stimulated emission, resulting in the strong dumping of the excited state.

The phase of the oscillation due to the vibrational wave packet is shifted by a quarter period in the Pt-Br complex. The negatively chirped pulses excite the sine-phase oscillation, while the positively chirped pulse induces cosine oscillation. The former corresponds to the excited-state wave packet and the latter to the ground-state wave packet.

Acknowledgments

The authors acknowledge collaborations with Prof. R. Lang at Tokyo University of Agriculture and Technology, Prof. J. Janszky at Research Laboratory for Crystal Physics in Hungary, and Prof. M. Yamashita at Tokyo Metropolitan University. TK was partly supported by the Research for the Future of Japan Society of the Promotion of Science (JSPS-RFTF-97P-00101).

References

1. *Femtochemistry*, M. Chergui, ed. (World Scientific, Singapore 1996)
2. A. Shirakawa, I. Sakane, M. Takasaka, and T.Kobayashi, *Appl. Phys. Lett.* **74**, 2268 (1999).
3. J. Janszky, P. Adam, A. V. Vinogradov, and T. Kobayashi, *Chem. Phys. Lett.* **213**, 368 (1993).
4. C. J. Bardeen, J. Che, K. R. Wilson, V. V. Yakovlev, V. A. Apkarian, C. C. Martens, R. Zadoyan, B. Kohler, and M. Messina, *J. Chem. Phys.* **106**, 8486 (1997)
5. *Nonlinear Fiber Optics*, G. P. Agrawal, (Academic Press, San Diego, 1989) Chap. 3.
6. N. Matsumoto, M. Yamashita, and S. Kida, *Bull. Chem. Soc. Jpn.* **51**, 2334 (1978).
7. W. P. de Boeij, M. S. Pshnichnikov, and D. A. Wiersma, *J. Phys. Chem.* **100**, 11806 (1996).
8. P. Cong, G. Roberts, J. L. Herek, A. Mohktari, and A. H. Zewail, *J. Phys. Chem.* **100**, 7832 (1996).
9. C. J. Bardeen, Q, Wang, and C. V. Shank, *Phys. Rev. Lett.* **75**, 3410 (1995).
10. G. Cerullo, C. J. Bardeen, Q. Wang, and C. V. Shank, *Chem. Phys. Lett.* **262**, 362 (1996).
11. M. D. Feit, J. A. Fleck, and A. Steiger, *J. Comput. Phys.* **47**, 42 (1982).
12. K. Mishima, M. Hayashi, J. T. Lin, K. Yamashita, and S. H. Lin, *Chem. Phys. Lett* **309**, 279 (1999).
13. K. Nasu, *J. Phys. Soc. Jpn.* **53**, 427 (1984).
14. M. Tanaka, S. Kurita, Y. Okada, T. Kojima, and Y. Yamada, *Chem. Phys.* **96**, 343 (1985).
15. Y. Wada, T. Mitani,, M. Yamashita, and T. Koda, *J. Phys. Soc. Jpn.* **54**, 3143 (1985).
16. H. Ooi, M. Yamashita, and T. Kobayashi, *Solid State Commun.* **86**, 789 (1993).

OPTICAL AND STRUCTURAL CONTROL OF QUANTUM INTERFERENCE IN SEMICONDUCTORS

W. PÖTZ AND T. KRIVOSHEEV

University of Illinois at Chicago, Physics Department, Chicago IL 60607, USA,

E-mail: wap@uic.edu

Electromagnetic and one structural coherent control schemes for semiconductors are discussed. These elementary schemes manipulate electronic inter(sub)band transitions and allow the coherent control of a variety of physical phenomena and processes, such as photo–absorption and emission, phonon emission, excitonic (many–body) effects, and THz emission. Here, we specifically address the use of sub–picosecond two–color pulses to control photo–absorption in bulk semiconductors and semiconductor quantum wells, as well as the initiation of coherent charge oscillations in double wells. It is shown that the phase sensitivity in absorption arises from the presence of the electron–electron interaction. It can be maintained as long as the pump pulse duration does not significantly exceed the inverse beating frequency associated with the two–color pump pulse.

1 Introduction and Theoretical Background

In the quantum–dynamical regime interference occurs between competing pathways which lead from a given initial state to a particular final state of the system. Competing pathways may arise by structural design of a system as, for example, in the case of Fano resonances, [1] or may be imposed upon a system by applying two or more "phase–locked" external perturbations. In the latter case, transition rates can be controlled externally by variation of the relative phase between these perturbations. This basic principle has been used in a variety of experimental and theoretical studies and predictions. [2,3]

In the last few years we have pursued a theoretical investigation of various coherent control schemes in semiconductors. While the basics of coherent control are well captured by Young's double slit experiment or, in the context of quantum mechanics, Fermi's golden rule, a detailed understanding of coherent phenomena and phase–breaking in semiconductors calls for a microscopic many–body approach. Two common approaches are non–equilibrium Green's function methods and the density matrix approach. [4–6] Both approaches require approximations to allow numerical evaluation. Hence, the most useful approach to the full many–body problem of interacting electrons and lattice ions depends on the specific application. The kinetic equations used in our work were originally derived within the Keldysh Green's function formalism but may as well be obtained within the density matrix approach. [7,8] They represent a generalization of the Boltzmann equation and account for the exact quantum–mechanical single–particle dynamics defined by the free–particle Hamiltonian and the external light field(s) used to manipulate the system. They have been termed Boltzmann–Bloch or semiconductor Bloch equations. Approximations are required to numerically account for many–body interactions. We have used the screened Hartree–Fock approximation to the self–energy to account for both mean–field and scattering contributions in the Boltzmann–Bloch equations. Thermal equilibrium of a non–interacting electron–phonon system can be characterized by single–particle distribution functions, which are the diagonal single–particle density matrix elements in the single–particle energy eigenfunction basis, *i. e.* $\rho_{\beta\alpha} = Tr\left\{\rho\hat{c}_{\alpha}^{\dagger}\hat{c}_{\beta}\right\}$, where Tr denotes the trace operation, ρ is the density operator, and $\hat{c}_{\alpha}^{\dagger}$ and $\hat{c}_{\beta}$ are single–particle operators for single–electron energy eigenstates α and β. Many–body interactions introduce non–diagonal density matrix elements into the kinetic equations which account for energy renormalization, quasi–particle formation, and the finite life–time of single–particle states provided by particle–particle "scattering". External perturbations introduce additional terms to both diagonal and off–diagonal single–particle density matrix elements. For example, the Hamiltonian for the electron–light–field coupling within the dipole approximation is of the form

$$H_{\text{e-LF}} = \sum_{i,f;k} \left\{a_{e-LF}(k,t)\cos(\omega_{LF}t + \phi_{LF})\hat{c}_f(k)^{\dagger}\hat{c}_i(k) + h.c.\right\} \quad . \tag{1}$$

Here $a_{e-LF}(k,t)$ are proportional to the dipole matrix elements between initial ($|i\rangle$) and final state ($|f\rangle$) electron (sub)bands. $H_{\text{e-LF}}$ leads to "interband polarization", *i. e.*, the build–up of non–diagonal density matrix elements ρ_{fi}. The latter account for the coherence initially present between photo–excited electron–hole pairs. Part of this interband polarization may disappear with the light field or decay due to many–body interactions (loss of coherence due to "scattering"). Yet another part of this interband polarization may persist to account for quasi–particle formation during the excitation process, such as excitons. [7]

As both external perturbations and many–body interactions initially provide a coherent contribution, adjustment of the phase of an external perturbation can be used to control the (coherent) dynamics of the system. Several elementary schemes which we and others have explored are shown in Fig. 1.

In semiconductors, energy levels are usually electron bands or subbands, rather than discrete levels as in atoms. Semiconductor microstructure design provides a powerful tool to structurally design a desired electronic structure. [9]

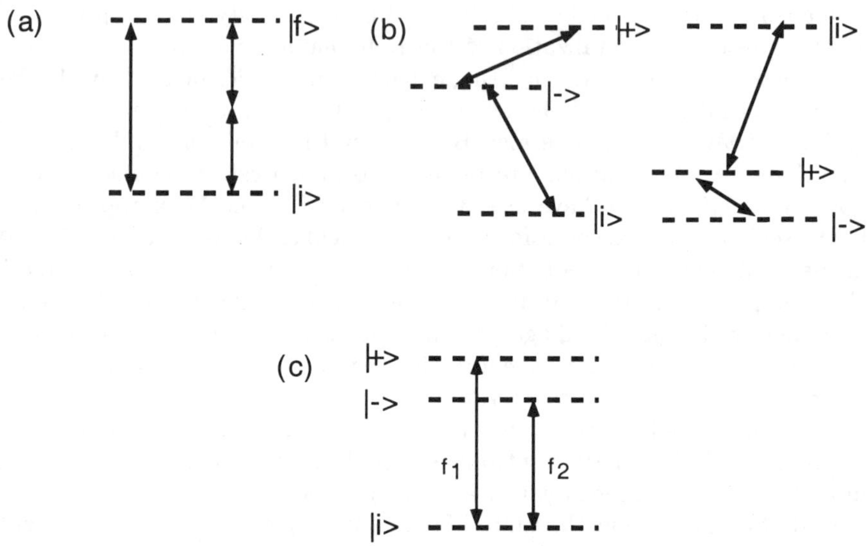

Figure 1. Schemes for electromagnetically initiated coherent control: (a) Conventional scheme based on interference between single– and multiple–photon absorption, (b) three–level scheme which uses one pump–field and one control field; (c) scheme based on two light fields of similar frequency and pulse shape with variable time–delay and polarization.

The most widely used electromagnetic control scheme exploits quantum interference between single– and multiple–photon absorption and is schematically presented in Fig. 1(a) for single– and two–photon absorption. As single– and two–photon absorption couple to final states of different symmetry, the symmetry of the final state may be controlled. This important control scheme has been discussed extensively in the literature and, therefore, will merely be mentioned here for completeness. [2,3] In semiconductors, this scheme has been used to control photocurrents. [10,11] We have proposed this scheme to control charge oscillations in double–quantum wells and the associated THz radiation. [12] It requires an extension of the standard semiconductor Bloch equations to account for $|s\rangle$ – and $|p\rangle$–mixing in valence and conduction bands. The rate of electron–hole formation controls the amplitude of charge oscillations.

The basic idea of control schemes based on a driven three–level system is shown in Fig. 1 (b). This scheme has been used to demonstrate stimulated Raman adiabatic processes in atomic systems, [13] as well as electromagnetically induced transparency in various media. [14,15] In semiconductor heterostructures, this scheme has been predicted to provide a variety of coherent control phenomena once high–intensity coherent microwave (mw) fields become available. These include coherent control of final–state population, optical absorption, optical gain, and LO phonon emission associated with electron intersubband transitions. [16–19] The physical mechanism behind this scheme is that the interband polarization ρ_{+-} between the doublet $|+\rangle/|-\rangle$ which is "generated" by the mw field enters the kinetic equations for interband polarization between the doublet and the third subband ($|i\rangle$), *i.e.*, $\rho_{\pm i}$. The latter govern transitions between $|i\rangle$ and the doublet. Hence, when the Rabi period associated with the driven doublet is *larger* than a pump (or probe) pulse coupling subband i to the subband doublet, the phase of the driving mw field matters. An interesting prediction of this scheme is the feasibility of coherent control of LO phonon emission in intersubband transitions. [19] In this case the transition rate of electrons from the doublet into the lower singlet via the electron–phonon coupling, which plays the role of a "probe field", is manipulated by the ρ_{+-} interband polarization. Similar to the case of coherent control of photo–emission, *i.e.* optical gain, the phonon emission rate can be controlled by the phase of the mw field. [18,19]

Unfortunately, intense coherent mw sources required for scheme Fig. 1 (b) do not seem to be available at this point in time. Therefore, it may be advantageous to pursue an approach by which slow interband oscillations are induced by two frequency–detuned coherent optical light pulses. Such a scheme is sketched in Fig. 1(c). In general, one may use two coherent light fields of variable time–delay, polarization, and/or frequency detuning. For example, Heberle and coworkers have used two linearly–polarized pulses of equal pulse shape and frequency to coherently control exciton formation (optical absorption) and Faraday rotation by variation of the time–delay. [20] Such a scheme works as long as the time delay is less than the characteristic phase–breaking ("decoherencing") time in the sample.

In this contribution we will use a special case of scheme Fig. 1(c) to predict coherent control of photo–absorption by using two short phase–matched pulses with frequency detuning $\delta f = f_2 - f_1$. The concurrent pulses have identical pulse shape and parallel linear polarization. Their amplitudes may be adjusted to give identical final carrier densities when applied individually. Variation of the relative phase between the two pulses allows coherent control of carrier absorption. The physical mechanism for this control phenomenon can be summarized as follows. Due to the different curvature of valence and conduction bands, frequency–detuned pulses tend to excite electron–hole pairs in different regions of k–space. In the absence of many–body interactions each photo–generated electron–hole pair acts like an isolated (driven) two–level system. As long as the spectral overlap between pulses is negligible, there is no control (or interference) effect. Due to the carrier–carrier Coulomb interaction, however, electron–hole pairs of different k–space regions communicate with one another via the exchange interaction. This, of course, is the well–known mechanism responsible for excitonic effects and Coulomb enhancement. Exciton formation is directly related to conduction–valence–band polarizations $\rho_{cv}(k)$. In case of two frequency–detuned pump pulses $\rho_{cv}(k)$ peaks at two different values of k. Due to the Coulomb interaction oscillations with frequencies near δf are induced. Hence, if the pulse duration τ_p is shorter than $\approx 1/\delta f$ the relative phase between the two pump pulses determines the magnitude of $\rho_{cv}(k)$ and hence the interaction strength between light and matter. One may also try to interpret this result as a consequence of pulse–shaping. In the regime $\tau_p < 1/\delta f$, variation of the relative phase leads to substantial pulse–shaping for the total light field. For longer pulses one merely recovers the standard beating behavior for which a change in relative phase merely changes the phase of the temporal evolution of $\rho_{cv}(k)$ but does not influence the net absorption as one averages over several δf oscillation cycles. However, this interpretation misses the important point that this effect is closely linked to many–body effects arising from the electron–electron Coulomb interaction.

2 Numerical Results

2.1 Optical control of photoabsorption

In what follows we present numerical results for two cases, limiting ourselves to the low–carrier–density and short–pulse regime in which scattering contributions from many–body interactions may be neglected. Excitation levels, temperature, and timescales are such that phonon processes may be neglected.

First we consider photo–excitation in bulk InP. One pulse is tuned in resonance with the direct single–electron energy gap, the second is detuned by +10 meV. This detuning corresponds to a period of 414 fs. The electronic structure near the Γ–point is well captured by a four–band $\mathbf{k} \cdot \mathbf{P}$ model. [21] Individual pulses provide approximately 1.6×10^{16} carriers per cm^3. Pulse duration is varied from 50 to 300 fs. Corresponding results for the total number of carriers are given in Figs. 2–4 for 100 fs, 200 fs, and 300 fs pulses, respectively. In each figure we show the results for individual pulses and two–color excitation with relative phase $\Delta\Phi = 0$, 0.5π, π, 1.5π. Our numerical results clearly demonstrate the conclusions drawn on the basis of the kinetic

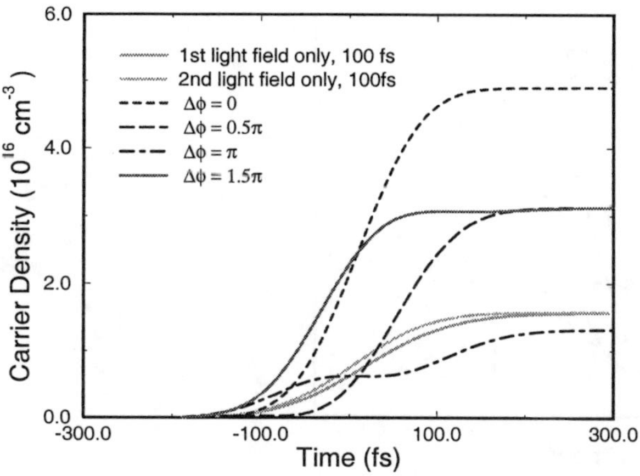

Figure 2. Density of photo–generated electron–hole pairs as a function of time for several phase difference $\Delta\phi$ for bulk InP. Thin lines for single pulses, thick lines for two pulses. Pulse duration is 100 fs.

equations. Significant phase control of the absorption can be achieved in the short–pulse regime. The effect becomes negligible once the pulse duration becomes significantly larger than the period of slow frequency oscillation of $\rho_{cv}(k)$, as shown in Fig. 5.

A second example is given for a biased asymmetric GaAs/AlGaAs double well. Here individual pulses provide 10^{10} carriers per cm^2. The double well is biased such that the lowest two electronic subbands are separated by 10 meV and the predominantly narrow–well associated level lies lower than the predominantly wide–well associated level. Furthermore, the structure is designed such that only the top heavy–hole subband(s) of the wide well participate in (resonant) optical transitions to the electron subband doublet. The total number of carriers for a 200 fs pulse and phase 0, 0.5π, π, and 1.5π is shown in Fig. 6. Again, clear phase sensitivity of the absorption process is evident. Indeed, the effect is even more pronounced than in bulk InP. This is consistent with the fact that the Coulomb interaction in two spatial dimensions is stronger than in three. In addition, the present three–subband scheme, Fig. 1(c) may be used to control coherent charge oscillations in this structure, an issue which will be discussed in detail in a future publication. In Fig. 7 we show the carrier density in the right (narrow) well versus time

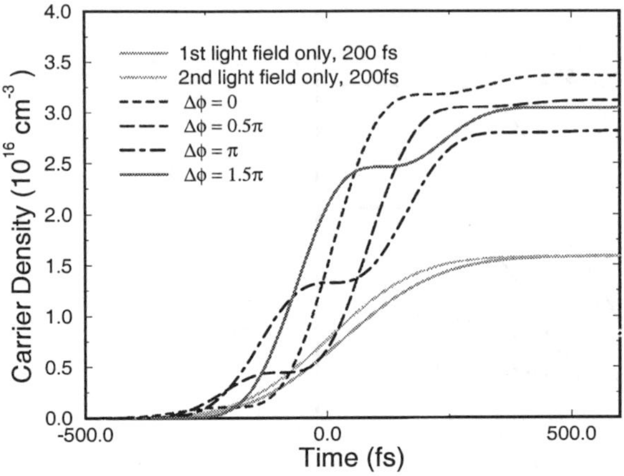

Figure 3. Density of photo–generated electron–hole pairs as a function of time and phase difference $\Delta\phi$ for bulk InP. Thin lines for single pulses, thick lines for two pulses. Pulse duration is 200 fs.

neglecting electron–electron scattering. It should be noted that somewhat lower carrier densities should be used to make this effect more robust against the effects of carrier scattering.

2.2 Structural Control of Photoabsorption

Asymmetric absorption lineshapes (Fano resonances) due to quantum interference in photo–ionization have found considerable attention in atoms and, more recently in semiconductors. [1, 22– 25] Maturing growth technologies for a large number of semiconductors have made feasible the design of heterostructures which can reveal Fano resonances. Young's double slit experiment teaches us that equal slits lead to optimum interference patterns. This also holds true for Fano resonances where it is important that the various pathways to ionization have approximately equal probability. [26] In the simplest case, a Fano resonance arises when ionization can occur either directly to the continuum or via a resonant level. [1] We have recently explored systematically various GaAs/AlGaAs multi–quantum well structures regarding their potential for "producing" Fano resonance features in optical absorption. [26] When inspecting semiconductor heterostructures it becomes clear that in many cases the direct coupling of the initial (bound) state to the continuum will be sig-

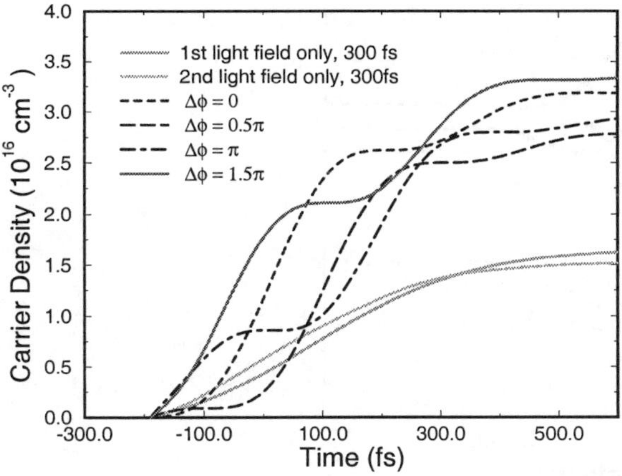

Figure 4. Density of photo–generated electron–hole pairs as a function of time and phase difference $\Delta\phi$ for bulk InP. Thin lines for single pulses, thick lines for two pulses. Pulse duration is 300 fs.

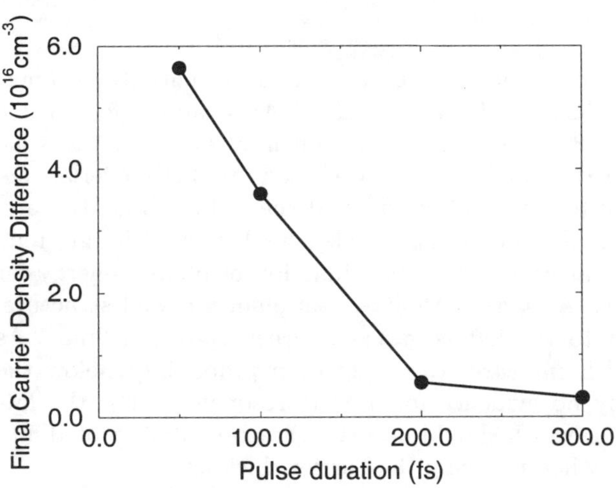

Figure 5. Difference in final carrier density for $\Delta\phi$=0 and π versus pulse duration for bulk InP. Detuning is 10 meV, corresponding to a period of 414 fs.

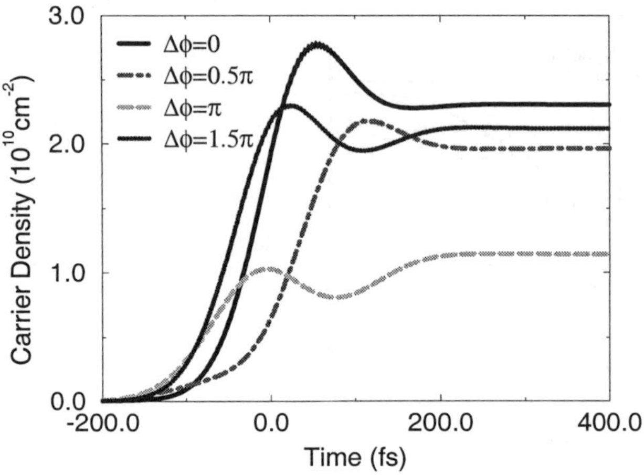

Figure 6. Density of photo–generated electron–hole pairs in an asymmetric GaAs/AlGaAs double well as a function of time for several phase differences $\Delta\phi$. Pulse duration is 200 fs. Pulse detuning is 10 meV.

nificantly weaker than to the resonant level, causing an essentially symmetric lineshape (except for density–of–state effects). An example for such a structure is given in Fig. 8(a). Due to the exponential decay of the scattering wave function into the barrier to the well, there is little overlap between the scattering state (of the continuum) and the initial (bound) state. Fig. 9 gives the square of the dipole matrix element between the ground–state wave function and scattering states as a function of photon energy for a $Al_{0.09}Ga_{0.91}As/Al_{0.2}Ga_{0.8}As/GaAs/Al_{0.2}Ga_{0.8}As$ quantum well structure. The $Al_{0.2}Ga_{0.8}As$ barrier to the left is varied between zero and 2nm. As reducing this barrier width *increases* the absorption probability below the resonance, there is clearly no evidence for a Fano resonance. For the latter, reduction of the barrier width should *decrease* the absorption probability below the resonance peak when reducing the barrier thickness.

Fano resonances can be found for structure type (b) in Fig. 8. [22] Here, the continuum wave function sweeps in "above" the initial–state wavefunction giving a good overlap with the ground state wave function. This is demonstrated in Fig. 10 for a $Al_{0.09}Ga_{0.91}As$ / $Al_{0.2}Ga_{0.8}As$ / GaAs / $Al_{0.2}Ga_{0.8}As$ / $Al_{0.09}Ga_{0.91}As$/ $Al_{0.2}Ga_{0.8}As$ double quantum well structure. Both well widths are 8nm. The separating $Al_{0.2}Ga_{0.8}As$ barrier width is varied between

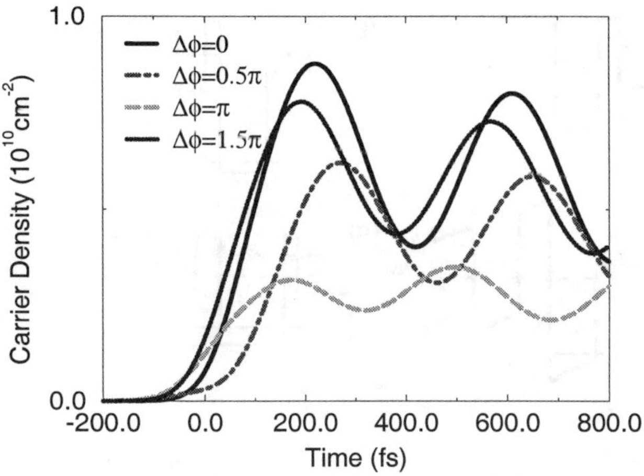

Figure 7. Density of photo–generated electron–hole pairs in the right (narrow) well of an asymmetric GaAs/AlGaAs double well as a function of time for several phase differences $\Delta\phi$. Pulse duration is 200 fs. Pulse detuning is 10 meV.

0.5nm and 2nm. It is observed that the minimum in absorption decreases with decreasing barrier width due to an increase in interference. In the absence of the shallow well (*i. e., the resonant level*), this minimum in absorption disappears (short dashed line), as it should.

Both data were obtained for *electron intersubband* transitions. This avoids a significant wash–out of resonance features due to effective–mass differences. However, we have qualitatively accounted for inhomogeneous broadening due to structural imperfections using a Lorentzian distribution function with a width of 3 meV in Fig. 10 to ensure that the feature survives some broadening. Fano resonances are quite sensitive to imperfections (as well as non–parabolicity in the energy bands).

Fano resonances may be tuned by an external electric field. [22] This is most effective when selected layers are heavily doped to leave band profiles flat, as indicated in Fig. 8(c). Moreover, the quantum–confined Stark effect should be minimized. As the resonant level in the double well moves below the energy continuum above the left band edge due to application of an electric field, resonant coupling between the excited state of the well and the continuum and, hence, the Fano resonance disappear. Fig. 11 shows the elimination of a Fano resonance as a function of applied electric field for the same structure

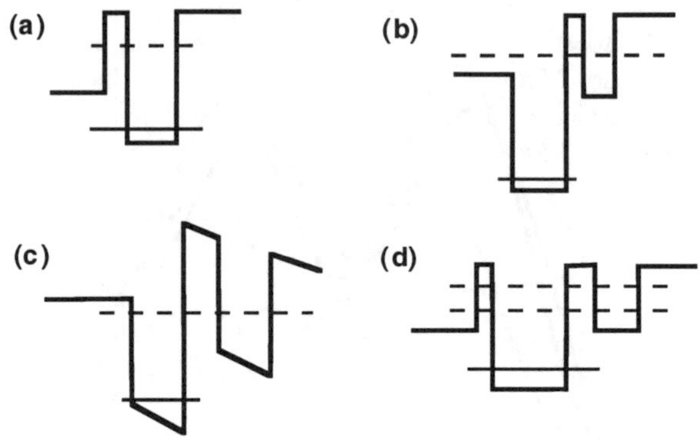

Figure 8. Examples for GaAs/AlGaAs heterostructures which allow photo–absorption via one or more resonances: (a) single quantum well structure, (b) asymmetric double well, (c) biased asymmetric double well, and (d) asymmetric double well with two (closely spaced) resonances.

as in Fig. 10, for a 1 nm wide separating barrier. Here the band edge of the left–most $Al_{0.09}Ga_{0.91}As$ layer was kept flat, assuming heavy doping in this region. At 10^6 V/m there is still a clear absorption minimum observable. However, at 2×10^6 V/m this minimum feature has disappeared.

Alternatively, Fano structural interference may occur between two or more (closely–spaced) resonances mediating two or more possible routes to the same final continuum state. [1] This situation may be achieved with a structure of type Fig. 8 (d). [23–25] Our calculations of dipole matrix elements show that the spacing of resonant levels should not exceed about 30 meV to lead to a measurable dip in absorption due to destructive quantum interference between the resonances. [26] Overall our study has not produced effects spectacular enough to encourage fabrication of "inversion–free" lasers as proposed previously. [27]

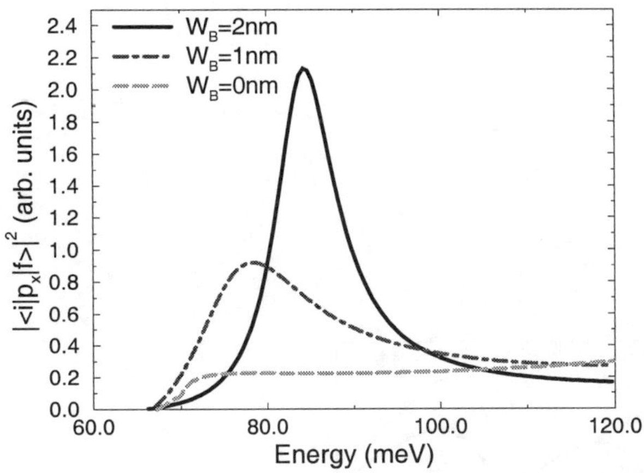

Figure 9. Square of dipole matrix element versus photon energy for structure (a) in Fig. 8.

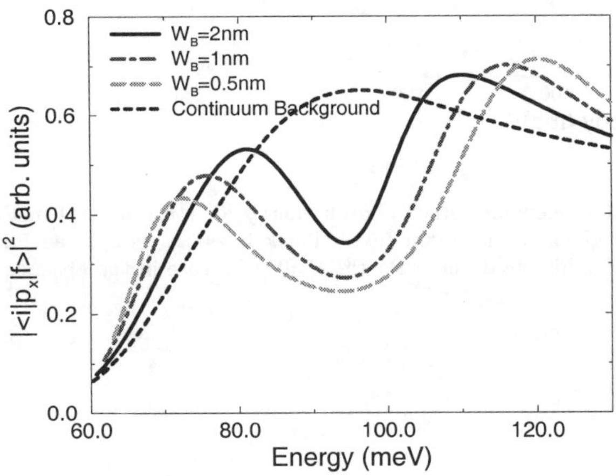

Figure 10. Square of dipole matrix element versus photon energy for structure (b) in Fig. 8. Details of the structure are given in the main text.

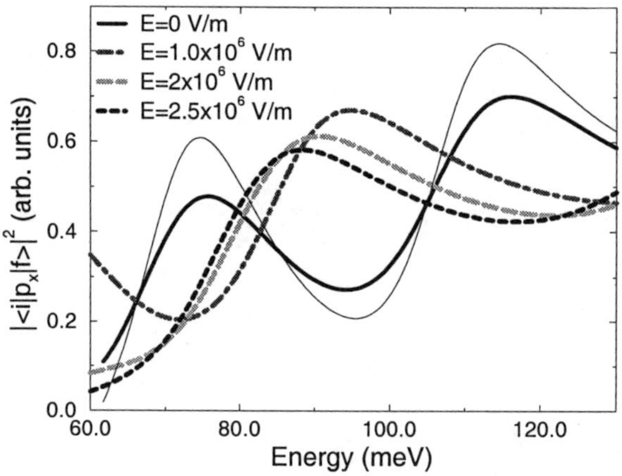

Figure 11. Square of dipole matrix element versus photon energy for structure (c) in Fig. 8. Details of the structure are given in the main text. Thick lines were obtained for a Lorentzian broadening of 3 meV. Thin solid line is for the E=0 V/m case in the absence of linewidth broadening.

3 Summary and Conclusions

We have briefly discussed aspects of coherent control in semiconductors based on structural design and manipulation by electromagnetic fields. A variety of control phenomena have been demonstrated by experimental groups and a number of further schemes have been proposed theoretically. Here we concentrated on two–color pump pulses to coherently control photo–absorption across the main energy gap of bulk semiconductors and GaAs/AlGaAs multi–quantum wells. Combined with modern growth techniques we believe that such control schemes may lead to the development of a new generation of ultrafast phase–sensitive optoelectronics. While heterostructures can be designed to give measurable structural quantum interference in absorption it is presently not clear whether the effect can be made strong enough to lead to a practical device.

Acknowledgments

Helpful discussions with W. A. Schroeder and P. Brumer are acknowledged.

References

[1] U. Fano, *Phys. Rev.* **124**, 1866 (1961).

[2] See, for example, R. J. Gordon, *Annual Reviews of Physical Chemistry* **48**, 595 (1997).

[3] See, for example, *Coherent Control in Atoms, Molecules, and Semiconductors*, edited by W. Pötz and W. A. Schroeder, (Kluwer, Dordrecht, 1999).

[4] J. Rammer and H. Smith, *Rev. Mod. Phys.* **58**, 323 (1986).

[5] T. Kuhn and F. Rossi, *Phys. Rev.* B **46**, 7496 (1992).

[6] H. Haug and A.-P. Jauho, *Quantum Kinetics in Transport and Optics of Semiconductors*, (Springer, Berlin, 1996).

[7] W. Pötz, *Phys. Rev.* B **54**, 5647–5664 (1996).

[8] U. Hohenester and W. Pötz, *Phys. Rev.* B **56**, 13177–13189 (1997).

[9] V. Narayanamurti, *Physics Today* **37**, 24 (1984).

[10] E. Dupont, P. B. Corkum, H. C. Liu, M. Buchanan, and Z. R. Wasilewski, *Phys. Rev. Lett.* **74**, 3596 (1995).

[11] A. Hachè, Y. Kostoulas, R. Atanasov, J. L. P. Hughes, J. E. Sipe, and H. M. van Driel, *Phys. Rev. Lett.* **78**, 306 (1997).

[12] W. Pötz, *Appl. Phys. Lett.* **72**, 3002–3004 (1998).

[13] U. Gaubatz, P. Rudecki, S. Schiemann, and K. Bergmann, *J. Chem. Phys* **92**, 5363 (1990).

[14] See, for example, S. E. Harris, *Phys. Today* **50** (7), 36 (1997).

[15] Y. Zhao, Optics & Photonics News, Dec. 1997, p. 30-31. *Appl. Phys. Lett.* **71**, 395–397 (1997).

[16] W. Pötz, *Appl. Phys. Lett.* **71**, 395–397 (1997).

[17] W. Pötz, *Phys. Rev. Lett.* **79**, 3262–3265 (1997).

[18] X. Hu and W. Pötz, *Appl. Phys. Lett.* **73**, 876–878 (1998).

[19] X. Hu and W. Pötz, *Phys. Rev. Lett.* **82**, 3116–3119 (1999).

[20] A. P. Heberle, J. J. Baumberg, and K. Köhler, *Phys. Rev. Lett.* **75**, 2598 (1995); A. P. Heberle, J. J. Baumberg, E. Binder, T. Kuhn, K. Köhler, and K. H. Ploog, *IEEE J. Select. Topics in Quantum Electronics* 2(3), 769–775 (1996).

[21] E. O. Kane, in *Semiconductors and Semimetals, Physics of III-V Compounds*, Vol. **1**, edited by R. K. Willardson and A. C. Beer, 75 (Academic, New York, 1966).

[22] J. Faist, C. Sirtori, F. Capasso, L. N. Pfeiffer, and K. W. West, *Optics Letters* **21**, 985–987 (1996).

[23] J. Faist, F. Capasso, C. Sirtori, A. L. Hutchinson, K. W. West, and L. N. Pfeiffer, *Appl. Phys. Lett.* **71**, 3477–3479 (1997).

[24] J. Faist, F. Capasso, C. Sirtori, K. W. West, and L. N. Pfeiffer, *Nature* **390**, 589 (1997).

[25] H. Schmidt, K. L. Campman, A. C. Gossard, and A. Imamoglu, *Appl. Phys. Lett.* **70**(25), 3455-3457 (1997).

[26] W. Pötz, *Superlatt. Microstruct.* **26** (2), 141–156 (1999).

[27] A. Imamoglu and R. J. Ram, *Opt. Lett.* **19**, 1744 (1994).

SPECTROSCOPIC APPLICATION OF THZ RADIATION GENERATED BY ULTRASHORT PULSES -STATIC FAR INFRARED ABSORPTION MEASUREMENTS IN CONDENSED PHASES-

KEISUKE TOMINAGA

Department of Chemisiry, Faculty, of Science, Kobe University, Nada, Kobe, 657-8501 Japan

E-mail: tominaga@kobe-u.ac.jp

HIDEYUKI OHTAKE, NOBUHIKO SARUKURA

Institute for Molecular Science, Myodaiji, Okazaki, 444-8585 Japan

E-mail: saru@ims.ac.jp

KEN-ICHI SAITOW

Graduate School of Science and Technology, Chiba University, Yayoi-cho Inage-ku Chiba, 263-8522 Japan

E-mail: saitoh@scichem.c.chiba-u.ac.jp

HIROAKI SASAKAWA, ATSUO TAMURA

Graduate School of Science and Technology, Kobe University, Nada, Kobe, 657-8501 Japan

E-mail: atamura@kobe-u.ac.jp

IGOR V. RUBTSOV, AND KEITARO YOSHIHARA

Japan Advanced lnstitute of Science and Technology, Tatsunokuchi, Ishikawa 923-1292 Japan

E-mail: yosihara@jaist.ac.jp

We have constructed a conventional far infrared spectrometer based on terahertz (THz) radiation as a spectral source. The THz radiation is generated by focusing a femtosecond pulse onto a semiconductor substrate placed in a strong magnetic field. Far infrared spectra of various condensed phase systems including neat liquids, supercritical fluids, a charge-transfer complex solution, and a protein solution have been measured by this spectrometer.

1 Introduction

Low-frequency vibrational spectra (far infrared absorption and depolarized Raman scattering) of condensed phases contain valuable information on cooperative effects of a number of intermolecular processes.[1-4] These processes arise from the different modes of molecular motions such as librations, hindered rotations, and quasi-lattice vibrations of the molecules. Understanding of these motions is essential for studies of chemical reactions and relaxation phenomena in condensed phases, and especially in liquids. Recent ultrafast laser spectroscopy has shown that chemical reactions in liquids are strongly influenced by the collective motions of surrounding liquid molecules.[5,6] Therefore, investigation of the low-frequency motion in a molecular level is necessary to clarify the solvent dynamical effect on reactions.

The Raman scattering technique normally requires high concentrations of the scattering molecules unless the incident light is resonant with an electronically excited state.[3,4] On the other hand, since the cross sections are greater for IR absorption than Raman scattering, far infrared absorption measurements are potentially more suitable for the study of the low frequency modes in the condensed phases.

A terahertz (THz, 1 THz = 33.3 cm^{-1}) pulse has its oscillating frequency in the low frequency region, and the first generation of subpicoseconds THz pulses was reported almost thirty years ago.[7] The recent advent of more stable and reliable solid state sources of femtosecond laser pulses has allowed us to develop new techniques in this frequency range.[8-19] Because of its short pulse duration, it is possible to perform time-resolved, pump-probe type experiments in the far IR region.[15-19] Examples of visible-pump, THz-probe experiments were performed on semiconductors,[17,18] superconductors,[19] and electron-transfer systems in solution.[15,16] These experiments tell us novel information about how the low frequency collective modes respond to a sudden change in the electronic state.

In order to generate a sub-picosecond THz pulse, a visible femtosecond pulse is focused onto an absorbing semiconductor, creating charge carriers and positive holes. Acceleration of the carriers and holes in an electric or magnetic field gives rise to a transient photocurrent that radiates electromagnetic waves. To allow an external electrical bias field to be applied, an anntena structure has been widely used in which two electrodes are laid down on a silicon or GaAs substrate with a separation in the order of 100 μm.[10-19] Generation of intense THz pulses in a magnetic field is more feasible and simple; application of a strong magnetic field to a semiconductor substrate enhances the intensity of the radiation.[8.9]

One of the final goals of this series of studies is to perform time-resolved, pump-probe experiments in the far IR region on various condensed phase systems. These studies will permit us to explore what kind of roles the low frequency collective motions play in various dynamical proceises in condensed phases. In the present study we have constructed a conventional far IR absorption spectrometer by using an intense THz pulse generated in a magnetic field. Steady-state far IR absorption spectra of a variety of condensed phase systems have been measured with this spectrometer.

2 Experimental

The experimental apparatus was described elsewhere and is shown in Figure 1.[8,9] A mode-locked Ti:sapphire laser delivers nearly-transform-limited ~70-fs pulses at a repetition rate of 80 MHz with an average power of 1 W. The laser beam is focused onto the emitter in a spot of about 2 mm in diameter. The THz radiation emitter is a non-doped bulk InAs with a (100) surface, and it is placed in a magnetic field. The conduction type of the InAs is slightly n, and the carrier density is 3.0×10^{16} cm^{-3}. Because the InAs sample itself is highly reflective in the THz region, the THz pulse radiation is totally generated in the reflection direction.[8,9] The InAs sample is placed at a angle of 60° to the excitation laser beam path. A 1.7 T magnetic field is applied in parallel to the <110> crystal axis by a water-cooled electro-magnet. In this geometry, the excitation laser and THz radiation are polarized horizontally, and the THz radiation power approaches a sub-mW level.[8,9]

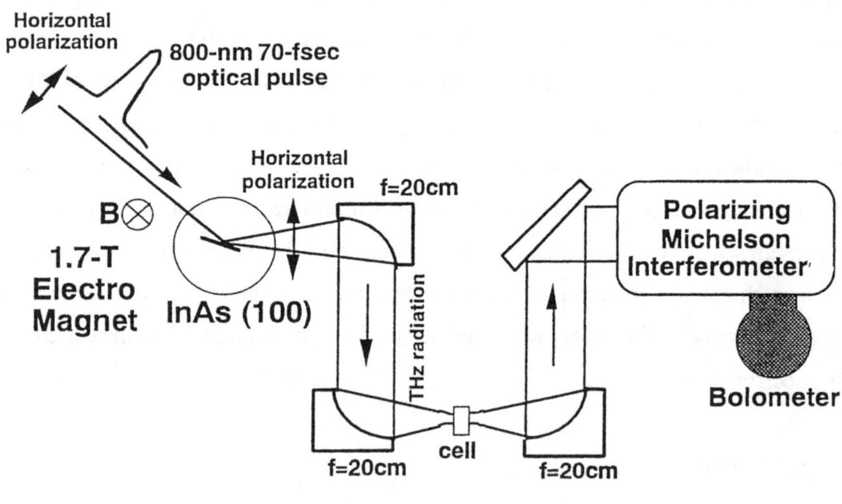

Figure 1. Schematic illustration of the transmission experiment in the terahertz region. The THz-radiation emitter is placed in the 1.7-T magnetic field. The beam size on the cell is about 2 mm in diameter.

The THz radiation is collimated and focused on the sample cell by a 20 cm off-axis parabolic mirrors. The transmitted signal is also collimated by a 20 cm off-axis parabolic mirror and is reflected onto a polarizing Michelson interferometers equipped with a liquid helium-cooled silicon bolometer. It takes about 30 minutes to measure one transmission spectrum.

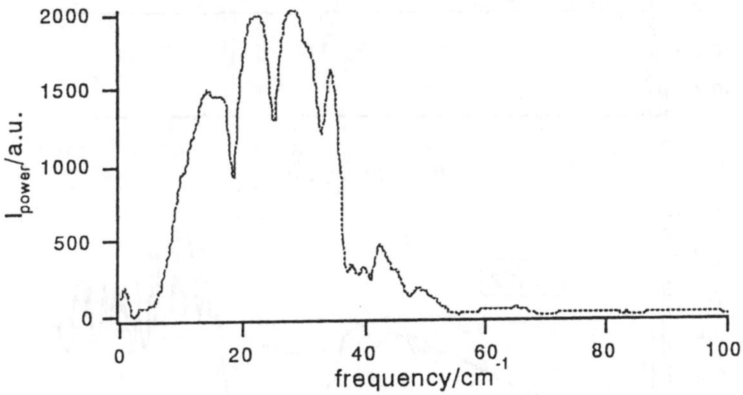

Figure 2. A power spectrum of the THz radiation.

A typical power spectrum of the THz radiation is shown in Figure 2. Structures observed in the spectrum result from absorption of water vapor in the air. The spectrum extends up to 100 cm^{-1}. The low frequency side is limited by the detector. Furthermore, since the interferometric method is employed to detect the signal, a slight change of the baseline in the interferrogram causes uncertainty in the very low frequency region. The high frequency side is limited by the spectral shape of the THz radiation or sensitivity of the detector. Intensity fluctuation in the region ν > 70 cm^{-1} is rather severe. By considering these limiting factor we concluded that we can resolve spectral features in the region of 4 to 70 cm^{-1}. The measurement was performed at room temperature (296 K).

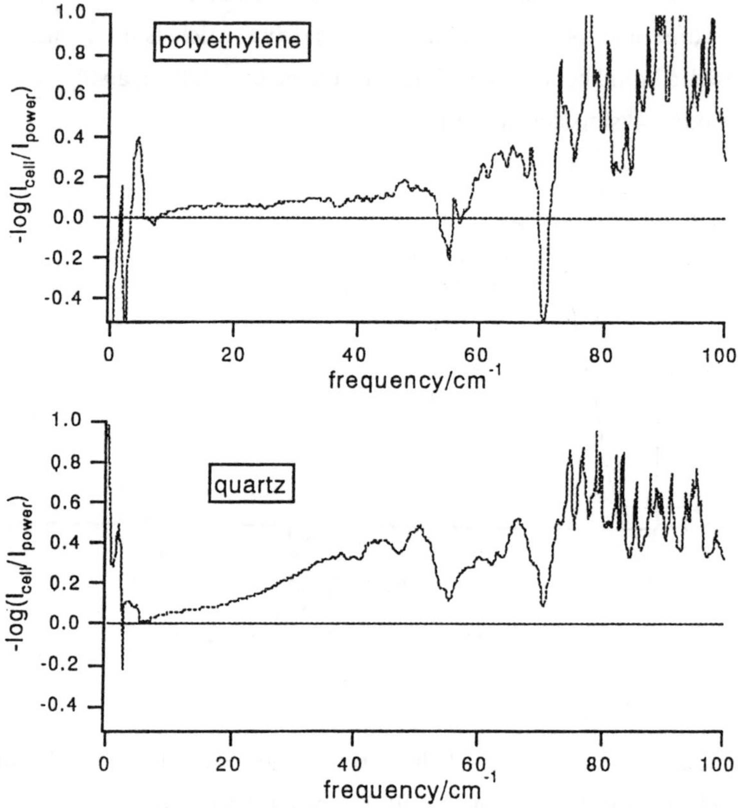

Figure 3. Far IR absorption spectra of the empty cell. The window is polyethylene (upper) and quartz (lower).

In this article we use two different quantities to express the absorption spectrum. One is the absorption coefficient at a frequency ν, $\alpha(\nu)$, which is defined as

$$\alpha(\nu) = -\frac{1}{l}\ln\left(\frac{I(\nu)}{I_0(\nu)}\right) , \qquad (1)$$

where $I(\nu)$ and $I_0(\nu)$ are the intensities of the radiation field after and before the sample, respectively, and l is the optical path length of the sample. The other is the extinction coefficient, $\varepsilon(\nu)$, defined as

$$\varepsilon(\nu) = -\frac{1}{cl}\log\left(\frac{I(\nu)}{I_0(\nu)}\right) , \qquad (2)$$

where c is the concentration of the sample solution. Normally, the absorption coefficient is used for neat liquids, and the extinction coefficient is used for solutions with a solute and solvent. We have to be careful when we measure absorption coefficient of neat liquids since loss of the intensities at interfaces complicates the measurement. To avoid any complication due to the effect at interfaces, we measure the transmitted intensities, I_1 and I_2, with two different optical path lengths, l_1 and l_2, respectively, and obtain the spectra by the following equation,

$$\alpha(\nu) = -\frac{1}{(l_2 - l_1)}\ln\left(\frac{I_2(\nu)}{I_1(\nu)}\right) . \qquad (3)$$

This method is also effective to minimize artificial oscillations in the spectrum caused by the interference of the THz radiation within the window glasses.

Figure 3 shows far IR absorption spectra of an empty cell with a polyethylene window (upper) and quartz window (lower). As a reference, the power spectrum of the radiation is used. It can be seen that the polyethylene window is more suitable for the far IR absorption measurement.

3 Results and discussion

3-1 Neat Liquids

We first measure far IR spectra of neat liquids (Figure 4). For nonpolar (CCl_4) and less polar (*m*-dichloro-benzene) liquids a sample cell with relatively long optical path lengths (l_1 = 1 mm and l_2 = 10 mm) is used. For a polar liquid such as $CHCl_3$, we used optical cells with $l_1 = 0.01$ and $l_2 = 0.1$ mm. All the spectra shown are consistent with the spectra reported before.[20-22]

In IR spectroscopy, the absorption coefficient at an angular frequency ω per unit path length, $\alpha(\omega)$, of a sample at temperature T is given by[23]

$$\alpha(\omega) \propto \frac{\omega \tanh(\hbar\omega / 2k_B T)}{\hbar n(\omega)} \int_{-\infty}^{+\infty} dt e^{-i\omega t} \langle M(t)M(0) \rangle \quad , \qquad (4)$$

where M is the total dipole moment of the sample, and $n(\omega)$ is the reftactive index. In the case of far IR spectroscopy one cannot neglect the term of the refractive index since it dramatically changes in this frequency range. The total dipole moment M can be expressed as a sum of the microscopic dipoles occurring in the liquid sample, namely,

$$M = \sum_i \mu_i + \sum_{ij} \mu_{ij} + \sum_{ijk} \mu_{ijk} + \cdots \qquad (5)$$

where μ_i takes into account the permanent dipole and the induced dipole at the molecule i, μ_{ij} is the two-body dipole coming from the short range interactions between molecules i and j, μ_{ijk} is the irreducible three-body dipole coming from nonelectrostatic forces, and so on. One can calculate the far IR absorption spectrum in terms of equation (4) and molecular dynamics simulations, which is now undertaken for the system of $CHCl_3$.[24]

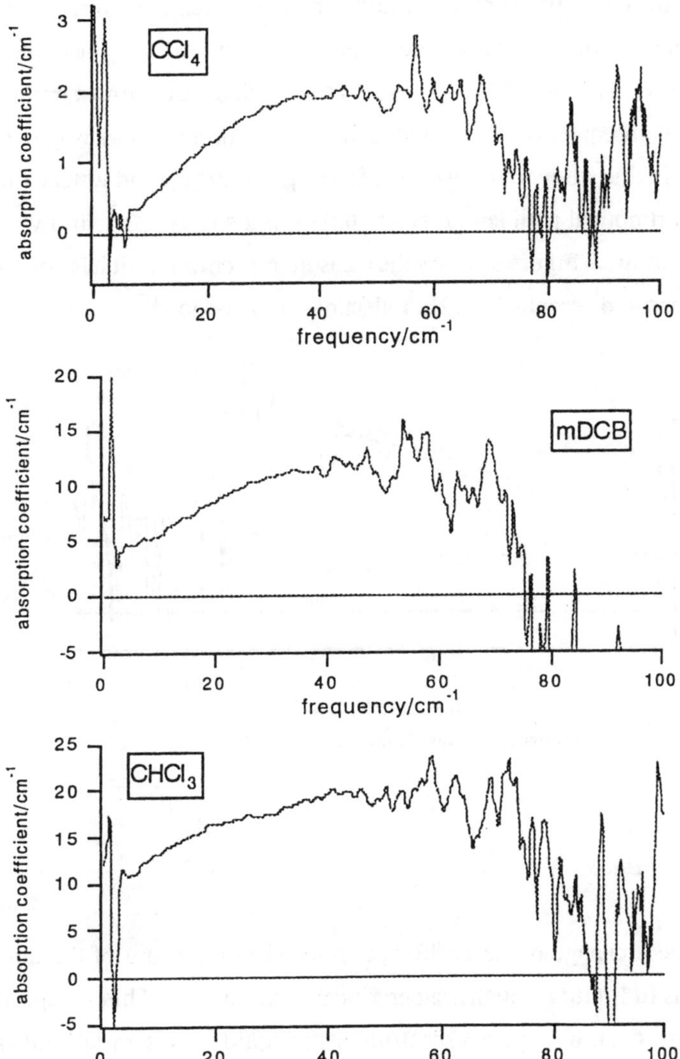

Figure 4. Far IR spectra of CCl_4 (upper panel), *m*-dichlorobenzene (middle panel), and $CHCl_3$ (lower panel).

We next measure the far IR spectrum of liquid water. In spite of its importarrce only few experimental far IR spectra of liquid water have been published.[11,12,14,25,26] Good broad band spectral sources are not available below 100 cm^{-1}, and the very large absorption coefficient of liquid water causes difficulties for trartsmission spectroscopy. Consequently, the reported far IR spectra of liquid water differ substantially. In this study we measure the far IR spectrum of liquid water by using a cell with a short optical path length of l_1 = 0.025 mm and l_2 = 0.1 mm with the polyethylene window. Figure 5 shows the measured spectrum, which is consistent with the one measured recently by THz reflection spectroscopy.[11]

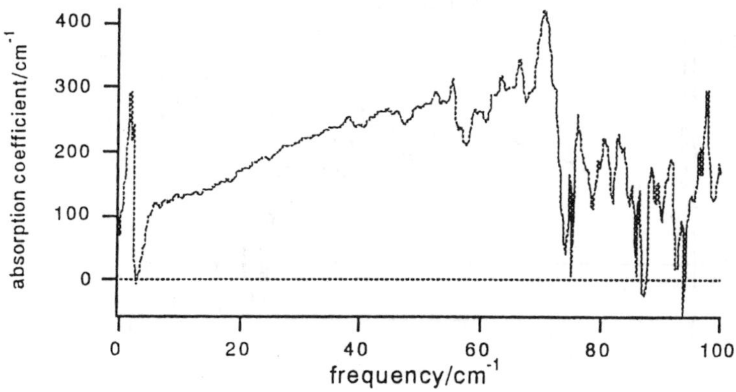

Figure 5. A far IR spectrum of water.

3-2 Supercritical Fluids

We next examine a change of the far IR spectrum when the phase of the fluid is changed from gas to liquid through a supercritical fluid state.[27] This is especially interesting because the low frequency spectrum of the liquid is substantially difirerent from that of the gas: in the liquid phase the spectrum has a broad and structureless feature caused by the intermolecular interaction, while in the gas phase progressions of the rotational state are clearly resolved. The fluid continuously changes its phase from gas to liquid through the supercritical state. Therefore, this experiment allows us to observe how the intermolecular interaction appears in the low frequency spectrum in the liquid state.

This experiment is also important for studies on structures and dynamics of the supercritical fluid. The specific character appears near a critical isochore, existing in the extension of a liquid-gas curve in the phase diagram,[27,28] and at this point physical[29] and chemical[30] properties show anomalous behaviors. The static structures of the supercritical fluid have been extensively studied by a small-angle X-ray-scattering technique.[27,28] Dynamical information such as rotations and intermolecular motions will be obtained from the far IR spectral investigation.

The fluid under investigation is CF_3H. A sample holder used for supercritical CF_3H was described in elsewhere.[28] The body and window are SUS 316 and diamond, respectively. The temperature is controlled by circulating water from a chiller to a water jacket and monitored by a thermocouple. Pressure is adjusted with an injector and monitored with a strain gauge and an amplifier. Temperature and pressure fluctuations were within 0.1%. CF_3H gas was used as received (99.5%).

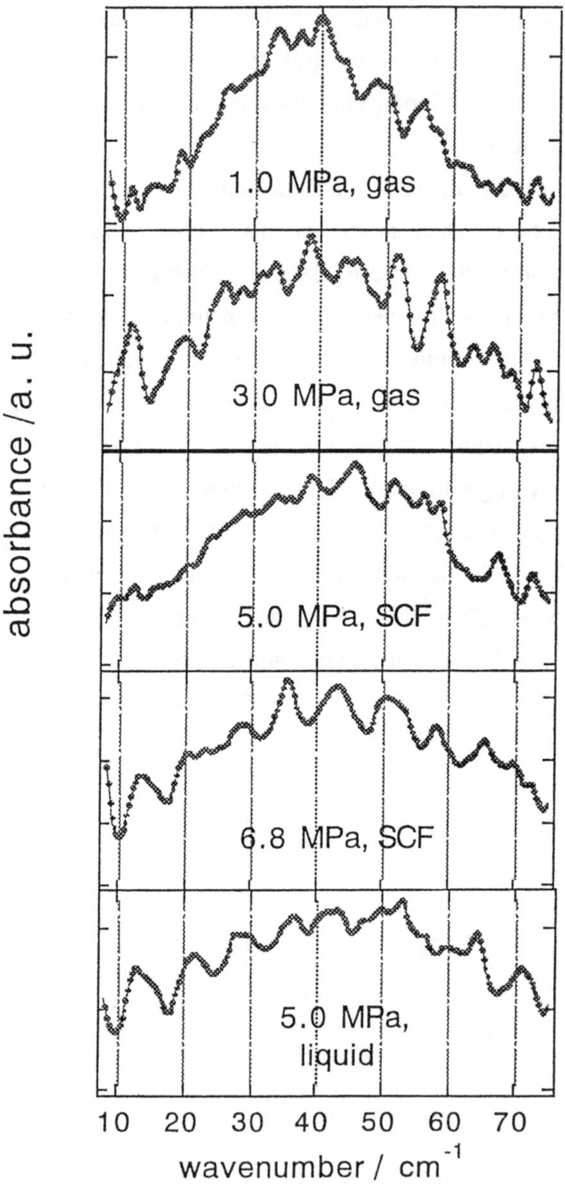

Figure 6. Far IR spectra of CF_3H. The pressure and its phase is shown in the figure (SCF; supercritical fluid).

Far IR absorption spectra for gases, supercritical, and liquid states are shown in Figure 6. We have used the power spectrum of the THz radiation as the reference spectrum to calculate the far IR spectrum in this experiments. Therefore, the spectrum is a product of the contributions from the fluid and window. However, since the spectrum of the window is not affected by a change of the fluid density, the observed density dependence of the spectrum reflects the nature of the fluid. The spectra were measured at the reduced temperature $T_r = T/T_c = 1.02$, except for the liquid state (the critical temperature $T_c = 299.06$ K), which was measured at $T_r = 0.97$ to form a liquid state. Structures observed in the spectra originate from a fringe in the windows. A newly designed cell can eliminate this fringe, which will be reported elsewhere.[31]

There are two features observed in the spectra; one is an increase of the spectral width as a function of pressure, and the other is a shift of the absorption peak toward higher energy as the sample pressure increases. The shift is not very clearly observed because of artificial oscillation. However, our recent measurement has unambiguously shown a shift of the peak.[31] There are two tentative interpretations for the broadening of the spectrum. One is a dynamical effect; the rotational relaxation time becomes shorter with increasing density since the collision frequency increases in a high density medium. The other is a static structural effect, namely the effect of the inhomogeneous distribution of the density in the fluid. The inhomogeneity causes a distribution of the rotational relaxation time, which broadens the spectrum. The mechanism of the peak shift is now under investigation, which will be discussed in detail elsewhere.[31,32]

3-3 Solute-Solvent System

It is interesting to see how a polar solute interacts with solvent molecules and what kind of effect can be seen in the low frequency spectrum caused by this interaction. We dissolved betaine, a highly polar molecule with a permanent dipole moment of 15 D in its ground state, in *m*-dichlorobenzene (mDCB) and measured the far IR spectrum. Recently, two groups performed a time-resolved experiment on this system, where the pump pulse excited betaine, and the response in the low frequency region was probed by a THz pulse.[15,16] Coherent beats were observed in the data, and they concluded that there were two components in the solvent response, one being a bulk diffusive dielectric solvation, and the other an oscillatory response tentatively assigned to librational modes localized close to the solute.[15] From a simple dielectric continuum theory the frequency of this oscillatory response was estimated to be 5.5 cm^{-1}.[15] At this frequency the solvent molecules oscillate around the reaction filed created by the solute dipole moment.

Figure 7 shows the far IR spectrum of the betaine/mDCB solution, where the contribution from the solvent was subtracted. In the region $\nu > 10$ cm^{-1} there is no prominent component observed with an extinction coefficient greater than 50 cm^{-1}. This indicates that the solute does not form a complex with the solvent molecules. On the other hand, in the very low frequency region $\nu < 10$ cm^{-1} there seems to be some band appearing. This region is very close to the limit of the sensitivity of the detector (4 cm^{-1}), and, therefore, careful attention should be paid to ascertain this band. A further study is needed to clarify this point.

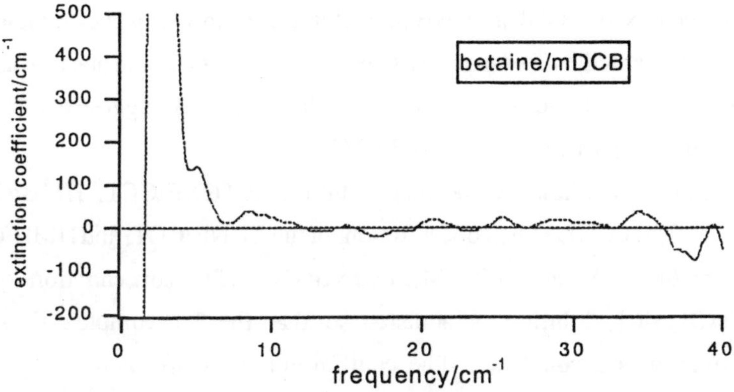

Figure 7. A far IR spectrum of a solution betaine/*m*-dichlorobenzene. The contribution from the solvent was subtracted.

3-4 Charge-Transfer Complex

It has been well known that tetracyanoethylene (TCNE) and hexamethylbenzene (HMB) form a complex in a relatively nonpolar solvent such as CCl_4.[34-39] The first electronically excited state of this complex corresponds to the charge-transfer state, where TCNE acts as an electron acceptor and HMB acts as a donor.

One long-standing and important question of this system is related to the intermolecular mode of the complex; what is the frequency of the intermolecular mode, and what kind of role does the intermolecular motion play in the charge-transfer dynamics? Recently, dynamieal aspects of this charge-transfer complex have been discussed by means of a femtosecond transient absorption[35] and a fluorescence up-conversion method.[37,38] A resonance Raman band at around 160 cm^{-1} was once assigned to be the intermolecular mode;[36] however, the recent ultrafast fluorescence upconversion experiment has shown that this band is an intramolecular mode of TCNE.[38] Moreover, an *ab initio* calculation study has suggested that the intermolecular stretching mode has a frequency of 59 cm^{-1}.[34] The transmission spectrum in the mixed crystal of TCNE and HMB for the low frequency region of 40 - 200 cm^{-1} was obtained by Rossi and Haselbach.[39] They observed a peak at around 80 cm^{-1} with a spectral width of 20 cm^{-1}.[39] This band was suggested to be the intermolecular mode between TCNE and HMB.

We have measured the far IR spectra of solutions of TCNE/CCl_4, HMB/CCl_4, and (TCNE+HMB)/CCl_4. The concentrations of the TCNE/CCl_4 and HMB/CCl_4 solutions are 0.006 M and 0.03 M, respectively. The concentration of the (TCNE+HMB)/CCl_4 solution is adjusted so that the 2:1 complex does not dominate over the 1:1 complex. The equilibrium constants for this system in CCl_4 were already reported; K_1 = 151 M^{-1} for D + A → D-A, and K_2 =7.3 M^{-1} for D-A+D → D-A-D, where D and A denote the electron donor and acceptor, respetively.[40,41] By using these values the amount of the 2:1 complex is estimated to be 18% of the 1:1 complex at this concentration. The far IR spectra of the TCNE/CCl_4 and HMB/CCl_4 solutions are similar to that of neat CCl_4, and there is not any effect of the solutes observed in the far IR spectra at these concentrations.

Figure 8 shows the far IR spectra of the (TCNE+HMB)/CCl_4 solutions. The concentrations of the complex (shown in the figure captions) were estimated from the reported value of the association constant of K_1 of 151 M^{-1} and were confirmed by the UV-VIS measurement. Increasing the complex concentration, a prominent band appears at around 60 cm^{-1}, though the signal-to-noise is not so high in this spectral range. We obtain the far IR absorption spectrum of the complex by subtracting the low concentration spectrum from the high concentration spectrum, as shown in Figure 8. The structures observed in the spectrum are due to noise. The extinction coefficient of this band is about 100 $M^{-1}cm^{-1}$. Since this band appears only in the mixed solvent with TCNE and HMB at a moderately high concentration and the frequency is in good agreement with the theoretical prediction, it is suggested to be the intermolecular band of the complex.

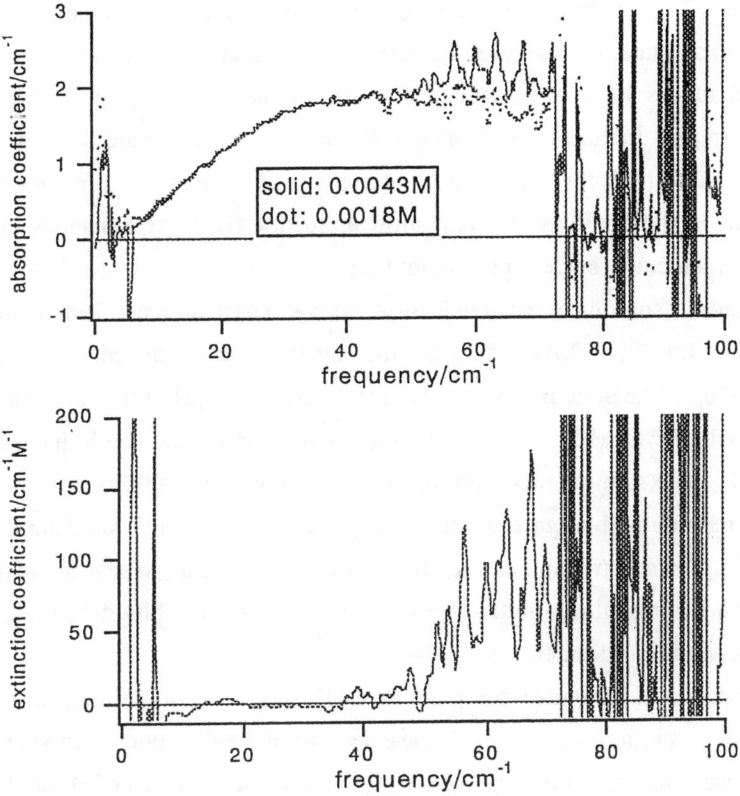

Figurc 8. Upper panel: Far IR spectra of the TCNE+HMB/CCl_4 solutions. The concentrations of the complex are 0.0043 M (solid line) and 0.0018 M (dotted line). Lower panel: a difference spectrum obtained by subtracting the lower concentration spectrtun from the higher concentration spectrum.

3-5 Proteins in Aqueous Solutions

There is an increasing interest in the low frequency collective vibrational modes of biomolecules.[42-45] These absorption lines may be used as a direct probe of the conformational state of the biomolecules. For example, protein collective vibrational modes have been calculated to lie in the far IR region for bovine pancreatic trypsin inhibitor and human lysozyme.[44,45] These modes are due to collective motion of the secondary and tertiary strucutures. Time-resolved experiments in the far IR region may allow us to directly probe dynamics of the higher order structures such as protein folding.

An aqueous solution of Ribonuclease A is chosen as a sample. Ribonuclease A (EC 3. I.27.5, Type XII-A) from bovine pancreas was purchased from Sigma Chemical Co. The protein was dissolved in H_2O and the pH of the solution was adjusted to 1.17 with HC1. If the pH were higher, the protein would have been aggregated. The solution was centrifuged for 10 minutes at 15,000 rpm to remove dust and present possible aggregation. The protein concentration was determined to be 68.2 mg/ml spectrophotometrically at pH 7.0 and 278 nm, with an absorbance of 0.71 for a 1.0 mg/ml solution. We use the same optical cell used for the far IR measurement on liquid water.

The upper panel in Figure 9 compares the far IR spectra of liquid water, $\alpha_{water}(\nu)$, and the protein solution, $\alpha_{protein}(\nu)$. A clear decrease of the absorption component is observed when the protein is dissolved in water. In the lower panel of Figure 9 the difference spectrum, $\alpha_{protein}(\nu)$ - $\alpha_{water}(\nu)$, is shown for reference.

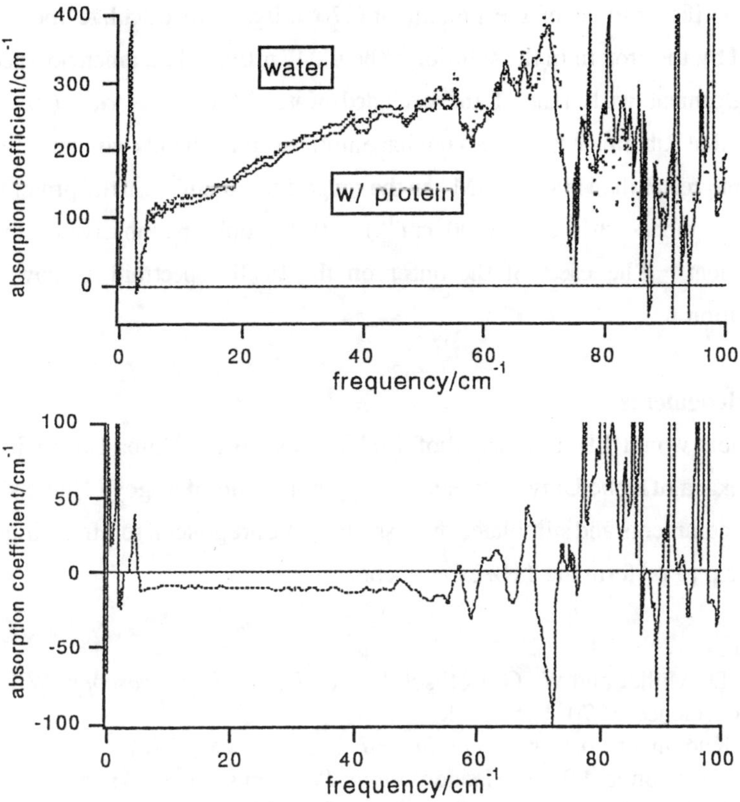

Figure 9. Upper panel: Far IR spectra of water (dotted line) and the protein solution (solid line). Lower panel: difference spectrum obtained by subtracting the water spectrum from the protein solution spectrum.

There are several possible contributions to this difference spectrum: low frequency motions of the proteins, interactions between the protein and water, effect of water molecules excluded by the presence of the protein, and effect of the buffers in the protein solution. A tentative explanation for the difference spectrum is the effect of excluded water molecules for the following reason. Using the reported value of the partial specific volume of the protein of 0.70 ml/g,[46] we calculate the volume occupied by the protein in the solution to be 0.047 ml/g. The difference spectrum is in agreement with that of the excluded water, -0.047 × $\alpha_{water}(\nu)$. This suggests that Ribonuclease A does not have any absorption bands with an extinction coefficient greater than 1000 $cm^{-1}M^{-1}$ in the range 4 to 70 cm^{-1}. If the protein had a vibrational mode with $\varepsilon > 1000$ $cm^{-1}M^{-1}$, that would be observable by this measurement. The effect of the buffer on the far IR spectrum is now under investigation.

Acknowledgments

The authors would like to thank Prof. Keiko Nishikawa of Chiba University, Dr. Shoji Takada of Kobe University, and Prof. Shinji Saito of Nagoya University for valuable comments and stimulated discussion. We are grateful to Mr. Shingo Ono for his help to perform the THz experiments.

References

1. K. D. Möller and W. G. Rothschild, *Far-Infrared Spectroscopy*, Wiley Interscience, 1970, New York.
2. E. Knözinger, *Angew. Chem. Int. Ed. Engl.* **15**, 25 (1976).
3. D. Kivelson and P. A. Madden, *Annu. Rev. Phys. Chem.* **31**, 523 (1980).
4. T. S. Perova, *Adv. Chem. Phys.* **87**, 427 (1994).
5. K. Yoshihara, K. Tominaga, and Y. Nagasawa, *Bull. Chem. Soc. Jpn.* **68**, 696 (1995).
6. J. C. Owrutsky, D. Raftry, and R. M. Hochstrasser, *Annu. Rev. Phys. Chem.* **45**, 519 (1994).
7. K. H. Yang, P. L. Richards, and Y. R. Shen, *Appl. Phys. Lett.* **19**, 285 (1971).
8. N. Sarukura, H. Ohtake, S. Izumida, and Z, Liu, *J. Appl. Phys.* **84**, 654 (1998).
9. S. Izumida, S. Ono, Z. Liu, H. Ohtake, and N. Sarukura, *Appl. Phys. Lett.* **75**, 451 (1999).
10. J. E. Pedersen and S. R. Keiding, *IEEE J. Quant. Elect.* **28**, 2518 (1992).
11. L. Thrane, R. H. Jacobsen, P. Uhd Jepsen, and S. R. Keiding, *Chem. Phys.*

Lett. **240**, 330 (1995).
12. C. Rønne, L. Thrane, P.-O. Åstrand, A. Wallqvist, K. V. Mikkelsen, and S. R. Keiding, *J. Chem. Phys.* **107**, 5319 (1997); C. Rønne, P.-O. Åstrand, and S. R. Keiding, *Phys. Rev. Lett.* **82**, 2888 (1999).
13. B. N. Flanders, R. A. Chevil1e, D. Grischkowsky, and N. F. Scherer, *J. Phys. Chem.* **100**, 11824 (1996).
14. J. T. Kindt and C. A. Schmuttenmaer, *J. Phys. Chem.* **100**, 10373 (1996); J. T. Kindt and C. A. Schmuttenmaer, *J. Chem. Phys.* **106**, 4389 (1997).
15. R. McElroy and K. Wynne, *Phys. Rev. Lett.* **79**, 3078 (1997).
16. G. Haran, W.-D. Sun, K. Wynne, and R. M. Hochstrasser, *Chem. Phys. Lett.* **274**, 365 (1997).
17. B. I. Greene, J. F. Fedrici, D. R. Dykaar, A. F. J. Levi, and L. Pfeifer, *Opt. Lett.* **16**, 48 (1991).
18. W. Fishler, P. Buchberger, R. A. Höpfel, and G. Zandler, *Appl. Phys. Lett.* **68**, 2778 (1996).
19. J. F. Federici, et al., *Phys. Rev. B* **46**, 11153 (1992).
20. S. K. Garg, J. E. Berite, H. Kilp, and C. P. Smyth, *J. Chem. Phys.* **49**, 2551 (1968).
21. M. Davies, G. W. F. Pardoe, J. E. Chamberlain, and H. A. Gebbie, *Trans. Faraday Soc.* **64**, 847 (1968).
22. J. Goulon and J. L. Rivail, *Chem. Phys. Lett.* **18**, 211 (1973).
23. B. Guillot, *J. Chem. Phys.* **95**, 1543 (1991).
24. S. Saito and K. Tominaga, manuscript in preparation.
25. J. K. Vij and F. Hufnagel, *Chem. Phys. Lett.* **155**, 153 (1989).
26. Z. Czumaj, *Mol. Phys.* **69**, 787 (1990).
27. *Special Issue of Supercritical Fluids* ed. by R. Noyori, *Chem. Rev.* **99**, (1999).
28. K. Nishikawa and T. Morita, *J. Supercritical Fluids.* **13**, 143 (1998).
29. K. Nishikawa and T. Morita, *J. Phys. Chem. B* **101**, 1413 (1997).
30. C. A. Eckert, D. H. Ziger, and K. P. Johnston, T. K. Ellison, *J. Phys. Chem.* **90**, 2738 (1986).
31. B. Otto, J. Schroeder, and J. Troe, *J. Chem. Phys.* **81**, 3931 (1984).
32. K. Saitow, H. Ohtake, H. Miyagi, Y. Shimokawa, H. Matsuo, K. Tominaga, N. Sarukura, and K. Nishikawa, manuscript in preparation.
33. K. Saitow, H. Ohtake, K. Tominaga, N. Sarukura, and K. Nishikawa, manuscript in preparation.
34. M. Hayashi, T.-S. Yang, J. Yu, A. Mebel, R. Chang, S. H. Lin, I. V. Rubtsov, and K. Yoshihara, *J. Phys. Chem.* **102**, 4256 (1998).

35. K. Wynne and R. M. Hochstrasser, *J. Chem. Phys.* **100**, 4707 (1994).
36. F. Markel, N. S. Ferris, I. R. Gould, A. B. Myers, *J. Amer. Chem. Soc.* **114**, 6208 (1992).
37. I. V. Rubtsov and K. Yoshihara, *J. Phys. Chem. A* **101**, 6138 (1997).
38. I. V. Rubtsov and K. Yoshihara, *J. Phys. Chem. A*, in press.
39. M. Rossi and E. Haselbach, *Helv. Chim. Acta* **62**, 140 (1979).
40. W. Liptay, T. Rehm, D. L. Schanne, W. Baumann,and W. Land, *Z. Naturforch. A* **37**, 1427 (1982).
41. M. L. Smith and J. L. McHale, *J. Phys. Chem.* **89**, 4002 (1985).
42. S. Krimm, *Biological Applications of Raman Spectroscopy 1*, ed. by T. G. Spiro, pp. 1-45, John Wiley & Sons, New York, 1987.
43. R. A. Friesner and R. M. Levy, *Biological Applications of Raman Spectroscopy 1*, ed. by T. G. Spiro, pp.263-305, John Wiley & Sons, New York, 1987.
44. A. Roitberg, R. Benny Gerber, R. Elber, and M. A. Ratner, *Science* **268**, 1319 (1995).
45. S. Hayward and N. Go, *Annu. Rev. Phys. Chem.* **46**, 223 (1995).
46. *Handbook of Biochemistry Selected Data for Molecular Biology 2nd ed.*, ed. by Sober, CRC Press, Cleveland (1970).